W9-CVA-642

Smoking Concerns
SOURCEBOOK

SECOND EDITION

Smoking Concerns
SOURCEBOOK

SECOND EDITION

Basic Consumer Health Information about Nicotine Addiction and Smoking Cessation, Featuring Facts about the Health Effects of Tobacco Use, Including Lung and Other Cancers, Heart Disease, Stroke, and Respiratory Disorders, Such as Emphysema and Chronic Bronchitis

Along with Information about Smoking Prevention Programs, Suggestions for Achieving and Maintaining a Smoke-Free Lifestyle, Statistics about Tobacco Use, Reports on Current Research Initiatives, a Glossary of Related Terms, and Directories of Resources for Additional Help and Information

OMNIGRAPHICS
615 Griswold, Ste. 520, Detroit, MI 48226

LONGWOOD PUBLIC LIBRARY

Bibliographic Note
Because this page cannot legibly accommodate all the copyright notices, the Bibliographic Note portion of the Preface constitutes an extension of the copyright notice.

* * *

OMNIGRAPHICS
Angela L. Williams, *Managing Editor*
* * *

Copyright © 2019 Omnigraphics

ISBN 978-0-7808-1689-3
E-ISBN 978-0-7808-1690-9

Library of Congress Cataloging-in-Publication Data

Names: Omnigraphics, Inc., issuing body.

Title: Smoking concerns sourcebook: basic consumer health information about nicotine addiction and smoking cessation, with facts about the health effects of tobacco use, including lung and other cancers, heart disease, stroke, and respiratory disorders such as emphysema and chronic bronchitis, a glossary of related terms, resources for additional help, and index / Angela Williams, managing editor.

Description: Second edition. | Detroit, MI: Omnigraphics, [2019] | Series: Health reference series | Includes bibliographical references and index.

Identifiers: LCCN 2019000723 (print) | LCCN 2019000835 (ebook) | ISBN 9780780816909 (ebook) | ISBN 9780780816893 (hard cover: alk. paper)

Subjects: LCSH: Smoking--Health aspects. | Smoking cessation. | Nicotine--Physiological effect.

Classification: LCC RA1242.T6 (ebook) | LCC RA1242.T6 S589 2019 (print) | DDC 613.85--dc23

LC record available at https://lccn.loc.gov/2019000723

Electronic or mechanical reproduction, including photography, recording, or any other information storage and retrieval system for the purpose of resale is strictly prohibited without permission in writing from the publisher.

The information in this publication was compiled from the sources cited and from other sources considered reliable. While every possible effort has been made to ensure reliability, the publisher will not assume liability for damages caused by inaccuracies in the data, and makes no warranty, express or implied, on the accuracy of the information contained herein.

This book is printed on acid-free paper meeting the ANSI Z39.48 Standard. The infinity symbol that appears above indicates that the paper in this book meets that standard.

Printed in the United States

Table of Contents

Part II: Tobacco-Related Health Hazards

Part III: Smoking Cessation

Part VI: Additional Help and Information

Preface

About This Book

According to the Centers for Disease Control and Prevention (CDC), around 480,000 Americans die annually from diseases attributed to cigarette smoking, with more than 41,000 of these deaths from exposure to secondhand smoke. It is estimated that smoking-related illness in the United States costs more than $300 billion a year. Cancer, heart disease, stroke, and emphysema are among the most well-known and feared effects of smoking. Other associated problems include chronic bronchitis, musculoskeletal disorders, diabetes, digestive disorders, erectile dysfunction, reproductive disorders, complications of pregnancy, and depression. Because of the health hazards associated with tobacco use and other concerns, nearly three-quarters of current smokers want to quit. Breaking free from nicotine addiction, however, is not easy. According to various studies, former smokers often make three, four, or many more attempts before they finally succeed in achieving smoking cessation goals.

Smoking Concerns Sourcebook, Second Edition provides basic facts about tobacco use, including how nicotine affects the body and how addiction develops. It offers facts about the health effects of smoking or using smokeless tobacco. A section on smoking cessation offers tips on preparing for, achieving, and sustaining a smoke-free lifestyle. Statistics about tobacco use, reports on current research initiatives, and information about public-health policies regarding tobacco control and use prevention are also included. Readers seeking additional help will find a glossary of related terms and directories of resources.

How to Use This Book

This book is divided into parts and chapters. Parts focus on broad areas of interest. Chapters are devoted to single topics within a part.

Part I: Smoking and Tobacco Products: An Overview provides facts about various tobacco products such as cigarettes, cigars, pipes etc. It describes the different types of chemicals present in the tobacco products and explains how addiction develops. Statistical information about trends in tobacco use among specific population and economic costs of smoking are also included.

Part II: Tobacco-Related Health Hazards discusses the diseases and disorders that can be caused by or worsen as a result of tobacco use. These include various types of cancer, respiratory disorders, cardio-vascular diseases, and mental problems. Hazards associated with smoking during pregnancy and the effects of smoking on the digestive system, musculoskeletal health, oral health, and sexual functioning are also explained.

Part III: Smoking Cessation provides information about how to quit smoking. It outlines the recommended steps for preparing to quit an various health benefits of quitting. It also elaborates what to expect during the first few days of a cessation program, and offers tips for coping with commonly experienced problems such as stress and depression. Medications that can help people achieve smoking cessation goals are also described.

Part IV: Tobacco-Related Research reports on areas of current investigation into how genetic factors are linked with tobacco use, the health-risks associated with tobacco products, the relation between nicotine dependency and lung cancer, and the strategies to lessen the impact of tobacco-related damage on body organs and systems.

Part V: Tobacco Control and Use Prevention explains various public-health policies used to reduce tobacco consumption, including programs to restrict minors' access to tobacco, clean indoor-air regulations, taxation policies, counter-advertising campaigns, and tobacco labeling requirements.

Part VI: Additional Help and Information includes a glossary of terms related to tobacco use and smoking cessation, a directory of resources for tobacco-related information, and a list of resources for help with smoking cessation.

Bibliographic Note

This volume contains documents and excerpts from publications issued by the following U.S. government agencies: Centers for Disease Control and Prevention (CDC); National Cancer Institute (NCI); National Center for Biotechnology Information (NCBI); National Heart, Lung, and Blood Institute (NHLBI); National Institute of Dental and Craniofacial Research (NIDCR); National Institute of Diabetes and Digestive and Kidney Diseases (NIDDK); National Institute on Drug Abuse (NIDA); National Institute on Drug Abuse (NIDA) for Teens; National Institutes of Health (NIH); *NIH News in Health*; NIH Osteoporosis and Related Bone Diseases—National Resource Center (NIH ORBD—NRC); Office of Disease Prevention and Health Promotion (ODPHP); Office of the Surgeon General (OSG); Substance Abuse and Mental Health Services Administration (SAMHSA); U.S. Department of Health and Human Services (HHS); U.S. Department of Veterans Affairs (VA); U.S. Environmental Protection Agency (EPA); U.S. Food and Drug Administration (FDA); and U.S. National Library of Medicine (NLM).

It may also contain original material produced by Omnigraphics and reviewed by medical consultants.

About the Health Reference Series

The *Health Reference Series* is designed to provide basic medical information for patients, families, caregivers, and the general public. Each volume takes a particular topic and provides comprehensive coverage. This is especially important for people who may be dealing with a newly diagnosed disease or a chronic disorder in themselves or in a family member. People looking for preventive guidance, information about disease warning signs, medical statistics, and risk factors for health problems will also find answers to their questions in the *Health Reference Series*. The *Series*, however, is not intended to serve as a tool for diagnosing illness, in prescribing treatments, or as a substitute for the physician/patient relationship. All people concerned about medical symptoms or the possibility of disease are encouraged to seek professional care from an appropriate healthcare provider.

A Note about Spelling and Style

Health Reference Series editors use *Stedman's Medical Dictionary* as an authority for questions related to the spelling of medical terms and the *Chicago Manual of Style* for questions related to grammatical structures, punctuation, and other editorial concerns. Consistent adherence is not always possible, however, because the individual volumes within the *Series* include many documents from a wide variety of different producers, and the editor's primary goal is to present material from each source as accurately as is possible. This sometimes means that information in different chapters or sections may follow other guidelines and alternate spelling authorities. For example, occasionally a copyright holder may require that eponymous terms be shown in possessive forms (Crohn's disease vs. Crohn disease) or that British spelling norms be retained (leukaemia vs. leukemia).

Medical Review

Omnigraphics contracts with a team of qualified, senior medical professionals who serve as medical consultants for the *Health Reference Series*. As necessary, medical consultants review reprinted and originally written material for currency and accuracy. Citations including the phrase "Reviewed (month, year)" indicate material reviewed by this team. Medical consultation services are provided to the *Health Reference Series* editors by:

Dr. Vijayalakshmi, MBBS, DGO, MD
Dr. Senthil Selvan, MBBS, DCH, MD
Dr. K. Sivanandham, MBBS, DCH, MS (Research), PhD

Our Advisory Board

We would like to thank the following board members for providing initial guidance on the development of this series:

- Dr. Lynda Baker, Associate Professor of Library and Information Science, Wayne State University, Detroit, MI

- Nancy Bulgarelli, William Beaumont Hospital Library, Royal Oak, MI

- Karen Imarisio, Bloomfield Township Public Library, Bloomfield Township, MI

- Karen Morgan, Mardigian Library, University of Michigan-Dearborn, Dearborn, MI

- Rosemary Orlando, St. Clair Shores Public Library, St. Clair Shores, MI

Health Reference Series *Update Policy*

The inaugural book in the *Health Reference Series* was the first edition of *Cancer Sourcebook* published in 1989. Since then, the *Series* has been enthusiastically received by librarians and in the medical community. In order to maintain the standard of providing high-quality health information for the layperson the editorial staff at Omnigraphics felt it was necessary to implement a policy of updating volumes when warranted.

Medical researchers have been making tremendous strides, and it is the purpose of the *Health Reference Series* to stay current with the most recent advances. Each decision to update a volume is made on an individual basis. Some of the considerations include how much new information is available and the feedback we receive from people who use the books. If there is a topic you would like to see added to the update list, or an area of medical concern you feel has not been adequately addressed, please write to:

Managing Editor
Health Reference Series
Omnigraphics
615 Griswold, Ste. 520
Detroit, MI 48226

Part One

Smoking and Tobacco Products: An Overview

Chapter 1

Cigarettes and Other Tobacco Products: The Facts

Chapter Contents

Section 1.1

Tobacco Timeline

This section contains text excerpted from the following sources:
Text in this section begins with excerpts from "Highlights: Tobacco
Timeline," Centers for Disease Control and Prevention (CDC), July
21, 2015. Reviewed February 2019; Text under the heading "Reports
of The Surgeon General: Brief History" is excerpted from "The
Reports of the Surgeon General Brief History," U.S. National Library
of Medicine (NLM), April 1, 2002. Reviewed February 2019.

Below is the timeline of tobacco products in the United States of
America:

- Cigarettes were first introduced in the United States in the early
  19th century. Before this, tobacco was used primarily in pipes
  and cigars, by chewing, and in snuff.

- By the time of the Civil War, cigarette use had become more
  popular. Federal tax was first imposed on cigarettes in 1864.
  Shortly afterward, development of the cigarette-manufacturing
  industry led to tobacco becoming a major U.S. product.

- Concurrently, a populist health-reform movement led to early
  antismoking activity. From 1880 to 1920, this activity was
  largely motivated by moral and hygienic concerns rather than
  health issues.

- The milder flue-cured tobacco blends used in cigarettes during
  the early 20th century made the smoke easier to inhale and
  increased nicotine absorption into the bloodstream.

- During World War I, U.S. Army surgeons praised cigarettes for
  helping the wounded relax and easing their pain.

- Smoking was first linked to lung cancer and other diseases in
  the late 1940s and early 1950s.

- In 1956, the U.S. Surgeon General's scientific study group
  determined that there was a causal relationship between
  excessive cigarette smoking and lung cancer.

- In England, the 1962 Royal College of Physicians report
  emphasized smoking's causative role in lung cancer.

- Antismoking messages had a significant impact on cigarette sales;
  however, when cigarette advertising on television and radio was
  banned in 1969, antismoking messages were discontinued.

- The 1972 Surgeon General's report became the first of a series of science-based reports to identify environmental tobacco smoke (ETS) as a health risk to nonsmokers.

- In 1973, Arizona became the first state to restrict smoking in a number of public places explicitly because ETS exposure is a public-health hazard.

- By the mid-1970s, the federal government began administratively regulating smoking within government domains. In 1975, the U.S. Army and Navy stopped including cigarettes in rations for service members. Smoking was restricted in all federal government facilities in 1979 and was banned in the White House in 1993.

- In 1988, Congress prohibited smoking on domestic commercial airline flights scheduled for two hours or less. By 1990, the ban was extended to all commercial U.S. flights.

- In 1992, the U.S. Environmental Protection Agency (EPA) classified ETS as a "Group A" carcinogen, the most dangerous class of carcinogens.

- In 1994, Mississippi became the first state to sue the tobacco industry to recover Medicaid costs for tobacco-related illnesses, settling its suit in 1997. A total of 46 states eventually filed similar suits. Three other states settled individually with the tobacco industry—Florida (1997), Texas (1998), and Minnesota (1998).

- On November 23, 1998, the tobacco industry approved a 46-state Master Settlement Agreement, the largest settlement in history, totaling nearly $206 billion to be paid through the year 2025. The settlement agreement contained a number of important public-health provisions.

- In April 1999, as part of the Master Settlement Agreement, the major U.S. tobacco companies agreed to remove all advertising from outdoor and transit billboards across the nation. The remaining time on at least 3,000 billboard leases, valued at $100 million, was turned over to the states for posting antitobacco messages.

- On March 21, 2000, the U.S. Supreme Court narrowly affirmed a 1998 decision of the U.S. Court of Appeals for the 4th Circuit and ruled that the U.S. Food and Drug Administration (FDA) lacks jurisdiction under the Federal Food, Drug, and Cosmetic Act (FDCA) to regulate tobacco products. As a result, the FDA's proposed rule to reduce access and appeal of tobacco products for young people became invalid.

Reports of the Surgeon General: Brief History

In a healthcare system chiefly directed toward treating disease and surgical intervention, the Surgeon General has pursued a complementary strategy: disease prevention and health promotion. Appointed by the President with the advice and consent of the Senate, the Surgeon General—whose title means "chief surgeon" and who is the federal government's principal spokesperson on matters of public health. The first Surgeon General was appointed in 1871 to head the Marine Hospital Service, itself established in 1798 to minister to sick and injured merchant seamen and reorganized as the U.S. Public Health Service (USPHS) in 1912. In recent decades, the Surgeon General has become the most widely recognized and respected voice on public-health issues, preventive medicine, and health promotion through public appearances, speeches, and, most influentially, the reports featured on this section.

The Surgeon General has often been called upon to deal with difficult and controversial issues, such as smoking and sexual health. In some cases, the public-health message has generated controversy, when it ran counter to the political beliefs of the time. But the Surgeon General's public statements often served to generate debate where there had been silence, to the benefit of the nation's health.

The role of the Surgeon General has changed much during the past four decades. As the head of the USPHS, for over half a century the Surgeon General oversaw infectious-disease eradication, rural sanitation, medical research, the provision of medical and hospital care to members of the Coast Guard and Merchant Marine, and other public-health activities. Until 1968, the Surgeon General's main responsibility was the day-to-day administration of the USPHS and its many programs, including directing the uniformed Commissioned Corps of physicians, dentists, nurses, pharmacists, sanitary engineers, and other health professionals that has been the institutional mainstay of USPHS.

In 1968, an organizational reform greatly reduced the Surgeon General's administrative role, abolished the Office of the Surgeon General (OSG) (though not the position of Surgeon General itself), and transferred line authority for the administration of USPHS to the

Assistant Secretary for Health within the U.S. Department of Health, Education, and Welfare (HEW) (since 1980, the U.S. Department of Health and Human Services (HHS)). Since 1968, the Surgeon General has not administered the USPHS, but focused instead on the primary official duty to advise the U.S. Secretary of Health and Human Services and the Assistant Secretary of Health on affairs of preventive health, medicine, and health policy. Left with few bureaucratic tasks, the Surgeons General since the 1960s have undertaken a more proactive role in informing the American public on health matters. They have relied on their professional credentials (all Surgeons General have been MDs) and political independence to make themselves into the most visible and, in the public's mind, impartial and therefore, trusted government spokespersons on health issues affecting the nation as a whole.

- 1912—The Marine Hospital Service reorganized as the USPHS

- 1913—R. J. Reynolds Tobacco Company launched Camel, the first modern mass-produced cigarette made from blended tobacco

- 1917—Cigarettes included in the field rations of American soldiers in World War I

- 1928—Herbert L. Lombard and Carl R. Doering offered the first detailed statistical data showing a higher proportion of heavy smokers among lung cancer patients than among controls

- 1929—U.S. Surgeon General Hugh S. Cumming (1920 to 1936) cautioned that smoking causes nervousness and insomnia, particularly among women

- 1938—Raymond Pearl demonstrated statistically that smoking shortens life expectancy

- 1941 to 45—Tobacco supplied to American servicemen in World War II

- 1942—In-vitro experiments established that tar, or solid particles of partially burnt tobacco, can act directly on cells to produce neoplasm, or new and abnormal growth

- 1953—Ernest Wynder, a researcher at Sloan-Kettering Cancer Center, painted smoke condensate on the skin of mice, producing cancerous tumors in 44 percent of the animals

- 1957—Surgeon General Leroy E. Burney (1956 to 1961) declared it to be the official position of the USPHS that a causal relationship exists between smoking and lung cancer (June 12)

- 1964—Surgeon General Luther L. Terry (1961 to 65) issued Smoking and Health, the first Surgeon General's report to receive widespread media and public attention (January 11)

- 1965—Congress mandated health warnings on cigarette packs

- 1968—The OSG was abolished and the position became that of an advisor to the Secretary of Health, Education, and Welfare, and to the Assistant Secretary of Health. The Surgeon General no longer directly administered the USPHS

- 1969—The Public Health Cigarette Smoking Act passed in Congress. It imposed a ban on cigarette advertising on TV and radio after September 30, 1970, and required that the Surgeon General produce an annual report on the latest scientific findings on the health effects of smoking

- 1973—Arizona passed the first state law designating separate smoking areas in public places

- 1979—Surgeon General Julius B. Richmond (1977 to 1981) issued Healthy People: The Surgeon General's Report on Health Promotion and Disease Prevention, emphasizing the role of nutrition, exercise, environmental factors, and occupational safety in advancing health

- 1980—With the report Maternal and Infant Health, the Surgeon General took up a subject that has been a focus of federal social policy since the creation of the Children's Bureau in 1912

- 1981—Acquired immune deficiency syndrome (AIDS) first diagnosed

- 1983—Lung cancer surpassed breast cancer as the leading cause of death from cancer in women

- 1986—Surgeon General C. Everett Koop (1981 to 89) released The Surgeon General's Report on Acquired Immune Deficiency Syndrome, describing AIDS as a preventable, manageable chronic disease

- 1987—The OSG was re-established with the Surgeon General as supervisor of the personnel system of the Commissioned Corps of the USPHS

- 1987—Congress banned smoking on all domestic flights of two hours or less; two years later smoking was banned on all domestic flights

- 1992—The EPA placed passive smoke on its list of major carcinogens, making it subject to federal workplace and other regulations

- 1999—Surgeon General David Satcher (1998–2002) published Mental Health, marking an expansion of the Surgeon General's concerns beyond a predominant focus on diseases of the body

- 2000—California became the first state to ban smoking in bars and restaurants

Section 1.2

What Is Tobacco?

This section includes text excerpted from "Cigarettes and Other Tobacco Products," National Institute on Drug Abuse (NIDA), June 2018.

Tobacco is a plant grown for its leaves, which are dried and fermented before being put in tobacco products. Tobacco contains nicotine, an ingredient that can lead to addiction, which is why so many people who use tobacco find it difficult to quit. There are also many other potentially harmful chemicals found in tobacco or created by burning it.

How Do People Use Tobacco?

People can smoke, chew, or sniff tobacco. Smoked tobacco products include cigarettes, cigars, bidis, and kreteks. Some people also smoke loose tobacco in a pipe or hookah (waterpipe). Chewed tobacco products include chewing tobacco, snuff, dip, and snus; snuff can also be sniffed.

How Does Tobacco Affect the Brain?

The nicotine in any tobacco product readily absorbs into the blood when a person uses it. Upon entering the blood, nicotine immediately stimulates the adrenal glands to release the hormone epinephrine (adrenaline). Epinephrine stimulates the central nervous system

(CNS) and increases blood pressure, breathing, and heart rate. As with drugs such as cocaine and heroin, nicotine activates the brain's reward circuits and also increases levels of the chemical messenger dopamine, which reinforces rewarding behaviors. Studies suggest that other chemicals in tobacco smoke, such as acetaldehyde, may enhance nicotine's effects on the brain.

What Are Other Health Effects of Tobacco Use?

Although nicotine is addictive, most of the severe health effects of tobacco use comes from other chemicals. Tobacco smoking can lead to lung cancer, chronic bronchitis, and emphysema. It increases the risk of heart disease, which can lead to stroke or heart attack. Smoking has also been linked to other cancers, leukemia, cataracts, and pneumonia. All of these risks apply to use of any smoked product, including hookah tobacco. Smokeless tobacco increases the risk of cancer, especially mouth cancers.

Pregnant women who smoke cigarettes run an increased risk of miscarriage, stillborn or premature infants, or infants with low birth weight (LBW). Smoking while pregnant may also be associated with learning and behavioral problems in exposed children.

People who stand or sit near others who smoke are exposed to secondhand smoke, either coming from the burning end of the tobacco product or exhaled by the person who is smoking. Secondhand smoke exposure can also lead to lung cancer and heart disease. It can cause health problems in both adults and children, such as coughing, phlegm, reduced lung function, pneumonia, and bronchitis. Children exposed to secondhand smoke are at an increased risk of ear infections, severe asthma, lung infections, and death from sudden infant death syndrome.

How Does Tobacco Use Lead to Addiction?

For many who use tobacco, long-term brain changes brought on by continued nicotine exposure result in addiction. When a person tries to quit, she or he may have withdrawal symptoms, including:

- Irritability
- Problems paying attention
- Trouble sleeping
- Increased appetite
- Powerful cravings for tobacco

10

Section 1.3

How Cigarettes Are Made

This section includes text excerpted from "How Cigarettes Are Made and How You Can Make a Plan to Quit," U.S. Food and Drug Administration (FDA), December 19, 2017.

Each day in the United States, more than 1,300 people die because of cigarette smoking, and nearly 400 kids under age 18 become daily smokers. In an effort to bring the number of smokers in the United States down to zero, on November 17, 2016, people across the country will join the Great American Smokeout: a movement spearheaded by the American Cancer Society (ACS), in which people across the country pledge to make a plan to quit smoking. But for many, quitting feels like an impossible task, and unfortunately, this may be by design.

How a Cigarette Is Engineered

The U.S. Food and Drug Administration (FDA) launched a new infographic, "How a Cigarette is Engineered," highlighting some of the reasons quitting smoking can be difficult. It's not only because cigarettes contain the addictive chemical nicotine—which keeps people smoking even when they don't want to be—but also because the design and content of cigarettes continue to make them addictive and attractive to consumers.

What exactly are you smoking when you smoke a cigarette? You inhale everything that is burned—the tobacco filler, the paper—even the chemicals that form when the cigarette is lit. While that may be an unappealing thought, the mix of more than 7,000 chemicals that smokers inhale in the smoking process is downright deadly.

Filter

Let's begin with the filter. Typically made from bundles of thin fibers, the filter is located at the holding end of the cigarette and is meant to minimize the amount of smoke inhaled. The design of modern cigarette filters only prevents a nominal portion of smoke from being inhaled.

Tipping Paper

Wrapped around the filter is the tipping paper, which contains small ventilation (vent) holes. The purpose of vent holes is to allow

11

fresh air into mix with smoke, diluting the toxic mix of chemicals inhaled. Unfortunately, vent holes are usually located where you would hold the cigarette, and often get blocked by your fingers or lips, making them largely ineffective. They may also lead you to inhale more deeply, pulling dangerous chemicals farther into your lungs.

Cigarette Paper and Tobacco Filler

Below the filter and the tipping paper is the cigarette paper, which contains added chemicals to control how quickly the cigarette burns.

Within the cigarette paper is the tobacco filler itself, which comprises of chopped tobacco leaves, stems, reprocessed pieces, and scraps. Dangerous chemicals can form in and be deposited on tobacco during processing. What's more is that when the tobacco filler is burned, other hazardous chemicals are created and breathed into your lungs.

Additives

Not only are chemicals created in the processing and the burning of tobacco filler, but manufacturers may also add hundreds of ingredients to a cigarette to make smoking more appealing and mask the harsh flavor and sensation of smoke. Flavor additives like menthol and sugar may be added to cigarettes to change the taste of smoke and make it easier to inhale. These and other additives may make cigarette smoke more palatable, but no less harmful. Cigarettes that are less harsh and easier to inhale may appeal to new smokers, especially adolescents, because they are easier to smoke.

Other chemicals may also be added to tobacco in an effort to optimize nicotine delivery and lung absorption. Ammonia—a chemical found in cleaning products—and other additives may be added to cigarette tobacco and may increase nicotine absorption, making cigarettes more addictive. Some additives are bronchodilators that can open the lungs and increase the amount of dangerous chemicals that are absorbed.

A Cigarette

Given this information, it becomes clear that a cigarette is not just tobacco wrapped in paper. Its design and content make it alluring and addictive. And when you inhale its smoke, you take in every part of the cigarette.

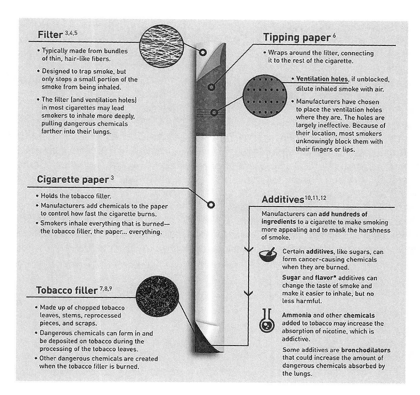

Filter [3,4,5]

- Typically made from bundles of thin, hair-like fibers.
- Designed to trap smoke, but only stops a small portion of the smoke from being inhaled.
- The filter (and ventilation holes) in most cigarettes may lead smokers to inhale more deeply, pulling dangerous chemicals farther into their lungs.

Tipping paper [6]

- Wraps around the filter, connecting it to the rest of the cigarette.
- **Ventilation holes**, if unblocked, dilute inhaled smoke with air.
- Manufacturers have chosen to place the ventilation holes where they are. The holes are largely ineffective. Because of their location, most smokers unknowingly block them with their fingers or lips.

Cigarette paper [3]

- Holds the tobacco filler.
- Manufacturers add chemicals to the paper to control how fast the cigarette burns.
- Smokers inhale everything that is burned— the tobacco filler, the paper... everything.

Additives [10,11,12]

Manufacturers can **add hundreds of ingredients** to a cigarette to make smoking more appealing and to mask the harshness of smoke.

Certain **additives**, like sugars, can form cancer-causing chemicals when they are burned.

Sugar and **flavor*** additives can change the taste of smoke and make it easier to inhale, but no less harmful.

Ammonia and other **chemicals** added to tobacco may increase the absorption of nicotine, which is addictive.

Some additives are **bronchodilators** that could increase the amount of dangerous chemicals absorbed by the lungs.

Tobacco filler [7,8,9]

- Made up of chopped tobacco leaves, stems, reprocessed pieces, and scraps.
- Dangerous chemicals can form in and be deposited on tobacco during the processing of the tobacco leaves.
- Other dangerous chemicals are created when the tobacco filler is burned.

Figure 1.1. *Design and Content of Cigarettes*

In 2009, The Family Smoking Prevention and Tobacco Control Act banned characterizing flavors in cigarettes, except for tobacco and menthol flavors.

(1) U.S. Department of Health and Human Services. A Report of the Surgeon General: How Tobacco Smoke Causes Disease (Fact Sheet). Atlanta, GA: U.S. Department of Health and Human Services, Centers for Disease Control and Prevention, National Center for Chronic Disease Prevention and Health Promotion, Office on Smoking and Health; 2010. (2) U.S. Department of Health and Human Services. The Health Consequences of Smoking—50 Years of Progress: A Report of the Surgeon General. Atlanta, GA: U.S. Department of Health and Human Services, Centers for Disease Control and Prevention, National Center for Chronic Disease Prevention and Health Promotion, Office on Smoking and Health; 2014. (3) Taylor MJ. The Role of Filter Technology in Reduced Yield Cigarettes. Filtrona. World Tobacco Exhibition Kunming. (4) Kiefer JE, Mumpower RC II. Parameters That Affect the Pressure Drop and Efficiency of Cellulose Acetate Cigarette Filters. Research Laboratories, Tennessee Eastman Company; 2004; Bates number: 81052204/2269. (5) U.S. Department of Health and Human Services. Let's Make the Next Generation Tobacco-Free: Your Guide to the 50th Anniversary Surgeon General's Report on Smoking and Health (Consumer Booklet). Atlanta, GA: U.S. Department of Health and Human Services, Centers for Disease Control and Prevention, National Center for Chronic Disease

Prevention and Health Promotion, Office on Smoking and Health; 2014. (6) Browne CL. The Design of Cigarettes. 3rd ed. Charlotte, NC: C Filter Products Division, Hoechst Celanese Corporation; 1990. (7) Spears AW. Effect of manufacturing variables on cigarette smoke composition. CORESTA Bulletin d'Information. 1974; 6:65–78. (8) Geiss O., Kotzias D. Tobacco, Cigarettes, and Cigarette Smoke: An Overview. European Commission, Directorate-General, Joint Research Centre; 2007. (9) Baker R. A Review of Pyrolysis Studies to Unravel Reaction Steps in Burning Tobacco. Journal of Analytical and Applied Pyrolysis. 1987; 11:555–573. (10) U.S. Department of Health and Human Services. How Tobacco Smoke Causes Disease: The Biology and Behavioral Basis for Smoking-Attributable Disease: A Report of the Surgeon General. Atlanta, GA: U.S. Department of Health and Human Services, Centers for Disease Control and Prevention, National Center for Chronic Disease Prevention and Health Promotion, Office on Smoking and Health; 2010. (11) Rabinoff M., Caskey N, Rissling A., Park, C. Pharmacological and Chemical Effects of Cigarette Additives. American Journal of Public Health. 2007; 97 (11): 1981–1991. (12) Talhout R., Opperhuizen A., Amsterdam J. Sugars as Tobacco Ingredient: Effects on Mainstream Smoke Composition. Food and Chemical Toxicology. 2006; 44(11):1789–1798.

Section 1.4

Other Tobacco Products Usage

This section includes text excerpted from "Tobacco, Nicotine, and E-Cigarettes," U.S. Food and Drug Administration (FDA), January 2018.

While cigarette smoking has declined significantly during the past 40 years, use of other tobacco products is increasing—particularly among young people. These include:

- **Cigars:** Tobacco wrapped in leaf tobacco or another tobacco-containing substance instead of paper, which can be bought individually

- **Cigarillos:** Small cigars that cost less and are also available for purchase individually

- **Hookahs or waterpipes:** Pipes with a long, flexible tube for drawing smoke from lit, flavored tobacco through water contained in a bowl

- **Smokeless tobacco:** Products like chewing tobacco and snuff that are placed in the mouth between the teeth and gums
- **Powder tobacco:** Mixtures that are inhaled through the nose

In 2014, almost one-quarter of high school students reported past month use of a tobacco product—with e-cigarettes (13.4%), hookahs (9.4%), cigarettes (9.2%), cigars (8.2%), smokeless tobacco (5.5%), and snus (moist powder tobacco) (1.9%) as the most popular.

Cigars

In 2016, an estimated 12 million people aged 12 or older (4.6% of the adolescent and adult population) smoked cigars during the past month. The majority of adolescents and young adults who smoked cigars also smoke cigarettes.

Cigarillos

Data from the Tobacco Use Supplement to the Current Population Survey (CPS) and National Survey on Drug Use and Health (NSDUH) suggest that younger and less economically advantaged males initiate tobacco use with cigarillos. From 2002 to 2011, past month cigarette smoking declined for males and females of all age groups. However, during this same period, rates of cigarillo use among males aged 18 to 25 remained constant (at approximately 9%).

Hookahs or Waterpipes

Between 2011 and 2014, use of hookah increased among middle and high school students, despite decreased use of cigarettes and cigars, according to the New York Theological Seminary (NYTS). Research also suggests that rates of hookah use for tobacco smoking increase during the first month of college. Nationally representative data from college students indicate that daily cigarette or cigar smokers (as well as marijuana users) were more likely to be frequent waterpipe users.

Hookah users may mistakenly believe that it is less addictive or dangerous than cigarettes; however, one session of hookah smoking exposed users to greater smoke volumes and higher levels of tobacco toxicants (e.g., tar) than a single cigarette. Additionally, hookah smoking is linked with nicotine dependence and its associated medical consequences. Reviews of the literature on waterpipe users suggest that like those who use other forms of tobacco, many have tried to quit but have been unsuccessful on their own. These findings indicate the need for tobacco control policies and prevention and treatment interventions for this form of nicotine delivery that are similar to those seen for cigarettes.

Smokeless Tobacco

In 2016, 8.8 million people aged 12 or older (3.3% of this population) used smokeless tobacco during the past month. Overall, use of smokeless tobacco among adults decreased from 1992 to 2003 but has held constant since. Longitudinal data suggest that people are more likely to switch from smokeless tobacco use to cigarette smoking than vice versa. Although smokers may attempt to use smokeless products to cut down or quit, research suggests that this approach is not effective. However, some argue that using smokeless tobacco in lieu of cigarettes may help reduce the harms associated with smoking traditional cigarettes.

Polytobacco Use

Some users of tobacco consume it in multiple forms (polytobacco use); this behavior is associated with greater nicotine dependence and the risk for other substance-use disorder (SUD). Analyses of a decade of data from NSDUH found steady rates of polytobacco use from 2002 to 2011 (8.7% to 7.4%) among people age 12 and older. However, use of some product combinations—such as cigarettes and smokeless tobacco, cigars and smokeless tobacco, and use of more than two products—increased over that period.

Among individuals younger than 26, rates of polytobacco use increased despite declines in overall tobacco use. Polytobacco use was associated with being male, having relatively low income and education, and engaging in risk-taking behaviors. In 2014, an estimated 2.2 million middle and high school students had used two or more types of tobacco products during the past month, according to the NYTS. Polytobacco use was common, even among students who used tobacco products five days or fewer during the past month. The 2012 NYTS had found that 4.3 percent of students used three or more types of tobacco. This study also observed that male gender, use of flavored products, nicotine dependence, receptivity to tobacco marketing, and perceived peer use were all associated with youth polytobacco use.

16

Chapter 2

Understanding Nicotine Addiction

Chapter Contents

Section 2.1

What Is Nicotine?

This section includes text excerpted from the following: Text in this section begins with excerpts from "Nicotine and Addiction," Smokefree.gov, U.S. Department of Health and Human Services (HHS), September 16, 2017; Text beginning with the heading "How Do People Use Tobacco and Nicotine?" is excerpted from "Mind Matters: The Body's Response to Nicotine," National Institute on Drug Abuse (NIDA) for Teens, January 30, 2019.

Nicotine is the chemical found in tobacco products that is responsible for addiction. When you use tobacco, nicotine is quickly absorbed into your body and goes directly to your brain. Nicotine activates areas of the brain that make you feel satisfied and happy. Whether you smoke, vape, or dip, the nicotine you are putting in your body is dangerously addictive and can be harmful to your developing brain.

How Do People Use Tobacco and Nicotine?

People can smoke, sniff, chew, or inhale the vapors of tobacco and nicotine products.

Some products that you smoke:

- Cigarettes
- Cigars
- E-cigarettes
- Hookahs

Smokeless products:

- Chewing tobacco
- Snuff (ground tobacco that can be sniffed or put between your cheek and gums)
- Dip (wet snuff that is chewed)
- Snus (small pouch of wet snuff)

How Does Nicotine Work?

Nicotine is absorbed into your bloodstream and goes to your adrenal glands just above your kidneys. The glands release adrenaline which increases your blood pressure, breathing, and heart rate. Adrenaline also gives you a lot of good feelings all at once.

Why Is Nicotine Dangerous?

Nicotine can lead to addiction, which puts you at risk of becoming a lifelong smoker and exposing you to the many harmful chemicals in tobacco. These chemicals cause cancer and harm almost every organ in your body. Teens are especially sensitive to nicotine's addictive effects because their brains are still developing. This makes it easier to get hooked. Using nicotine during your teen years can also rewire your brain to become more easily addicted to other drugs. Nicotine can have other long-lasting effects on your brain development, making it harder for you to concentrate, learn, and control your impulses.

Section 2.2

Is Nicotine Addictive?

This section contains text excerpted from "Is Nicotine Addictive," National Institute on Drug Abuse (NIDA), January 2018.

Yes. Most smokers use tobacco regularly because they are addicted to nicotine. Addiction is characterized by compulsive drug-seeking and use, even in the face of negative health consequences. The majority of smokers would like to stop smoking, and each year about half try to quit permanently. Yet, only about six percent of smokers are able to quit in a given year. Most smokers will need to make multiple attempts before they are able to quit permanently. Medications including varenicline, and some antidepressants (e.g., bupropion), and nicotine replacement therapy (NRT) can help in many cases.

A transient surge of endorphins in the reward circuits of the brain causes a slight, brief euphoria when nicotine is administered. This surge is much briefer than the "high" associated with other drugs. However, like other drugs of abuse, nicotine increases levels of the neurotransmitter dopamine in these reward circuits which reinforces the behavior of taking the drug. Repeated exposure alters these circuits' sensitivity to dopamine and leads to changes in other brain circuits involved in learning, stress, and self-control. For many tobacco users, the long-term brain changes induced by continued nicotine exposure result in addiction, which involves withdrawal symptoms when not smoking, and difficulty adhering to the resolution to quit.

The pharmacokinetic properties of nicotine, or the way it is processed by the body, contribute to its addictiveness. When cigarette smoke enters the lungs, nicotine is absorbed rapidly in the blood and delivered quickly to the brain, so that nicotine levels peak within 10 seconds of inhalation. But the acute effects of nicotine also dissipate quickly, along with the associated feelings of reward; this rapid cycle causes the smoker to continue dosing to maintain the drug's pleasurable effects and prevent withdrawal symptoms.

Withdrawal occurs as a result of dependence when the body becomes used to having the drug in the system. Being without nicotine for too long can cause a regular user to experience irritability, craving, depression, anxiety, cognitive and attention deficits, sleep disturbances, and increased appetite. These withdrawal symptoms may begin within a few hours after the last cigarette, quickly driving people back to tobacco use.

When a person quits smoking, withdrawal symptoms peak within the first few days of the last cigarette smoked and usually subside within a few weeks. For some people, however, symptoms may persist for months, and the severity of withdrawal symptoms appears to be influenced by a person's genes.

In addition to its pleasurable effects, nicotine also temporarily boosts aspects of cognition, such as the ability to sustain attention and hold information in memory. However, long-term smoking is associated with cognitive decline and risk of Alzheimer disease (AD), suggesting that short-term nicotine-related enhancement does not outweigh long-term consequences for cognitive functioning. In addition, people in withdrawal from nicotine experience neurocognitive deficits such as problems with attention or memory. These neurocognitive withdrawal symptoms are increasingly recognized as a contributor to continued smoking. A small research study also suggested that withdrawal may impair sleep for severely dependent smokers and that this may additionally contribute to relapse.

In addition to the drug's impact on multiple neurotransmitters and their receptors, many behavioral factors can affect the severity of withdrawal symptoms. For many people who smoke, the feel, smell, and sight of a cigarette and the ritual of obtaining, handling, lighting, and smoking the cigarette are all associated with the pleasurable effects of smoking and can make withdrawal or craving worse. Learning processes in the brain associate these cues with nicotine-induced dopamine surges in the reward system—similar to what occurs with other drug addictions. Nicotine replacement therapies such as gum, patches, and inhalers, and other medications approved for the treatment of nicotine addiction may help alleviate the physiological aspects of withdrawal; however, cravings often persist because of the power of these cues. Behavioral therapies can help smokers identify environmental triggers of craving so they can use strategies to avoid these triggers and manage the feelings that arise when triggers cannot be.

Are There Other Chemicals That May Contribute to Tobacco Addiction?

Research is showing that nicotine may not be the only ingredient in tobacco that affects its addictive potential. Smoking is linked with a marked decrease in the levels of monoamine oxidase (MAO), an important enzyme that is responsible for the breakdown of dopamine, as well as a reduction in MAO binding sites in the brain. This change is likely caused by some as-yet-unidentified ingredient in tobacco smoke other than nicotine because we know that nicotine itself does not dramatically alter MAO levels.

Animal research suggests that MAO inhibition makes nicotine more reinforcing, but more studies are needed to determine whether MAO inhibition affects human tobacco dependence. Animal research has also shown that acetaldehyde, another chemical in tobacco smoke created by the burning of sugars added as sweeteners, dramatically increase the reinforcing properties of nicotine and may also contribute to tobacco addiction.

Section 2.3

Triggers for Addiction

This section includes text excerpted from "Tobacco Cessation: An Abbreviated Mini-Workbook," U.S. Department of Veterans Affairs (VA), July 2015. Reviewed February 2019.

Tobacco Use Is Linked to Other Habits

Tobacco use is associated with several different behaviors that are very closely related. Tobacco use is a learned behavior, meaning you learned this from your family, friends, magazine ads, television, movies, or during your military service. When you were in the military, you may have heard someone say "Smoke 'em if you got 'em."

Who Helped You Learn Your Tobacco Habit

Tobacco use is also a triggered behavior, meaning certain activities or times of day may make you think about having a cigarette or chewing tobacco.

A Challenge for You: Addressing Your Triggers

Try not smoking during one of your triggers. For example, you might pick after eating breakfast as the trigger. Starting tomorrow, try not smoking for 10 minutes after breakfast. Wait 20 minutes to smoke after breakfast the next day. Wait 30 minutes the day after that. Add another 10 minutes each day. By the end of the week, you will be waiting for one hour between the trigger and smoking. Once you have tackled one trigger, try adding another trigger until you have a few of them managed.

By adding time between the trigger and the action of using tobacco, the trigger becomes weaker. This will help when you get to your quit day. If you use chewing tobacco, you can do the same thing by looking at your triggers to put in a new dip. To help you avoid using tobacco, come up with a few things you can do rather than use tobacco.

Some examples would be:

- Going for a walk

- Using a substitute such as chewing gum or candy

- Doing deep breathing exercises

Finally, tobacco use is also an automatic behavior, meaning you may find you smoke without thinking about it. You may recall times when you lit a cigarette and then noticed you already had one lit. There are a few ways to help reduce this automatic behavior:

Tip 1: Move Your Tobacco to a Different Location

- If you carry your cigarettes in your front pocket, put them on the kitchen counter. If you smoke outside, take one cigarette with you instead of the whole pack. You will need to physically walk to the pack to smoke the next cigarette. This leads you to think about smoking that cigarette and possibly consider whether you really need it or if you can wait longer to have the next one.

- If you use chewing tobacco, try placing the can on the kitchen counter or another location in the house, rather than in your pocket. You will need to walk to the location of the tobacco can before putting in a new plug of chew.

Tip 2: Keep Track of Your Cigarettes as You Smoke Them

- Track your tobacco use on a sheet of paper you keep in your pocket or in your pack of cigarettes. Each time you smoke or dip, note on your tracker the time of day, your mood at that time (e.g., happy, bored, angry, etc.,) and your level of need for tobacco (e.g., low, medium, high).

- Keeping track of the cigarettes helps you work on the automatic behavior of smoking. You can see if changes in your mood cause you to smoke more or less. Many people find that they smoke more when they are unhappy, while others might find they smoke more when they are bored. Times of celebration may be a situation when your tobacco use might increase. You may find that you smoke more when you are around certain friends who also smoke.

- If you use chewing tobacco, you can use the same method to track the number of dips you take in a day.

Section 2.4

Nicotine Addiction and Quitting

This section includes text excerpted from "Tobacco Cessation: An Abbreviated Mini-Workbook," U.S. Department of Veterans Affairs (VA), July 15, 2015. Reviewed February 2019.

Nicotine is found naturally in tobacco that causes feelings of pleasure, relaxation or stimulation, and stress reduction. Nicotine is the addictive part of tobacco but is not by itself harmful to your body. The other substances found in tobacco and substances formed when tobacco is burned to harm your body. There are more than 7,000 chemicals in tobacco smoke and at least 69 that are harmful to humans and can cause cell damage, cell death, and cancer. Some of the more harmful chemicals include the following:

- Carbon monoxide
- Hydrogen cyanide
- Ammonia
- Lead
- Cadmium
- Polonium-210
- Arsenic
- Benzene
- Formaldehyde

Even though it is not burned, smokeless tobacco also contains some of these harmful chemicals. Other products and devices that deliver nicotine, including the electronic cigarette, may also be harmful to you.

Some products with nicotine can help you quit using tobacco, like nicotine replacement therapy (nicotine patch, nicotine gum, and nicotine lozenges). It is important to only use nicotine products to quit that is approved by the U.S. Food and Drug Administration (FDA).

Coping with Nicotine Withdrawal

Nicotine is one of the most addictive substances on earth and this is why it is so hard to quit. You feel a need for a cigarette when the level of nicotine in your body starts to drop. If you go for long periods of time between cigarettes, such as sleeping through the night, you will have a strong craving to smoke. You might feel the following effects when you are low on nicotine:

- Irritability, frustration, anger
- Anxiety
- Difficulty concentrating
- Restlessness
- Depressed mood
- Difficulty sleeping
- Increased appetite
- Coughing
- Runny nose
- Cravings/urges

Most of these symptoms start the first or second day you are off tobacco. They are at their worst in the first week and get better with time. Most symptoms disappear after two to four weeks, but the urge to smoke can stay with you for a long time. The urge to use tobacco will be stronger when you first quit and seem to last minutes. However, after the first two to four weeks, the urges will become shorter. For most people, the urge lasts only seconds after they have been off tobacco for a month or longer. Nicotine withdrawal symptoms can be managed by certain medications and behavioral coping strategies.

Table 2.1. Behavioral Strategies for Coping with Nicotine Withdrawal Symptoms

Withdrawal Symptoms	What Can I Do about It?
Irritability	• Avoid stress • Practice relaxation techniques • Exercise
Depressed mood	• Do something fun • Get support from family and friends • Discuss with your medical provider
Difficulty concentrating	• Avoid stress • Plan your work accordingly
Dizziness	• Get up slowly from sitting position

Table 2.1. Continued

Withdrawal Symptoms	What Can I Do about It?
Chest tightness	• Practice relaxation techniques
Fatigue	• Get more sleep • Take naps • Don't push yourself
Hunger	• Drink lots of water • Eat low-calorie snacks
Stomach pain, constipation, gas	• Drink fluids • Eat fruits and vegetables
Cough, dry throat, runny nose	• Drink fluids • Eat sugar-free candy • Use cough drops
Difficulty sleeping	• Reduce caffeine consumption (e.g., reduce daily intake by 50%)

How Tobacco Affects Your Body

Tobacco can be harmful to almost every part of your body. Many of these problems can be completely or almost completely reversed if you stop tobacco. Here is a list of the most common problems of tobacco use going from your head to your toes:

Head

- Stroke (blockage or breaking of a blood vessel in the brain)
- Mouth and throat cancers
- Cavities and loss of teeth
- Bad breath
- Decreased night vision
- Yellow staining of skin and teeth
- Nose congestion and infections
- Wrinkles

Lungs

- Cancer (up to 85% of all lung cancers are from smoking)
- Emphysema and chronic bronchitis

- Worsening of asthma
- Lung infections

Heart

- Congestive heart failure
- Heart attacks
- Increased blood pressure and heart rate

Stomach / Intestines

- Cancers
- Ulcers
- Heartburn

Pancreas

- Cancer

Circulation in Arms, Legs, and Feet

- Reduced circulation in arms, legs, and feet that sometimes leads to amputations in severe cases

Bones

- Increased bone thinning leading to a higher risk of broken bones

Genitals / Urinary System

- Cancers in kidneys, bladder, and reproductive organs
- Erectile dysfunction in men
- Sexual dysfunction in women

Recovery of Your Body after Stopping Tobacco

After you quit using tobacco products, your body will start to heal and you will start seeing many improvements. It is never too late to quit. You will start seeing benefits the very first day you quit using tobacco and these benefits will increase the longer you remain tobacco free.

- **20 minutes after you quit**—Reduction in your heart rate and blood pressure; the temperature of your hands and feet will start returning to normal.

- **12 hours after you quit**—Carbon monoxide level in your blood drops.

- **24 hours after you quit**—Anxiety and irritability may start due to withdrawal from nicotine. These symptoms get better the longer you are off tobacco.

- **2 to 3 days after you quit**—Nerve endings in your body start to regenerate and you may notice a return in your taste and smell. Anger, anxiety, and irritability from nicotine withdrawal may be at the worst level during this time. Nicotine replacement with nicotine gum or lozenges may help this. Breathing may be easier now.

- **1 week after you quit**—Tobacco cravings and urges may be less frequent and shorter in duration.

- **2 weeks after you quit**—Blood circulation in your gums and teeth are similar to a nonsmoker. You should no longer have anger, anxiety, and irritability from nicotine withdrawal. Cravings and urges should be shorter and less frequent.

- **1 to 3 months after you quit**—Your heart attack risk has started to drop and your lung function is improving. The blood circulation in your body has improved and walking might be easier. Give walking a try and see if you can go farther than when you were smoking. If you had a cough when you smoked, the cough should be gone now.

- **1 to 9 months after you quit**—Smoking-related nasal congestion, fatigue, and shortness of breath should be improving. Cilia (little hairs in the lungs, throat, and nose) have regrown in your lungs and can clean your lungs to remove irritants and mucus and reduce infections.

- **1 year after you quit**—The risk of cardiovascular disease, heart attack, and stroke has dropped to less than half that of a smoker.

- **10 to 15 years after you quit**—Your risk of having a stroke or heart attack has dropped to a similar rate as a nonsmoker. Your risk of lung cancer is 30 to 50 percent less than a continuing smoker's risk. Your risk of death from lung cancer is one-half of the risk if you were an average smoker (one pack per day). Your risk of pancreatic cancer is similar to a person who has not smoked and your risk of mouth, throat, and esophageal cancer has reduced significantly. Your risk of tooth loss has decreased to a rate similar to someone who has never smoked.

- **20 years after you quit (women)**—Your risk of death from smoking-related causes, including cancer and lung disease, is the same as a person who never smoked.

Section 2.5

Electronic Nicotine Delivery System

This section contains text excerpted from the following sources: Text beginning with the heading "Vaporizers, E-Cigarettes, and Other Electronic Nicotine Delivery Systems" is excerpted from "Vaporizers, E-Cigarettes, and Other Electronic Nicotine Delivery Systems (ENDS)," U.S. Food and Drug Administration (FDA), January 3, 2019; Text beginning with the heading "What's the Bottom Line?" is excerpted from "About Electronic Cigarettes (E-Cigarettes)," Centers for Disease Control and Prevention (CDC), November 15, 2018.

Vaporizers, E-Cigarettes, and Other Electronic Nicotine Delivery Systems

Vapes, vaporizers, vape pens, hookah pens, electronic cigarettes (e-cigarettes or e-cigs), and e-pipes are some of the many terms used to describe electronic nicotine delivery systems (ENDS). ENDS are noncombustible tobacco products.

These products use an "e-liquid" that may contain nicotine, as well as varying compositions of flavorings, propylene glycol, vegetable glycerin, and other ingredients. The liquid is heated to create an aerosol that the user inhales.

ENDS may be manufactured to look like conventional cigarettes, cigars, or pipes. Some resemble pens or Universal Serial Bus (USB) flash drives. Larger devices, such as tank systems or mods, bear little or no resemblance to cigarettes.

Statistics about E-Cigarette Use among U.S. Youth

- Among middle and high school students, 3.62 million were users of e-cigarettes in 2018.

- E-cigarette use, from 2017 to 2018, increased 78 percent among high school students (11.7% to 20.8%) and 48 percent among middle school students (3.3% to 4.9%) from 2017 to 2018.

- According to a 2013 to 2014 survey, 81 percent of youth e-cigarette users cited the availability of appealing flavors as the primary reason for use.

U.S. Food and Drug Administration Regulation of Electronic Nicotine Delivery System

In 2016, U.S. Food and Drug Administration (FDA) finalized a rule extending FDA's Center for Tobacco Products (CTP) regulatory authority to cover all tobacco products, including electronic nicotine delivery systems (ENDS) that meet the definition of a tobacco product. FDA regulates the manufacture, import, packaging, labeling, advertising, promotion, sale, and distribution of ENDS, including components and parts of ENDS but excluding accessories. Examples of components and parts of ENDS include:

- E-liquids
- A glass or plastic vial container of e-liquid
- Cartridges
- Atomizers
- Certain batteries
- Cartomizers and clearomizers
- Digital display or lights to adjust settings
- Tank systems
- Drip tips
- Flavorings for ENDS
- Programmable software

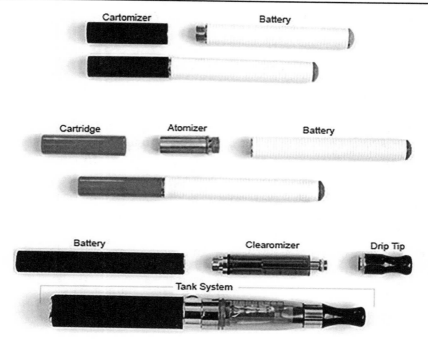

Figure 2.1. *Components of Electronic Nicotine Delivery System*

- However, products marketed for therapeutic purposes (for example, marketed as a product to help people quit smoking) are regulated by FDA through the Center for Drug Evaluation and Research (CDER). FDA published a rule clarifying the jurisdiction over tobacco products, drugs, and devices.

What's the Bottom Line?

- E-cigarettes have the potential to benefit adult smokers who are not pregnant if used as a complete substitute for regular cigarettes and other smoked tobacco products.

- E-cigarettes are not safe for youth, young adults, pregnant women, or adults who do not currently use tobacco products.

- While e-cigarettes have the potential to benefit some people and harm others, scientists still have a lot to learn about whether e-cigarettes are effective for quitting smoking.

- If you've never smoked or used other tobacco products or e-cigarettes, don't start.

- Additional research can help understand long-term health effects.

What Are E-Cigarettes?

- E-cigarettes come in many shapes and sizes. Most have a battery, a heating element, and a place to hold a liquid.

- E-cigarettes produce an aerosol by heating a liquid that usually contains nicotine—the addictive drug in regular cigarettes, cigars, and other tobacco products—flavorings, and other chemicals that help to make the aerosol. Users inhale this aerosol into their lungs.

- Bystanders can also breathe in this aerosol when the user exhales into the air.

- E-cigarettes are known by many different names. They are sometimes called "e-cigs," "e-hookahs," "mods," "vape pens," "vapes," "tank systems," and "electronic nicotine delivery systems (ENDS)."

- Some e-cigarettes are made to look like regular cigarettes, cigars, or pipes. Some resemble pens, USB sticks, and other everyday items. Larger devices such as tank systems, or "mods," do not resemble other tobacco products.

- Using an e-cigarette is sometimes called "vaping."

- E-cigarettes can be used to deliver marijuana and other drugs.

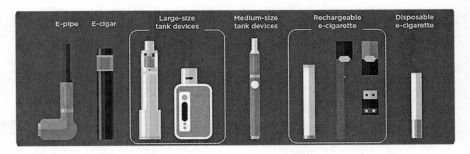

Figure 2.2. *E-Cigarettes*

Some e-cigarettes are made to look like regular cigarettes, cigars, or pipes. Some resemble pens, USB sticks, and other everyday items.

What Is in E-Cigarette Aerosol?

The e-cigarette aerosol that users breathe from the device and exhale can contain harmful and potentially harmful substances, including:

- Nicotine

- Ultrafine particles that can be inhaled deep into the lungs

- Flavoring such as diacetyl, a chemical linked to a serious lung disease

- Volatile organic compounds

- Cancer-causing chemicals

- Heavy metals such as nickel, tin, and lead

It is difficult for consumers to know what e-cigarette products contain. For example, some e-cigarettes marketed as containing zero percent nicotine have been found to contain nicotine.

What Are the Health Effects of Using E-Cigarettes?

E-cigarettes are still fairly new, and scientists are still learning about their long-term health effects.

- Most e-cigarettes contain nicotine, which has known health effects.

- Nicotine is highly addictive.

- Nicotine is toxic to developing fetuses.

- Nicotine can harm adolescent brain development, which continues into the early to mid-twenties.

- Nicotine is a health danger for pregnant women and their developing babies.

- Besides nicotine, e-cigarette aerosol can contain substances that harm the body.

- This includes cancer-causing chemicals and tiny particles that reach deep into the lungs. However, e-cigarette aerosol generally contains fewer harmful chemicals than smoke from burned tobacco products.

- E-cigarettes can cause unintended injuries.

- Defective e-cigarette batteries have caused fires and explosions, some of which have resulted in serious injuries. Most explosions happened when the e-cigarette batteries were being charged.

- The U.S. Food and Drug Administration (FDA) collects data to help address this issue. You can report an e-cigarette explosion, or any other unexpected health or safety issue with an e-cigarette.

- In addition, acute nicotine exposure can be toxic. Children and adults have been poisoned by swallowing, breathing, or absorbing e-cigarette liquid through their skin or eyes.

What Are the Risks of E-Cigarettes for Youth, Young Adults, and Pregnant Women?

Most e-cigarettes contain nicotine, which is addictive and toxic to developing fetuses. Nicotine exposure can also harm adolescent brain development, which continues into the early to mid-twenties. E-cigarette aerosol can contain chemicals that are harmful to the lungs. And youth e-cigarette use is associated with the use of other tobacco products, including cigarettes.

Are E-Cigarettes Less Harmful than Regular Cigarettes?

Yes—but that doesn't mean e-cigarettes are safe. E-cigarette aerosol generally contains fewer toxic chemicals than the deadly mix of 7,000 chemicals in smoke from regular cigarettes. However, e-cigarette aerosol is not harmless. It can contain harmful and potentially harmful substances, including nicotine, heavy metals like lead, volatile organic compounds, and cancer-causing agents.

Can E-Cigarettes Help Adults Quit Smoking Cigarettes?

E-cigarettes are not currently approved by the FDA as a quit smoking aid. The U.S. Preventive Services Task Force (USPSTF), a group of health experts that makes recommendations about preventive healthcare, has concluded that evidence is insufficient to recommend e-cigarettes for smoking cessation in adults, including pregnant women.

However, e-cigarettes may help nonpregnant adult smokers if used as a complete substitute for all cigarettes and other smoked tobacco products.

- To date, the few studies on the issue are mixed. A Cochrane Review found evidence from two randomized controlled trials that e-cigarettes with nicotine can help smokers stop smoking in the long-term compared with placebo (nonnicotine) e-cigarettes. However, there are some limitations to the existing research, including the small number of trials, small sample sizes, and wide margins of error around the estimates.

- A Centers for Disease Control and Prevention (CDC) study found that many adults are using e-cigarettes in an attempt to quit smoking. However, most adult e-cigarette users do not stop smoking cigarettes and are instead continuing to use both products (known as "dual use"). Dual use is not an effective way to safeguard your health, whether you're using e-cigarettes, smokeless tobacco, or other tobacco products in addition to regular cigarettes. Because smoking even a few cigarettes a day can be dangerous, quitting smoking completely is very important to protect your health.

Who Is Using E-Cigarettes?

E-cigarettes are the most commonly used tobacco product among youth.

- In the United States, youth are more likely than adults to use e-cigarettes.

- In 2018, more than 3.6 million U.S. middle- and high-school students used e-cigarettes in the past 30 days, including 4.9 percent of middle school students and 20.8 percent of high school students.

- In 2017, 2.8 percent of U.S. adults were e-cigarette users.

- In 2015, among adult e-cigarette users overall, 58.8 percent also were regular cigarette smokers, 29.8 percent were former regular cigarette smokers, and 11.4 percent had never been regular cigarette smokers.

- Among e-cigarette users aged 45 years and older in 2015, most were either newly started or former regular cigarette smokers, and 1.3 percent had never been cigarette smokers.

- In contrast, among e-cigarette users aged 18 to 24 years, 40.0 percent had never been regular cigarette smokers.

Section 2.6

Genes and Nicotine Addiction

This section includes text excerpted from "Studies Link Family
of Genes to Nicotine Addiction," National Institute on Drug
Abuse (NIDA), December 1, 2009. Reviewed February 2019.

One person reaches for a cigarette soon after waking, smokes a
pack a day, and cannot seem to quit. Another smokes a few cigarettes
now and then but never feels driven by the need for nicotine. A third
person smoked for a while in youth and then stopped. According to
several National Institute on Drug Abuse (NIDA) funded studies,
such contrasting smoking patterns, and responses may arise because
individuals inherit different forms of half a dozen genes that dictate
the features of the brain receptor to which nicotine binds.

Scientists have long known that nicotine produces many of its
effects by attaching to receptors for acetylcholine, a neurochemical
that influences memory, arousal, attention, and mood. This nicotinic acetylcholine (nACh) receptors comprise five subunits arranged
around a central pore, like sections of an orange. Each of the genes
identified by the studies provides the blueprint for one of a dozen

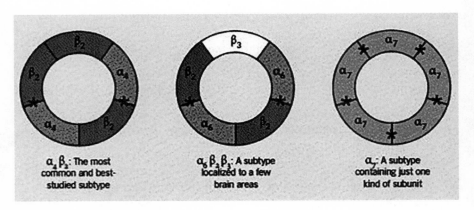

Figure 2.3. *Nicotinic Receptors*

*Nicotinic receptors vary in component proteins and activity. Nicotine initiates its
effects by binding to nicotinic acetylcholine (nACh) receptors, each consisting of five
proteins arranged in a circle around a central pore. The receptors occur in subtypes,
which differ in their constituent proteins and physiological and pharmacological characteristics. Asterisks indicate where nicotine and acetylcholine bind to each receptor
subtype.*

proteins, labeled α2-10 and β2-4, that serve as subunits in nACh receptors. Variations in the deoxyribonucleic acid (DNA) that encodes these genes may alter the structure or amount of the proteins produced, which in turn can modify what happens when nicotine molecules attach to the receptors.

Initially, research examining the influence of nACh receptor proteins on nicotine addiction focused on the α4 and β2 subunits. These are the most abundant and widely distributed nACh subunit proteins in the brain. Animal and human imaging studies have shown that nACh receptors consisting of two α4 and three β2 subunits are critical for the rewarding effects of nicotine.

The section highlight genes that code for less common nACh receptor proteins (see Table 2.2 below). Researchers have implicated the genes—located on chromosome 15—for the α3, α5, and β4 proteins in early initiation of smoking, the transition to dependence, and two smoking-related diseases: lung cancer and peripheral arterial disease (PAD). Investigators have also found that whether or not a person experiences extreme dizziness upon first trying cigarettes, as well as his or her risk of addiction, depends in part on the genes—on chromosome 8—for the α6 and β3 proteins. Taken together, the results suggest that genes for several nACh receptor proteins drive different aspects of the multistep process of nicotine addiction.

Development of an Addiction

The studies link genes for subunits of the nicotinic acetylcholine receptor to early smoking, initial responses to tobacco smoke, and vulnerability to addiction.

A study linked another aspect of risk with one of the α5 SNPs associated with nicotine dependence in the Weiss study. Drs. Bierut, Ovide Pomerleau, and colleagues found that regular smokers who had the higher-risk allele were more likely to recollect having had a pleasurable response the first time they smoked.

From the First Cigarette to Addiction

Drs. Bierut, Scott Saccone, and colleagues found that the genes for the α5 and β3 nACh receptor proteins affect an individual's risk of progressing from casual smoking to addiction. The 1,900 individuals who contributed DNA to their study had each smoked at least 100 cigarettes in their lifetime; about half were moderate to severely

Table 2.2. Development of an Addiction

Aspect of Smoking	Subunit Gene(s)	Type of Study	Participants	Researchers
Dizziness from first cigarette	β3	Gene association*	1,075 adolescent smokers and nonsmokers	Marissa A. Ehringer et al.
Pleasure from initial cigarette	α5	Gene association*	435 adult smokers	Laura Jean Bierut, Ovide Pomerleau, Richard Sherva, et al.
Age of smoking initiation	α5, β4	Gene association*	1,075 adolescent smokers and nonsmokers	M. A. Ehringer et al.
Increased risk of dependence among early smokers	α5	Candidate-gene**	2,827 long-term smokers	Robert B. Weiss et al.
Transition to dependence	β3	Genome-wide association***	1,929 smokers, Collaborative Genetic Study of Nicotine Dependence	L.J. Bierut, Scott Saccone, et al.
Transition to dependence	α3, α5, β3	Candidate-gene**	1,929 smokers, Collaborative Genetic Study of Nicotine Dependence	L.J. Bierut, S. Saccone, et al.
Transition to dependence	α5	Genome-wide association***	Approximately 15,000 adults	Wade Berrettini et al.
Transition to dependence	α3	Genome-wide association***	Approximately 14,000 smokers, 16,000 nonsmokers	Kári Stefánsson et al.
Lung cancer and peripheral arterial disease	α3	Genome-wide association***	Approximately 14,000 smokers, 16,000 nonsmokers	K. Stefánsson et al.

* Links genes with smoking by comparing the genetic markers of participants with and without the condition.
** Compares genes, selected on the basis of a demonstrated or hypothesized link, from individuals with and without the condition.
*** Considers genetic markers across the entire genome to compare participants with and without the condition.

dependent on nicotine, while the others had not developed a dependence on the drug, according to the Fagerstrom Test for Nicotine Dependence. The researchers looked for correlations between these divergent smoking histories and 4,000 SNPs within 348 genes that previous research had linked to nicotine dependence. The strongest associations were with five SNPs: two in the gene for the α3 nACh receptor subunit, two in the β3 subunit gene, and one in the α5 subunit gene. One of the α5 alleles carried double the risk of its alternative allele.

The trial participants who were not dependent on nicotine were unusually resistant to the drug's addictive effects, Dr. Bierut says. These are smokers who can quit at any time. Identifying the ways that the trial participants' genetic makeup protected them, even past the 100-cigarette threshold, could provide powerful clues to prevention and treatment.

"Our findings point to a role for the α3 and α5 proteins, as well as β4, in nicotine dependence," says Dr. Bierut. "The α5 proteins are expressed in the brain's reward areas, which makes our findings particularly intriguing."

Dr. Saccone says, "We are very excited to discover that genes encoding nicotinic acetylcholine receptor proteins confer risk for nicotine dependence because they have strong biological relevance to the addiction. My colleagues and I are using samples of DNA from different groups of people to confirm these findings; the next step is to determine exactly how alleles of these genes cause someone to keep smoking." Both St. Louis studies also found a link between the gene for the β3 protein—which was highlighted in the results of the Colorado studies—and nicotine dependence, bolstering confidence in this gene's involvement in the addiction.

Other researchers have corroborated and extended the St. Louis team's findings. For example, Dr. Wade Berrettini and colleagues at the University of Pennsylvania found that SNPs in the gene for the α5 protein influence the risk of smoking a pack of cigarettes a day as compared with smoking fewer than five cigarettes a day. A study led by Dr. Kari Stefansson of deCODE Genetics, a biopharmaceutical company based in Reykjavik, Iceland, also implicated an α3 gene SNP that influences nicotine dependence but not smoking initiation. Both of these studies employed a genome-wide association (GWA) scan methodology, which analyzes hundreds of thousands of SNPs simultaneously. Although GWA scans sometimes produce spurious associations, the convergence of these results with those of the St. Louis group provides strong evidence for the positive findings.

From Gene Discovery to Treatment

The SNPs identified in these studies may themselves affect nACh responses to nicotine by altering the form of their product proteins or their patterns of expression in different regions of the nervous system. Alternatively, further investigations may reveal that some of the SNPs are genetic bystanders that correlate with smoking behaviors only because they are inherited along with nearby, as yet unidentified DNA variants. Scientists plan to sort through these possibilities, pinpointing which alleles affect receptor responses to nicotine singly and which work in concert with others and determining how those altered responses promote or protect against nicotine addiction. A similar agenda applies to research on several genes that code for proteins that are unrelated to the nACh receptor but have recently been linked to addiction.

"Once scientists determine how these genetic variants affect nicotinic receptor function and behavioral responses to nicotine, they can develop pharmacotherapy interventions," says Dr. David Shurtleff of NIDA's Division of Basic Neuroscience and Behavioral Research. "Of the nicotinic receptors identified as novel targets in these studies, the α5 protein stands out. It seems to influence severe nicotine addiction," he says.

NIDA geneticist Dr. Joni Rutter agrees that the α5 receptor, in particular, is an interesting target. "This protein is not as abundant in the brain as other nicotinic receptor subtypes, so medications that target it might have few side effects and higher efficacies," she adds.

For people who have a genetic predisposition to various aspects of smoking addiction, the solution is simple, says Dr. Stefansson." It is only smoking that converts the risk into addiction and disease," he notes. "The ultimate preventive measure for these conditions is to never start smoking."

Chapter 3

What's in Tobacco Products?

Chapter Contents

41

Section 3.1

Overview of Tobacco Products

This section includes text excerpted from "Highlights: Tobacco Products," Centers for Disease Control and Prevention (CDC), July 21, 2015. Reviewed February 2019.

Fast Facts

- More than 4,000 chemical compounds have been identified in tobacco smoke. Of these, at least 43 are known to cause cancer.

- Current tobacco product regulation requires cigarette manufacturers to disclose levels of magnify tar and nicotine. Smokers receive very little information regarding chemical constituents in tobacco smoke, however, and the use of terms such as "light" and "ultra light" on packaging and in advertising may be misleading.

- Cigarettes with low tar and nicotine contents are not substantially less hazardous than higher-yield brands. Consumers may be misled by the implied promise of reduced toxicity underlying the marketing of such brands.

- Early data showed a lower cancer risk from low-tar cigarettes; however, more recent data suggest otherwise. Lower-yield cigarettes may be somewhat better than very high-yield cigarettes; but, when comparing full flavor cigarettes and current light cigarettes, there is no evidence to suggest a lower cancer risk from the low-tar cigarettes.

Cigarette Additives

- Federal law (the Comprehensive Smoking Education Act of 1984 and the Comprehensive Smokeless Tobacco Health Education Act of 1986) requires cigarette and smokeless tobacco manufacturers to submit a list of ingredients added to tobacco to the Secretary of Health and Human Services.

- Hundreds of ingredients are used in the manufacture of tobacco products. Additives make cigarettes more acceptable to the consumer—they make cigarettes milder and easier to inhale, improve taste, and prolong burning and shelf life.

- In 1994, six major cigarette manufacturers reported 599 ingredients that were added to the tobacco of manufacture cigarettes. Although these ingredients are regarded as safe when ingested in foods, some may form carcinogens when heated or burned.

- Knowledge about the impact of additives in tobacco products is negligible and will remain so as long as brand-specific information on the identity and quantity of additives is unavailable.

Smokeless Additives

- In 1994, 10 manufacturers of smokeless tobacco products released a list of additives used in their products. The additives list contained 562 ingredients approved for foods by the U.S. Food and Drug Administration (FDA).

- Moist snuff products with low nicotine content and pH levels have a smaller proportion of free nicotine. In contrast, moist snuff products with high nicotine content and pH levels have a higher proportion of free nicotine.

- The epidemiology of moist snuff use among teenagers and young adults indicates that most novices start with brands having low levels of free nicotine and then "graduate" to brands with higher levels.

- Sweeteners and flavorings, such as cherry juice concentrate, apple juice, chocolate liqueur, or honey are used in various smokeless tobacco products. As with manufactured cigarettes, these additives increase palatability and may increase the use of smokeless tobacco, at least among novices.

Section 3.2

Harmful Chemicals in Tobacco

This section contains text excerpted from the following sources: Text under the heading "Harmful and Potentially Harmful Constituents" is excerpted from "Harmful and Potentially Harmful Constituents," U.S. Food and Drug Administration (FDA), October 31, 2018; Text under the heading "Harmful and Potentially Harmful Constituents in Tobacco Products and Tobacco Smoke: Established List" is excerpted from "Harmful and Potentially Harmful Constituents in Tobacco Products and Tobacco Smoke: Established List," U.S. Food and Drug Administration (FDA), May 1, 2018; Text beginning with the heading "What Harmful Chemicals Does Tobacco Smoke Contain?" is excerpted from "Harms of Cigarette Smoking and Health Benefits of Quitting," National Cancer Institute (NCI), December 19, 2017.

Harmful and Potentially Harmful Constituents

The Food, Drug and Cosmetic Act (FD&C Act) requires tobacco manufacturers and importers to report the levels of harmful and potentially harmful constituents (HPHCs) found in their tobacco products and tobacco smoke. HPHCs are chemicals or chemical compounds in tobacco products or tobacco smoke that cause or could cause harm to smokers or nonsmokers.

U.S. Food and Drug Administration (FDA) must publish HPHC quantities in each brand and subbrand of tobacco product, in a way that people find understandable and not misleading. There are several efforts underway at the FDA to make progress toward that goal.

Preliminary Harmful and Potentially Harmful Constituents List

The FDA published a preliminary list of 93 HPHCs in March 2012. This HPHC list focuses on chemicals that are linked to the five most serious health effects of tobacco use (cancer, cardiovascular disease, respiratory effects, reproductive problems, and addiction).

Tobacco Industry Reporting Requirements

The FDA issued draft guidance in 2012 that identified a subset of 20 HPHCs for which manufacturers and importers are to test and report to the FDA. The FDA chose these 20 because testing methods were well established and widely available. The FDA is now evaluating the quality and reliability of the data submitted by manufacturers.

Communicating to the Public

Making sure the public can clearly understand the real and potential risks of tobacco use is an important goal. Presently, the FDA is conducting research about how best to ensure that the public is made aware of the dangers of the chemicals and chemical compounds in tobacco products and smoke and to communicate the levels of HPHCs in each brand and subbrand of tobacco product. In the meantime, the FDA is including messages about HPHCs in its ongoing public health campaigns. The FDA has also created three videos and interactive tools to lay the foundation for an important public health goal: we aim to publish a list of the levels of harmful and potentially harmful chemicals in tobacco, in a way that is easy for the public to understand. As an important step toward that goal, we invite you to explore the chemicals in tobacco in three stages of cigarettes, from plant to product to puff.

Harmful and Potentially Harmful Constituents in Tobacco Products and Tobacco Smoke: Established List

The FDA has established a list of harmful and potentially harmful constituents (HPHCs) in tobacco products and tobacco smoke (the established HPHC list) as required by the FD&C Act.

The established list of 93 HPHCs is included in the notice and in table 3.1 below.

Table 3.1. Established List of the Chemicals and Chemical Compounds Identified by the FDA as Harmful and Potentially Harmful Constituents in Tobacco Products and Tobacco Smoke

Constituent	Carcinogen (CA), Respiratory Toxicant (RT), Cardiovascular Toxicant (CT), Reproductive or Developmental Toxicant (RDT), Addictive (AD)
Acetaldehyde	CA, RT, AD
Acetamide	CA
Acetone	RT
Acrolein	RT, CT
Acrylamide	CA
Acrylonitrile	CA, RT

Table 3.1. Continued

Constituent	Carcinogen (CA), Respiratory Toxicant (RT), Cardiovascular Toxicant (CT), Reproductive or Developmental Toxicant (RDT), Addictive (AD)
Aflatoxin B1	CA
4-Aminobiphenyl	CA
1-Aminonaphthalene	CA
2-Aminonaphthalene	CA
Ammonia	RT
Anabasine	AD
o-Anisidine	CA
Arsenic	CA, CT, RDT
A-α-C (2-Amino-9H-pyrido[2,3-b]indole)	CA
Benz[a]anthracene	CA, CT
Benz[j]aceanthrylene	CA
Benzene	CA, CT, RDT
Benzo[b]fluoranthene	CA, CT
Benzo[k]fluoranthene	CA, CT
Benzo[b]furan	CA
Benzo[a]pyrene	CA
Benzo[c]phenanthrene	CA
Beryllium	CA
1,3-Butadiene	CA, RT, RDT
Cadmium	CA, RT, RDT
Caffeic acid	CA
Carbon monoxide	RDT
Catechol	CA
Chlorinated dioxins/furans	CA, RDT
Chromium	CA, RT, RDT
Chrysene	CA, CT
Cobalt	CA, CT
Coumarin	Banned in food
Cresols (o-, m-, and p-cresol)	CA, RT
Crotonaldehyde	CA
Cyclopenta[c,d]pyrene	CA

Table 3.1. Continued

Constituent	Carcinogen (CA), Respiratory Toxicant (RT), Cardiovascular Toxicant (CT), Reproductive or Developmental Toxicant (RDT), Addictive (AD)
Dibenz[a,h]anthracene	CA
Dibenzo[a,e]pyrene	CA
Dibenzo[a,h]pyrene	CA
Dibenzo[a,i]pyrene	CA
Dibenzo[a,l]pyrene	CA
2,6-Dimethylaniline	CA
Ethyl carbamate (urethane)	CA, RDT
Ethylbenzene	CA
Ethylene oxide	CA, RT, RDT
Formaldehyde	CA, RT
Furan	CA
Glu-P-1 (2-Amino-6-methyldipyrido[1,2-a:3',2'-d]imidazole)	CA
Glu-P-2 (2-Aminodipyrido[1,2-a:3',2'-d]imidazole)	CA
Hydrazine	CA, RT
Hydrogen cyanide	RT, CT
Indeno[1,2,3-cd]pyrene	CA
IQ (2-Amino-3-methylimidazo[4,5-f]quinoline)	CA
Isoprene	CA
Lead	CA, CT, RDT
MeA-α-C (2-Amino-3-methyl)-9H-pyrido[2,3-b]indole)	CA
Mercury	CA, RDT
Methyl ethyl ketone	RT
5-Methylchrysene	CA
4-(Methylnitrosamino)-1-(3-pyridyl)-1-butanone (NNK)	CA
Naphthalene	CA, RT
Nickel	CA, RT
Nicotine	RDT, AD

Table 3.1. Continued

Constituent	Carcinogen (CA), Respiratory Toxicant (RT), Cardiovascular Toxicant (CT), Reproductive or Developmental Toxicant (RDT), Addictive (AD)
Nitrobenzene	CA, RT, RDT
Nitromethane	CA
2-Nitropropane	CA
N-Nitrosodiethanolamine (NDELA)	CA
N-Nitrosodiethylamine	CA
N-Nitrosodimethylamine (NDMA)	CA
N-Nitrosomethylethylamine	CA
N-Nitrosomorpholine (NMOR)	CA
N-Nitrosonornicotine (NNN)	CA
N-Nitrosopiperidine (NPIP)	CA
N-Nitrosopyrrolidine (NPYR)	CA
N-Nitrososarcosine (NSAR)	CA
Nornicotine	AD
Phenol	RT, CT
PhIP (2-Amino-1-methyl-6-phenylimidazo[4,5-b]pyridine)	CA
Polonium-210	CA
Propionaldehyde	RT, CT
Propylene oxide	CA, RT
Quinoline	CA
Selenium	RT
Styrene	CA
o-Toluidine	CA
Toluene	RT, RDT
Trp-P-1 (3-Amino-1,4-dimethyl-5H-pyrido[4,3-b]indole)	CA
Trp-P-2 (1-Methyl-3-amino-5H-pyrido[4,3-b]indole)	CA
Uranium-235	CA, RT
Uranium-238	CA, RT
Vinyl acetate	CA, RT
Vinyl chloride	CA

What Harmful Chemicals Does Tobacco Smoke Contain?

Tobacco smoke contains many chemicals that are harmful to both smokers and nonsmokers. Breathing even a little tobacco smoke can be harmful.

Of the more than 7,000 chemicals in tobacco smoke, at least 250 are known to be harmful, including hydrogen cyanide, carbon monoxide, and ammonia.

Among the 250 known harmful chemicals in tobacco smoke, at least 69 can cause cancer. These cancer-causing chemicals include the following:

- Acetaldehyde
- Aromatic amines
- Arsenic
- Benzene
- Beryllium (a toxic metal)
- 1,3-Butadiene (a hazardous gas)
- Cadmium (a toxic metal)
- Chromium (a metallic element)
- Cumene
- Ethylene oxide
- Formaldehyde
- Nickel (a metallic element)
- Polonium-210 (a radioactive chemical element)
- Polycyclic aromatic hydrocarbons (PAHs)
- Tobacco-specific nitrosamines
- Vinyl chloride

What Are Some of the Health Problems Caused by Cigarette Smoking?

Smoking is the leading cause of premature, preventable death in this country. Cigarette smoking and exposure to tobacco smoke cause about 480,000 premature deaths each year in the United States. Of those premature deaths, about 36 percent are from cancer, 39 percent are from heart disease and stroke, and 24 percent are from lung disease. Mortality rates among smokers are about three times higher than among people who have never smoked.

Smoking harms nearly every bodily organ and organ system in the body and diminishes a person's overall health. Smoking causes cancers of the lung, esophagus, larynx, mouth, throat, kidney, bladder, liver, pancreas, stomach, cervix, colon, and rectum, as well as acute myeloid leukemia (AML).

Smoking also causes heart disease, stroke, aortic aneurysm (a balloon-like bulge in an artery in the chest), chronic obstructive pulmonary disease (COPD) (chronic bronchitis and emphysema), diabetes, osteoporosis, rheumatoid arthritis, age-related macular degeneration (AMD), and cataracts, and worsens asthma symptoms in adults. Smokers are at higher risk of developing pneumonia, tuberculosis, and other airway infections. In addition, smoking causes inflammation and impairs immune function.

Since the 1960s, a smoker's risk of developing lung cancer or COPD has actually increased compared with nonsmokers, even though the number of cigarettes consumed per smoker has decreased. There have also been changes over time in the type of lung cancer smokers develop—a decline in squamous cell carcinomas but a dramatic increase in adenocarcinomas. Both of these shifts may be due to changes in cigarette design and composition, in how tobacco leaves are cured, and in how deeply smokers inhale cigarette smoke and the toxicants it contains.

Smoking makes it harder for a woman to get pregnant. A pregnant smoker is at higher risk of miscarriage, having an ectopic pregnancy, having her baby born too early and with an abnormally low birth weight (LBW), and having her baby born with a cleft lip and/or cleft palate. A woman who smokes during or after pregnancy increases her infant's risk of death from Sudden Infant Death Syndrome (SIDS). Men who smoke are at greater risk of erectile dysfunction.

The longer a smoker's duration of smoking, the greater their likelihood of experiencing harm from smoking, including earlier death. But regardless of their age, smokers can substantially reduce their risk of disease, including cancer, by quitting.

What Are the Risks of Tobacco Smoke to Nonsmokers?

Secondhand smoke (also called environmental tobacco smoke, involuntary smoking, and passive smoking) is the combination of "sidestream" smoke (the smoke given off by a burning tobacco product) and "mainstream" smoke (the smoke exhaled by a smoker).

The U.S. Environmental Protection Agency (EPA), the U.S. National Toxicology Program (NTP), the U.S. Surgeon General, and the International Agency for Research on Cancer (IARC) have classified secondhand smoke as a known human carcinogen (cancer-causing agent). Inhaling secondhand smoke causes lung cancer in nonsmoking adults. Approximately 7,300 lung cancer deaths occur each year among adult nonsmokers in the United States as a result of exposure to secondhand smoke. The U.S. Surgeon General estimates that living with a smoker increases a nonsmoker's chances of developing lung cancer by 20 to 30 percent.

Secondhand smoke causes disease and premature death in non-smoking adults and children. Exposure to secondhand smoke irritates the airways and has immediate harmful effects on a person's heart and blood vessels. It increases the risk of heart disease by an estimated 25 to 30 percent. In the United States, exposure to secondhand smoke is estimated to cause about 34,000 deaths from heart disease each year. Exposure to secondhand smoke also increases the risk of stroke by 20 to 30 percent. Pregnant women exposed to secondhand smoke are at increased risk of having a baby with a small reduction in birth weight.

Children exposed to secondhand smoke are at an increased risk of SIDS, ear infections, colds, pneumonia, and bronchitis. Secondhand smoke exposure can also increase the frequency and severity of asthma symptoms among children who have asthma. Being exposed to second-hand smoke slows the growth of children's lungs and can cause them to cough, wheeze, and feel breathless.

Is Smoking Addictive?

Smoking is highly addictive. Nicotine is the drug primarily responsible for a person's addiction to tobacco products, including cigarettes. The addiction to cigarettes and other tobacco products that nicotine causes is similar to the addiction produced by using drugs such as heroin and cocaine. Nicotine is present naturally in the tobacco plant. But tobacco companies intentionally design cigarettes to have enough nicotine to create and sustain addiction.

The amount of nicotine that gets into the body is determined by the way a person smokes a tobacco product and by the nicotine content and design of the product. Nicotine is absorbed into the bloodstream through the lining of the mouth and the lungs and travels to the brain in a matter of seconds. Taking more frequent and deeper puffs of tobacco smoke increases the amount of nicotine absorbed by the body.

Are Other Tobacco Products, Such as Smokeless Tobacco or Pipe Tobacco, Harmful and Addictive?

Yes. All forms of tobacco are harmful and addictive. There is no safe tobacco product.

In addition to cigarettes, other forms of tobacco include smokeless tobacco, cigars, pipes, hookahs (waterpipes), bidis, and kreteks.

- **Smokeless tobacco:** Smokeless tobacco is a type of tobacco that is not burned. It includes chewing tobacco, oral tobacco, spit or spitting tobacco, dip, chew, snus, dissolvable tobacco, and snuff. Smokeless tobacco causes oral (mouth, tongue, cheek, and gum), esophageal, and pancreatic cancers and may also cause gum and heart disease.

- **Cigars:** These include premium cigars, little-filtered cigars (LFCs), and cigarillos. LFCs resemble cigarettes, but both LFCs and cigarillos may have added flavors to increase appeal to youth and young adults. Most cigars are composed primarily of a single type of tobacco (air-cured and fermented), and have a tobacco leaf wrapper. Studies have found that cigar smoke contains higher levels of toxic chemicals than cigarette smoke, although unlike cigarette smoke, cigar smoke is often not inhaled. Cigar smoking causes cancer of the oral cavity, larynx, esophagus, and lung. It may also cause cancer of the pancreas. Moreover, daily cigar smokers, particularly those who inhale, are at increased risk for developing heart disease and other types of lung disease.

- **Pipes:** In pipe smoking, the tobacco is placed in a bowl that is connected to a stem with a mouthpiece at the other end. The smoke is usually not inhaled. Pipe smoking causes lung cancer and increases the risk of cancers of the mouth, throat, larynx, and esophagus.

- **Hookah or waterpipe (other names include argileh, shelton, hubble bubble, shisha, booty, goza, and narghile):** A hookah is a device used to smoke tobacco (often heavily flavored) by passing the smoke through a partially filled water bowl before being inhaled by the smoker. Although some people think hookah smoking is less harmful and addictive than cigarette smoking, research shows that hookah smoke is at least as toxic as cigarette smoke.

- **Bidis:** A bidi is a flavored cigarette made by rolling tobacco in a dried leaf from the tendu tree, which is native to India. Bidi use is associated with heart attacks and cancers of the mouth, throat, larynx, esophagus, and lung.

- **Kreteks:** A kretek is a cigarette made with a mixture of tobacco and cloves. Smoking kreteks is associated with lung cancer and other lung diseases.

Is It Harmful to Smoke Just a Few Cigarettes a Day?

There is no safe level of smoking. Smoking even just one cigarette per day over a lifetime can cause smoking-related cancers (lung, bladder, and pancreas) and premature death.

What Are the Immediate Health Benefits of Quitting Smoking?

The immediate health benefits of quitting smoking are substantial:

- Heart rate and blood pressure, which are abnormally high while smoking, begin to return to normal.

- Within a few hours, the level of carbon monoxide in the blood begins to decline. (Carbon monoxide reduces the blood's ability to carry oxygen.)

- Within a few weeks, people who quit smoking have improved circulation, produce less phlegm, and don't cough or wheeze as often.

- Within several months of quitting, people can expect substantial improvements in lung function.

- Within a few years of quitting, people will have lower risks of cancer, heart disease, and other chronic diseases than if they had continued to smoke.

What Are the Long-Term Health Benefits of Quitting Smoking?

Quitting smoking reduces the risk of cancer and many other diseases, such as heart disease and COPD, caused by smoking.

Data from the U.S. National Health Interview Survey (NHIS) show that people who quit smoking, regardless of their age, are less likely to die from smoking-related illness than those who continue to smoke. Smokers who quit before age 40 reduce their chance of dying prematurely from smoking-related diseases by about 90 percent, and those who quit by age 45 to 54 reduce their chance of dying prematurely by about two-thirds.

Regardless of their age, people who quit smoking have substantial gains in life expectancy, compared with those who continue to smoke. Data from the U.S. National Health Interview Survey also show that those who quit between the ages of 25 and 34 years live about 10 years longer; those who quit between ages 35 and 44 live about 9 years longer; those who quit between ages 45 and 54 live about 6 years longer; and those who quit between ages 55 and 64 live about 4 years longer.

Also, a study that followed a large group of people age 70 and older found that even smokers who quit smoking in their 60s had a lower risk of mortality during followup than smokers who continued smoking.

Does Quitting Smoking Lower the Risk of Getting and Dying from Cancer?

Yes. Quitting smoking reduces the risk of developing and dying from cancer and other diseases caused by smoking. Although it is never too late to benefit from quitting, the benefit is greatest among those who quit at a younger age.

The risk of premature death and the chances of developing and dying from a smoking-related cancer depend on many factors, including the number of years a person has smoked, the number of cigarettes smoked per day, and the age at which the person began smoking.

Is It Important for Someone Diagnosed with Cancer to Quit Smoking?

Quitting smoking improves the prognosis of cancer patients. For patients with some cancers, quitting smoking at the time of diagnosis may reduce the risk of dying by 30 to 40 percent. For those having surgery, chemotherapy, or other treatments, quitting smoking helps improve the body's ability to heal and respond to therapy. It also lowers the risk of pneumonia and respiratory failure. In addition, quitting smoking may lower the risk that the cancer will recur, that a second cancer will develop, or that the person will die from the cancer or other causes.

Where Can I Get Help to Quit Smoking?

The National Cancer Institute (NCI) and other agencies and organizations can help smokers quit:

- Visit Smokefree.gov for access to free information and resources, including Create My Quit Plan, smartphone apps, and text message programs

- Call the NCI Smoking Quitline at 877-44U-QUIT (877-448-7848) for individualized counseling, printed information, and referrals to other sources

Section 3.3

Flavors in Tobacco Products

This section includes text excerpted from "Menthol and Other Flavors in Tobacco Products," U.S. Food and Drug Administration (FDA), July 20, 2018.

Menthol and Other Flavors in Tobacco Products

Flavors are added to tobacco products to improve flavor and taste by reducing the harshness, bitterness, and astringency.

However, the use of flavors in tobacco products raises important public health questions. For example, U.S. Food and Drug Administration (FDA) is aware of early reports that some flavors could help adult cigarette smokers switch to potentially less harmful tobacco products. On the flip side, research has shown that sweet-tasting flavors are particularly appealing to youth and young adults.

The FDA has already banned cigarettes with certain kid-appealing flavors and is examining options for regulating other flavored tobacco products, including menthol cigarettes. The FDA is dedicated to understanding how flavors influence tobacco use and addiction, with the goal of identifying regulatory actions that will best protect the public's health based on their net impact.

Menthol

Menthol is a flavor additive with a minty taste and aroma that is widely used in consumer and medicinal products due to its reported cooling or painkilling properties. When used in cigarettes, menthol may reduce the irritation and harshness of smoking. However, research suggests menthol cigarettes may be harder to quit than nonmenthol cigarettes, particularly among African American smokers.

In the United States:

- 19.7 million people are current smokers of menthol cigarettes.

- 84.6 percent of African American smokers, 44.4 percent of Hispanic smokers, 37.5 percent of Asian smokers, and 28.5 percent of White smokers smoke menthol cigarettes.

- Youth who smoke are more likely to smoke menthol cigarettes than older smokers. More than half of smokers ages 12 to 17 smoke menthols.

Menthol is also used in other tobacco products, such as cigars, hookah (waterpipe) tobacco, smokeless tobacco (dip, chew, snuff, and snus), and e-cigarettes and other electronic nicotine delivery systems (ENDS).

Other Flavors in Tobacco Products

In 2009, the FDA banned cigarettes with characterizing flavors other than menthol (e.g., cherry, chocolate), which are known to appeal to youth and young adults. This ban was an important first step for responsible tobacco regulation to protect the American public, particularly children, from the dangers of cigarettes—the product most responsible for tobacco-related death and disease in the United States.

Currently, no flavors are banned from other tobacco products, although research suggests flavors may also make these products more enticing to youth and young adults. Data from the FDA's Population Assessment of Tobacco and Health found that nearly 80 percent of youth ages 12 to 17 and nearly 75 percent of young adults ages 18 to 25 who were current tobacco users in 2014 reported that the first tobacco product they ever used was flavored. Alternatively, the FDA is aware of self-reported information suggesting that the availability of flavors in some noncombusted tobacco products such as e-cigarettes and other ENDS may help some adult users reduce cigarette use or switch to potentially less harmful products.

Supporting Research to Understand the Role of Menthol and Other Flavors in Tobacco Use

The FDA is committed to a science-based approach that addresses the public health questions and issues raised by menthol and other tobacco product flavors. The FDA supports a wide range of research to understand the differences between menthol and nonmenthol cigarettes, as well as the full spectrum of other flavored tobacco products, such as cigars, e-cigarettes and other ENDS, and hookah (waterpipe) tobacco.

New scientific research, as well as information received through public comments on these topics, will help the FDA make informed decisions about appropriate regulatory actions.

Section 3.4

Tobacco Ingredient Reporting

This section includes text excerpted from "Tobacco Ingredient and Nicotine Reporting," Centers for Disease Control and Prevention (CDC), February 20, 2018.

Description of the Laws
The Federal Cigarette Labeling and Advertising Act and Comprehensive Smokeless Tobacco Health Education Act
Cigarettes

The Federal Cigarette Labeling and Advertising Act (FCLAA), 15 U.S.C. Section 1335a(a), Public Law 89–92, in part requires that each person who manufactures, packages, or imports cigarettes annually submit to the Department of Health and Human Services (HHS) a list of ingredients added to tobacco in the manufacture of cigarettes (Ingredient Report). The Centers for Disease Control and Prevention (CDC), Office on Smoking and Health (OSH), has been delegated the responsibility of implementing these provisions. The Ingredient Report must include all additives and flavors. Submissions are due to CDC, OSH by

March 31; and for importers, the Ingredient Report is also due upon initial importation into the United States. The report submitted by March 31st each year must represent the ingredients added to tobacco in the manufacture of cigarettes during the previous calendar year.

Under FCLAA, one may submit information which does not identify the company that uses the ingredients or the brand of cigarettes that contain the ingredients. A person or group of persons required to provide the ingredient list may designate an individual or entity to provide the list. In the event that another individual or entity, such as an attorney or counsel, is designated to submit this information on your behalf, this individual or entity should clearly state the name(s) of the importer(s) or company(ies) for whom they are submitting the information.

Smokeless Tobacco

Comprehensive Smokeless Tobacco Health Education Act (CSTHEA), 15 U.S.C. §4403(a)(A), Public Law 99–252, in part requires each manufacturer, packager, or importer of smokeless tobacco products to annually submit to HHS the list of ingredients added to tobacco in the manufacture of smokeless tobacco products (Ingredient Report). CSTHEA, 15 U.S.C. §4403(a)(1)(B), further requires the submission of the quantity of nicotine contained in each smokeless tobacco product (Nicotine Report). CDC, OSH has been delegated the responsibility of implementing these provisions. Submissions are due to CDC, OSH by March 31; and for importers, the Ingredient Report is also due upon initial importation into the United States. The reports submitted by March 31st each year must represent the ingredients added to tobacco in the manufacture of smokeless tobacco during the previous calendar year; and the specification of the quantity of nicotine contained in smokeless tobacco products manufactured or imported during the previous calendar year.

Under CSTHEA, one may submit information which does not identify the company that uses the ingredients or the brand of smokeless tobacco that contains the ingredients. A person or group of persons required to provide the nicotine data and ingredient list may designate an individual or entity to provide the list. In the event that you designate another individual or entity, such as an attorney or counsel, to submit this information on your behalf, this individual or entity should clearly state the name(s) of the company(ies) for whom they are submitting the information.

Reporting Instructions

The requirements for submission and the format for submitting the Ingredient Report for cigarettes and the Ingredient and Nicotine

Reports for smokeless tobacco, and a specification of the quantity of nicotine for smokeless tobacco, are available in the Federal Register.

Certification
Deadline

As detailed in 64 FR 14086, March 23, 1999; and 66 FR 17559, April 2, 2001, all submissions required under the FCLAA and CSTHEA for cigarettes and smokeless tobacco are due upon initial importation and annually thereafter by March 31. Submissions to the CDC's Office on Smoking and Health are reflective of ingredient information required by FCLAA and CSTHEA during the previous calendar year.

Once an accurate submission of an Ingredient Report and Nicotine Report (where applicable) have been received, the CDC will issue a Certificate of Compliance valid until March 31 of the following year.

Note: All faxed submissions should be immediately followed with a mailed original.

Report Format

The CDC requests all submissions be on letterhead of the manufacturer, packager, importer, respective counsel, or designated individual or entity.

Because the CDC cannot ensure the confidentiality of information submitted via e-mail, that is not an acceptable format. However, submission of data by way of mailing a CD, 3-inch floppy disk, or thumb drive is acceptable.

Reports may also be submitted via facsimile, but all faxed lists should be followed-up with a mailed original.

To the best of our knowledge, laboratory analysis is not available that will provide a complete representation of the ingredients added to tobacco in the manufacture of cigarettes or smokeless tobacco. Laboratory analysis in lieu of the Ingredient Report is not acceptable.

Note: If no ingredients are added to tobacco in the manufacture of cigarettes or smokeless tobacco, a statement to that effect must be submitted in writing.

Categories

The reporting status of manufacturers, packagers, and importers will be coded by the CDC as either "compliant," "noncompliant," or "inactive."

Compliant

Importers of cigarettes and smokeless tobacco products must submit a list of ingredients added to tobacco in the manufacture of the product upon initial importation of said product. Additionally, all manufacturers, packagers, and importers must annually submit by March 31 a list of ingredients added to tobacco in the manufacture of cigarettes and smokeless tobacco products during the previous calendar year.

A Certificate of Compliance will be issued for submissions that meet all of the following requirements:

The submission clearly states on whose behalf the submission is made.

The list of ingredients, including chemical names and corresponding Chemical Abstract Service (CAS) registry numbers, added to tobacco in the manufacture of cigarettes and/or smokeless tobacco products is complete and without error.

Example:

Chemical Name: Menthol
CAS Number: 89-78-1

The submission is signed and certified as correct by the submitter.

The submission for smokeless tobacco products contains a specification of the quantity of nicotine through the reporting of the amount of total nicotine, amount of unionized nicotine and percentage of unionized nicotine, total moisture, and pH for each smokeless product.

Noncompliant

Failure to provide the annual Ingredient Report and Nicotine Report (for smokeless tobacco) by the March 31 deadline and failure to correct inadequacies or errors in a submission within 60 days of notification will result in the CDC deeming the manufacturer, packager, or importer noncompliant.

In addition, if a report is submitted by a designated individual, the manufacturer, packager, or importer on whose behalf the report is submitted must be identified with the submission. Otherwise, the company will be deemed noncompliant.

Noncompliant status will be changed upon receipt of required information.

Inactive

Companies are encouraged to inform the CDC if they are no longer manufacturing, packaging, or importing tobacco products. The CDC communicates with other federal agencies involved in the regulation of tobacco products and will share the information of a company's status as appropriate.

Other Requirements

In addition to the requirements detailed here, manufacturers, packagers, and importers of tobacco products may have additional legal obligations to consider. Although not an exhaustive list, other federal agencies that may have applicable laws include the Alcohol and Tobacco Tax and Trade Bureau (TTB), U.S. Customs (Customs), and the Federal Trade Commission (FTC).

Frequently Asked Questions

What Is the Federal Cigarette Labeling and Advertising Act (FCLAA)?

The FCLAA, Public Law 89–92, is a statute that was designed to "establish a comprehensive federal program to deal with cigarette labeling and advertising with respect to any relationship between smoking and health." In 1984, this Act was amended by PL 98–474, the Comprehensive Smoking Education Act, 15 U.S.C. § 1335a(a) to in part provide the Department of Health and Human Services with a list of the ingredients added to tobacco in the manufacture of cigarettes.

What Is the Comprehensive Smokeless Tobacco Health Education Act (CSTHEA)?

The CSTHEA, Public Law 98–252, was passed in 1986 to inform the public of any dangers to human health resulting from the use of smokeless tobacco products. CSTHEA requires, in part, that each person who manufactures, packages, or imports smokeless tobacco products to annually submit to the Department of Health and Human Services the list of ingredients added to tobacco in the manufacture of smokeless tobacco products as well as a specification of the quantity of nicotine contained in each smokeless tobacco product.

What Products Are Covered by FCLAA and CSTHEA?

The FCLAA defines a cigarette as "any roll of tobacco wrapped in paper or in any substance not containing tobacco, and any roll of tobacco wrapped in any substance containing tobacco which, because

61

of its appearance, the type of tobacco used in the filler, or its packaging and labeling, is likely to be offered to, or purchased by, consumers as a cigarette."

The CSTHEA defines smokeless tobacco as "any finely cut, ground, powdered, or leaf tobacco that is intended to be placed in the oral cavity."

What Tobacco Products Are Not Covered by FCLAA or CSTHEA?

The FCLAA and CSTHEA do not apply to cigars or cut rag tobacco unless the cut rag tobacco is packaged as a final product for consumption (i.e., no further manipulation is required in order to be consumed). Other items not covered by FCLAA or CSTHEA include little cigars; or roll-your-own, hookah, or pipe tobacco.

To Whom Does FCLAA and CSTHEA Apply?

Manufacturers, packagers, and importers of cigarettes and smokeless tobacco products.

Who Receives the Ingredient Report and the Nicotine Report?

Both FCLAA and CSTHEA require manufacturers, packagers, and importers to annually report ingredients added to tobacco in the manufacture of cigarettes and smokeless tobacco to the Secretary of Health and Human Services. In turn, the Secretary has delegated this responsibility to the Centers for Disease Control and Prevention, Office on Smoking and Health (OSH), as stated in 50 FR 49617, December 3, 1985; 59 FR 4717, February 1, 1994, respectively.

Are the Submission Requirements the Same for Cigarettes and Smokeless Tobacco Products?

Both FCLAA and CSTHEA require manufacturers, packagers, and importers to annually report ingredients added to tobacco in the manufacture of cigarettes and smokeless tobacco.

However, CSTHEA additionally requires manufacturers, packagers, and importers of smokeless tobacco products to annually report on the specific quantity of nicotine in these products by submitting data on total nicotine, unionized nicotine, total moisture, and pH.

What Does a Nicotine Report Consist Of?

The Nicotine Report is a specification of the quantity of nicotine (total nicotine, unionized nicotine, total moisture, and pH) contained in smokeless tobacco products from the previous calendar year. A uniform analytical protocol was developed that consists of standard laboratory methods to measure nicotine, moisture, and pH in smokeless tobacco products, and an equation to calculate unionized nicotine. It includes standardized parameters for pH, moisture, and nicotine determination (sample size, sample preparation, quantity and purity of standards and reagents, instrumentation, measurement time and conditions, etc.,). It is the responsibility of the manufacturer, packager, or importer to submit the Nicotine Report according to the specifications set forth in the protocol.

When Are Submissions Due?

All submissions are due annually by March 31 and upon initial import. These submissions reflect ingredients added to cigarettes and smokeless tobacco products which are manufactured, packaged, or imported in the previous calendar year; as well as a specification of the quantity of nicotine contained in smokeless tobacco products manufactured or imported during the previous calendar year.

Can I Submit My Ingredient Report and Nicotine Report Electronically?

Because the CDC cannot ensure the confidentiality of information submitted via E-mail, this is not an acceptable format. However, submission of data by way of mailing a CD, 3-inch floppy disk, or thumb drive is acceptable. Reports may also be submitted via facsimile, but all faxed lists should be followed-up with a mailed original.

What Are Chemical Abstract Service (CAS) Numbers?

Each chemical compound has a unique CAS number for purposes of identification, assigned by the American Chemical Society. A CAS registry is available from the National Institutes of Health National Library of Medicine at www.chem.sis.nlm.nih.gov/chemidplus/chemid-heavy.jsp. Initially natural additives and flavors were not given CAS registry numbers; however, because these ingredients are now regularly used in manufacturing, many have been assigned CAS registry numbers.

Flavors made of multiple chemicals will have a separate CAS registry number for each chemical in that compound.

Why Must CAS Numbers Be Included with the Ingredient Report?

CAS numbers are required to properly identify and reduce ambiguity among ingredients used in tobacco products. This is particularly important, as an ingredient with a common name may have different chemical names and thus, different CAS assignments. An example of this is sugar, whose chemical names can be fructose, glucose, or sucrose.

Do I Have to Report Information That Is Trade Secret or Confidential?

Yes, the U.S. Congress gave full weight and consideration to the sensitive nature of tobacco ingredients in drafting FCLAA and CSTHEA. The list of ingredients added to tobacco in the manufacture of cigarettes and smokeless tobacco products and the specification of the quantity of nicotine contained in each smokeless tobacco product received by the CDC, OSH under 15 U.S.C. Section 1335a(a) of FCLAA and 15 U.S.C. Section 4403(a)(1)(A) and (B) of CSTHEA are generally trade secret or confidential information subject to section 552(b)(4) of Title 5. However, 15 U.S.C. Section 1335a(2)(B) of FCLAA and 15 U.S.C. Section 4403(b)(2)(B) of CSTHEA do not authorize the withholding of the list of ingredients from any duly authorized subcommittee or committee of the Congress. If a subcommittee or committee of the Congress requests the Secretary of HHS to provide it such a list, the Secretary of HHS or a representative of the CDC will make the list available to the subcommittee or committee and shall, at the same time, notify in writing the person who provided the list of such request.

What Steps Does CDC Take to Secure Confidential Information?

A limited number of staff has access to confidential information. Each of these persons has signed a pledge of confidentiality and these statements along with a list of their names are located in a locked safe. Additionally, the list of ingredients and the specification of the quantity of nicotine are secured in a locked safe located in a locked office. Electronic information is stored on a removable hard drive also kept in the safe. Further, OSH redacts privileged, proprietary, and confidential information from public files.

How Does a Manufacturer, Packager, or Importer Submit an Ingredient Report or Nicotine Report?

The CDC requests all submissions be provided on letterhead of the manufacturer, packager, importer, or designated individual or entity. They may be mailed or faxed; however, a faxed copy should be accompanied by a mailed original. Because the CDC cannot ensure the confidentiality of information submitted via E-mail, this is not an acceptable format. However, submission of data by way of mailing a CD, 3-inch floppy disk, or thumb drive is acceptable.

Must the Designated Individual Disclose the Identity of Its Clients When Submitting an Ingredient Report and / or Nicotine Report?

Yes. Under 15 U.S.C. Section 1335a(a) of FCLAA and 15 U.S.C. Section 4403(a)(2) of CSTHEA a person or group of persons required to provide a list of ingredients and a specification of the quantity of nicotine may designate an individual or entity to provide the above-referenced lists required by these federal laws. In the event that another individual or entity is designated to submit the required information, such as an attorney acting on behalf of a manufacturer or importer; or a manufacturer acting on behalf of an importer, that individual or entity should clearly state the name of the company for whom the information is being submitted.

Must an Ingredient Report Be Submitted If There Have Been No Changes in the List of Ingredients Added to the Tobacco Products?

If there have been no changes to the ingredient list in the previous calendar year, the manufacturer, packager, importer, or designated individual or entity as addressed above may submit a letter stating that there have been no changes in the list of ingredients added to the tobacco products along with a copy of the original submission.

Must an Ingredient Report Be Submitted If No Ingredients Have Been Added?

In order to confirm whether or not a manufacturer, packager, or importer is compliant with FCLAA or CSTHEA, a statement should be submitted in writing if no ingredients are added to tobacco in the manufacture of cigarettes or smokeless tobacco.

What If I Am an Importer and the Manufacturer Will Not Provide Me with the List of Ingredients?

The manufacturer of the tobacco product or of the flavoring added to the tobacco products may submit a list of ingredients and/or a specification of the quantity of nicotine contained in the smokeless tobacco product directly to the CDC on behalf of the importer. As with any submission, the list of ingredients and the specification of the quantity of nicotine shall be treated as trade secret or confidential information subject to section 552(b)(4) of Title 5, United States Code and "shall not be revealed... to any person other than those authorized by the Secretary in carrying out their official duties under this section."

What Happens after I Submit the Ingredient Report?

Once the accuracy of the submitted information is verified, the CDC will issue a certificate of compliance for that calendar year. If missing data or errors are noted (e.g., an invalid or missing CAS registry number), a letter will be sent requesting clarification of the submission within 60 days.

The process of verifying an Ingredient Report and providing a response may take several weeks. Frequently, however, this process only takes a few days, although this can vary depending on the complexity and completeness of submissions. If the Ingredient Report contains many errors, the process may become more lengthy.

Are There Other Federal Agencies Responsible for Collecting Information under FCLAA?

Yes. While this is not an exhaustive list of federal agencies that are responsible for tobacco-related matters, the Federal Trade Commission's (FTC) Division of Advertising Practices is responsible for reviewing and approving health warning label plans for all cigarette and smokeless tobacco products, whether manufactured domestically or abroad, that enter the U.S. stream of commerce.

Are There Any Other Federal Agencies with Whom I May Need to Interact?

Yes. While this is not an exhaustive list, other agencies with whom you may need to interact include the Federal Trade Commission (FTC); the Bureau of Alcohol, Tobacco, Firearms and Explosives (ATF); Alcohol and Tobacco Tax and Trade Bureau (TTB); and U.S. Customs.

Are There Penalties for Failing to Comply with FCLAA or CSTHEA?

Yes. According to 15 U.S.C. §1338 "Any person who violates the provisions of this chapter shall be guilty of a misdemeanor and shall on conviction thereof be subject to a fine of not more than $10,000." 15 U.S.C. Section 4404 has a similar penalty provision.

Does FCLAA or CSTHEA Preempt State and Local Initiatives?

Yes, both laws contain some preemptive provisions, however, they do not preempt stricter state-level requirements regarding the submission of ingredient information. 15 U.S.C. 1334(a)-(b), of FCLAA reads:

(a) Additional statements

No statement relating to smoking and health, other than the statement required by section 1333 of this title, shall be required on any cigarette package.

(b) State regulations

No requirement or prohibition based on smoking and health shall be imposed under State law with respect to the advertising or promotion of any cigarettes the packages of which are labeled in conformity with the provisions of this chapter.

And the CSTHEA, 15 U.S.C. §4406(a)-(c), states:

(a) Federal action

No statement relating to the use of smokeless tobacco products and health, other than the statements required by section 4402 of this title, shall be required by any Federal agency to appear on any package or in any advertisement (unless the advertisement is an outdoor billboard advertisement) of a smokeless tobacco product.

(b) State and local action

No statement relating to the use of smokeless tobacco products and health, other than the statements required by section 4402 of this title, shall be required by any state or local statute or regulation to be included on any package or in any advertisement (unless the advertisement is an outdoor billboard advertisement) of a smokeless tobacco product.

(c) Effect on liability law

Nothing in this chapter shall relieve any person from liability at common law or under state statutory law to any other.

Chapter 4

Types of Smoked Tobacco Products

Chapter Contents

Section 4.1

Types of Tobacco Products

This section includes text excerpted from "Recognize
Tobacco in Its Many Forms," U.S. Food and Drug
Administration (FDA), January 17, 2018.

Tobacco use is the single largest preventable cause of disease and death in the United States, but can you recognize all the different forms of a tobacco product? The marketplace includes an array of new products, with many looking very different from the traditional tobacco products you may know about.

To attract users, tobacco companies regularly modify their products and introduce novel tobacco products to the market. "Parents should stay updated on the various products available and, discuss the dangers of tobacco use with their children," says Ii-Lun Chen, M.D., a pediatrician and medical branch chief in the Office of Science at U.S. Food and Drug Administration (FDA) Center for Tobacco Products (CTP).

Cigarettes

The basic components of most cigarettes are tobacco, a filter, and paper wrapping. Although smokers use cigarettes to get nicotine, they are exposed to toxic and cancer-causing chemicals that are created when the cigarette is burned.

Cigars, Little Cigars, and Cigarillos

Generally, cigars are cured tobacco wrapped in leaf tobacco or a substance containing tobacco. Cigars vary in size—with smaller sizes sometimes referred to as little cigars or cigarillos. Large cigars can deliver as much as 10 times the nicotine, 2 times more tar, and more than 5 times the carbon monoxide than a filtered cigarette. Although cigarettes with characterizing flavors are illegal, there are products available on the market that look like cigarettes but are labeled as "little cigars," and some include candy and fruit flavors that appeal to adolescents and youth adults. Cigars also may appeal to youth because they may be less expensive than cigarettes. In addition, young adults may think that cigars are less addictive and present fewer health risks than cigarettes.

70

Dissolvable Products

In the past, smokeless tobacco products have required spitting or discarding the product remains. There are new tobacco products that are not smoked and are often called "dissolvables." These products can be more easily concealed as no product disposal is needed. They are sold as lozenges, strips, or sticks, and may look like candy. The advertised appealing flavor and discreet forms of these products may encourage young people to take them up, but the nicotine content can lead to addiction and may also present an accidental poisoning risk for children.

Electronic Cigarettes (Also Referred to As: Vape Pen, E-Hookah, Hookah Pen)

Electronic cigarettes often resemble traditional cigarettes but they use a heat source, usually powered by a battery, to turn "e-liquid," a liquid that usually contains nicotine from tobacco and flavorings, into an aerosol that is inhaled by the user. The amount of nicotine in the aerosol may vary by brand. Little information about the safety of electronic cigarettes exists.

Traditional Smokeless Tobacco Products

There are two main types of smokeless tobacco that have been traditionally marketed in the United States: chewing tobacco and moist snuff. Chewing tobacco is cured tobacco in the form of loose leaf, plug, or twist. Snuff is finely cut or powdered, cured tobacco that can be dry, moist, or packaged in sachets. Snus is a finely ground moist snuff that can be loose or packaged. Most smokeless tobacco use involves placing the product between the cheek or lip and the gum.

The availability of flavored, lower-nicotine, smokeless tobacco products lacking harsh attributes promoted by manufacturers may allow for experimentation by new users, but may also lead to nicotine addiction and continued use of smokeless or other tobacco products. Over time, smokeless tobacco users may switch from lower-nicotine smokeless tobacco products to products that deliver more nicotine.

Waterpipes (Also Referred to As: Hookah, Shisha, Narghile, Argileh)

Waterpipes (also known as hookah, shisha, narghile, or argileh) are used to smoke specially made tobacco that comes in a variety of flavors like mint, cherry, and licorice. Waterpipe smoking delivers the addictive drug nicotine and the smoke from a waterpipe is at least as toxic as, or more toxic than cigarette smoke. In fact, research shows that waterpipe smokers may absorb even more of the harmful components found in cigarette smoke because smoking sessions are longer. A typical one-hour hookah session involves inhaling 100 to 200 times the volume of smoke from a single cigarette. Waterpipe tobacco flavoring, exotic paraphernalia, and social use at hookah bars have increased its popularity with people who don't already smoke cigarettes and younger people in the United States.

Section 4.2

E-Cigs, Menthol, and Dip

This section includes text excerpted from "E-Cigs, Menthol and Dip," Smokefree.gov, U.S. Department of Health and Human Services (HHS), May 31, 2018.

What We Know about Electronic Cigarettes

E-cigarettes usually contain nicotine and may have other harmful substances too. There's a lot of conflicting information about them.

E-cigarettes are known by many different names. They are often called e-cigs, e-hookahs, vapes, vape pens, tank systems, or mods. They come in many different shapes and sizes—some look like a regular cigarette, some look very different. There is a lot of talk about e-cigarettes. Some of it is true, but some of it is not.

You may have seen ads or stories on the Internet that say e-cigarettes are not harmful, or are a good way to help smokers quit smoking. However, doctors and researchers still have a lot to learn about the health effects of e-cigarettes. While e-cigarettes may be less harmful than regular cigarettes, this does not mean that they are harmless.

What Are E-Cigarettes?

E-cigarettes are battery powered devices that work by heating a liquid into an aerosol that the user inhales and exhales. The e-cigarette liquid typically contains nicotine, propylene glycol, glycerin, flavorings, and other chemicals. Nicotine is the addictive drug found in regular cigarettes and other tobacco products. Research shows that e-cigarette aerosol often contains substances that can be harmful, including flavoring chemicals (like diacetyl, which is linked to lung disease), metals (like lead), and other cancer-causing chemicals.

Are E-Cigarettes Regulated?

Companies that make or sell e-cigarettes must follow certain U.S. Food and Drug Administration (FDA) regulations. For example, only people age 18 and over are allowed to buy e-cigarettes. Researchers are working hard to gather more information about e-cigarettes and how they are used. This information may lead to additional regulations and could be helpful for informing the public about what's in e-cigarettes and the potential health risks of using them.

What Are the Known Health Risks of E-Cigarettes?

Compared with regular cigarettes, e-cigarettes have been on the market a short time—about 11 years. Scientists are studying e-cigarettes to understand how using them affects people's health. Here's what doctors and researchers know right now:

- E-cigarettes usually contain nicotine. Nicotine is what makes tobacco products addictive. Be aware that some e-cigarettes that claim to be nicotine-free have been found to contain nicotine.

- E-cigarettes are harmful for youth, young adults, and pregnant women. The nicotine in e-cigarettes is harmful for developing babies, and can lead to addiction and harm brain development in children and young adults into their early 20s. Although there is still much to learn about e-cigarettes, the evidence is clear that the harmful health effects of using e-cigarettes means teens and young adults should not use them.

- E-cigarettes may contain other harmful substances. While e-cigarettes typically have fewer chemicals than regular cigarettes, they may still contain heavy metals like lead, flavorings linked to lung disease, small particles that can be inhaled deep into the lungs, and cancer-causing chemicals. Being near someone using an e-cigarette can expose you to the aerosol and the chemicals in it. This is similar to secondhand smoke from regular cigarettes.

Can E-Cigarettes Help People Quit Smoking?

E-cigarettes are not approved by the FDA as a quit smoking aid. So far, the research shows there is limited evidence that e-cigarettes are effective for helping smokers quit. There are other proven, safe, and effective methods for quitting smoking. One way to start is to talk with your doctor, nurse, or a trained quitline counselor to figure out the best strategies for you.

Many people use quit smoking medication, like nicotine replacement therapy (NRT), in the form of a patch or gum, which doctors and other experts agree is one of the most helpful tools smokers can use to quit. You can also get free, effective support from a quitline counselor by calling 800-QUIT-NOW (800-784-8669), or by enrolling in SmokefreeTXT or using a Smokefree.gov smartphone app. The combination of medication and support is known to increase the chance of quitting for good. Explore your options and find a quit method that's right for you.

Know More about Menthol Cigarettes

Menthol is a substance naturally found in mint plants such as peppermint and spearmint. It gives a cooling sensation. It is often used to relieve minor pain and irritation and prevent infection.

Menthol is added to many products. These include lozenges, syrups, creams and ointments, nasal sprays, powders, and candy. But none of these products are lighted or smoked when used. That makes them different from menthol cigarettes.

Many smokers think menthol cigarettes are less harmful. There is no evidence that cigarettes, cigars, or smokeless tobacco products that have menthol are safer than other cigarettes.

Like other cigarettes, menthol cigarettes harm nearly every organ in the body. They cause many diseases, including cancer and heart disease. Some research shows that menthol cigarettes may be more addictive than nonmenthol cigarettes.

Menthol Marketing

Menthol was first added to cigarettes in the 1920s. In the past, the tobacco industry marketed menthol cigarettes as being healthier and safer. Advertisements emphasized their cool and refreshing taste. The ads often showed nature, coldness, springtime, water, and other refreshing qualities. The tobacco industry also targeted "beginner" smokers, smokers with health concerns, and certain population groups.

Many people chose menthol cigarettes because they believed they were safer than nonmenthol cigarettes. They are not.

The Dangers of Dip

There are at least 30 chemicals in dip that are linked to cancer, including lead, uranium, and arsenic. Also, dip has more nicotine than cigarettes. Dipping two cans a week gives you as much nicotine as smoking 3½ packs of cigarettes a day for one week. Nicotine from dip stays in your blood longer than nicotine from smoking, and that can make it harder to quit.

Dip Can Harm Your Mouth

It may cause:

- Cancer of the mouth

- White patches and red sores in the mouth that can turn into cancer over time

- The roots of your teeth to decay (break down)

- Your teeth to fall out

Using Dip Can Cause Problems in Other Parts of the Body

For example:

- Studies show that using dip may cause pancreatic cancer

- It's possible that dip plays a role in causing heart disease and stroke

Why Do You Want to Quit Dip?

There are a lot of good reasons to quit dip. Do you want to be healthier? Look your best? Save money? Whatever your reasons are, quitting dip is the right choice.

Think about what's most important to you right now. Does dipping help or get in the way of what's important? You can also try asking yourself these questions:

- What do I miss out on when I dip?

- How does dip affect my health?

- Is dip stopping me from looking the way I want?

- What do my friends and family think about dip?

- What will happen to me if I keep using dip?

- How will my life get better when I quit?

- Remembering why you want to quit can inspire you to stop using dip for good.

Need More Reasons to Quit Dip?

There may be benefits of quitting that you haven't even thought of yet, including some that kick in as soon as you stop. For example:

- You'll look better without dip in your mouth

- You'll have more money to spend on things you enjoy

- No more worrying about when or where you can dip

- Food will taste better

   Over time, quitting dip also means:

- No more tobacco stains on your teeth and gums

- No more sores or white patches in your mouth

- You'll be less likely to get sick (like with a cold)

- Lower risk of cancer, heart disease, and other serious health problems

   The bottom line: Quitting will make you feel better and improve your health.

Make a List of All the Reasons You Want to Quit Dip

Keep the list in a place where you will see it often, like your car or where you used to keep your dip. It might also help to keep the list on your phone. When you have a craving for dip, take a look at the list to remind yourself why you want to quit.

Get through Cravings

For many people, dealing with cravings is one of the hardest parts of quitting dip. That's why it's important to have a plan to beat that urge to dip.

Cravings typically last 5 to 10 minutes. It might be uncomfortable, but try to wait it out. Here are a few ideas to help you handle cravings when they hit. Try one or more of these ideas until you find what works for you to beat the urge to dip. It's also helpful to keep track of things that have worked for you. When you find something that helps, write it down. You can use that same approach if another craving comes along.

- Do something else. Stop what you're doing right away and do something different. Sometimes, just changing your routine helps you shake off a craving.

- Keep your mouth busy. Make sure you have sugar-free gum or snacks with you. Drinking more water can also help.

- Get active. Try taking a quick walk or go up and down the stairs a few times. Physical activity, even in short bursts, can help boost your energy and beat a craving.

- Take some deep breaths. Breathe in slowly through your nose and breathe out slowly through your mouth. Repeat this 10 times to help you relax.

- Think about your reasons for quitting. They'll remind you why quitting is important to you and help you manage the craving until it passes.

You don't have to do this alone. You can reach out for help:

- Call or text a friend. Lean on people you trust.

- Use the quitline. Call 877-44U-QUIT (877-448-7848) to talk with a trained counselor for free.

- Chat with an online counselor.

- Get free real-time help from a trained counselor from the National Cancer Institute (NCI).

Dealing with Triggers

Triggers are the things that make you want to dip, like dealing with a stressful situation or seeing someone else dip. Different people may have different triggers, but one thing is true for everyone: knowing your triggers and making a plan to manage your cravings will help you quit dip for good. Here are some common triggers and tips for how to deal with them.

Do You Dip When You're Really Happy? What about When You're Feeling Down?

Many dip users would answer "yes" to both of those questions. It's very common to want to use dip to escape a bad mood or to make a good mood better.

For example, you may notice an urge to dip when you're having negative emotions, like when you feel:

- Stressed or anxious

- Lonely

- Bored

- Sad

- Amped up after an argument

Positive emotions can also trigger the urge to dip, like when you're feeling:

- Happy

- Excited

- Relieved

How to Deal

You can learn how to manage your feelings without turning to dip. Try these tips:

- Get active. Getting your body moving is a great way to handle emotions. When you exercise, your brain releases chemicals that make you feel good.

- Listen to music. Music can relax you by slowing your heart rate, lowering your blood pressure, and lowering the amount of stress hormones in your body.

- Take some deep breaths. Slow, deep breathing helps slow your body down, quiet your mind, and make cravings less intense.

Are There Things You Can't Imagine Doing without Dip?

Many people connect using dip with certain activities. Plan ahead with ways to stay dip-free when you're:

- Watching TV or playing video games
- Driving
- Working out
- Talking on the phone
- Finishing a meal
- Drinking coffee
- Taking a study or work break
- Doing chores
- Working outside

How to Deal

Try to break the link between dip and the activity that triggers you by replacing it with something else. Here are some ways:

- Keep your mouth busy. If you're used to dipping while you drive, keep sugar-free gum or toothpicks in your car.
- Change your routine. Try working out at a different time or brushing your teeth right after you eat a meal instead of before bed.
- Get active. Go for a walk or hop on your bike. Physical activity can help distract you from triggers.

Do You Crave Dip in Social Situations?

Social situations or events are another common trigger. For example, you may have cravings when you:

- Go to a party
- See someone else dip or smoke
- Spend time with friends who dip or smoke
- Go hunting or fishing

How to Deal

Many people who have quit dip find it helpful to make some changes to their social lives, at least for a while. Once you've decided to quit, you may want to:

79

- Avoid places where people dip
- Ask friends not to dip or smoke around you
- Spend time with friends and family who don't dip

You might not be able to stay away from all your triggers. But you can be ready with ways to handle them. Prepare ahead to deal with increased cravings you might have when you're in places where other people are using dip. Over time, it will get easier to handle social situations that make you want to dip. Try to stick with it and continue to ask friends and family for their support.

Handling Dip Withdrawal

Nicotine withdrawal is different for every person who uses dip. Changing the things you do can help you manage withdrawal symptoms.

The most common symptoms include:

- Having cravings for dip
- Feeling down or sad
- Having trouble sleeping
- Feeling irritable' on edge' or grouchy
- Having trouble thinking clearly and concentrating
- Feeling restless and jumpy
- Having a slower heart rate
- Feeling more hungry or gaining weight

Over time, these symptoms and cravings will fade if you stay away from dip.

Stress

Withdrawal can be uncomfortable and some people may feel high levels of symptoms. Some people feel increased sadness after they quit. If you are having extreme sadness, let people know.

Cravings

For many people who use dip, cravings last longer than other symptoms of withdrawal. Cravings can be set off by reminders of dipping.

These reminders are often called triggers. People, places, and things can trigger a craving. This means it's important to have a plan for how you'll handle a craving when it hits.

How to Deal with Stress

Feeling stressed or sad after quitting dip? It's normal. But stress or bad moods can make it harder to stay off dip. That's why it's important to learn how to manage your feelings without dip.

Many people use dip to help cope with stress. It's also common to dip when you're in a bad mood. One of the biggest challenges of being tobacco-free is learning new ways to deal with stress and other emotions.

Dip Isn't the Answer

It can be helpful to know that dip isn't a good way to deal with stress and bad moods. Check out some reasons why:

- The relief you might feel from dip only lasts a short time. As soon as you start to feel stressed or down again, you'll want to dip.

- Dip doesn't solve your problem—it just hides it. The cause of your stress or bad mood isn't going away because of dip.

- Using dip causes more stress than it relieves. Studies show that a person's stress levels tend to go down after quitting tobacco.

The Good News Is There Are Lots of Things That Can Help You Deal with Your Emotions.

Here are a few ideas to get you started:

Move your body. If you're feeling down, think about getting active. Any kind of physical activity can help. Try taking a walk, going to the gym, or joining a team sport. It might be hard to get motivated at first because feeling down can drain your energy. But if you stick with it, physical activity can help you feel better.

Spend time with people you care about. Getting support from the important people in your life can be key to helping you feel better. Focus on spending time with people who make you feel good about yourself and want to help you stay tobacco-free. If you're feeling down, you might want to spend more time alone. That's normal, but just talking with someone you trust can help boost your mood.

Build healthy habits. Being physically run down can make it harder to deal with a bad mood. Take care of yourself—eat regularly, get enough sleep, and build in time for fun and healthy activities, like lifting weights or going fishing with a friend.

Prepare yourself for stressful situations. If something is coming up that's making you nervous, doing a little homework in advance can make a big difference. Nervous about talking to your teacher or coach? Practice what you're going to say in front of a mirror. Got a big game or performance coming up? Picture yourself nailing it.

When you're really upset, try using the stop-breathe-think method. Take a timeout—stop, take a deep breath, and think about what's going on. A short break from a stressful or upsetting situation can help you think more clearly and make a healthy decision about what to do next.

Reward yourself with fun. If you're feeling really stressed or sad, it can be tough to build fun activities into your life. Find ways to reward yourself—even small things like picking up your favorite magazine or listening to music can improve your mood. Choose an activity that you stopped doing when you started dipping, or try something new that you've always wanted to do.

Look out for Signs of Depression

Feeling sad after you quit dip is normal. If you are feeling extreme sadness, you may need help from a professional. It is common for people who are feeling depressed to think about hurting themselves or dying. If you or someone you know is having these feelings, you can get help now. Call a 24-hour crisis center at 800-273-TALK (800-273-8255) or 800-SUICIDE (800-784-2433) for free, private help or dial 911. The Substance Abuse and Mental Health Services Administration (SAMHSA)—a part of the U.S. Department of Health and Human Services (HHS)—runs both crisis centers.

Ditch Dip for Good

Quitting dip is hard, but there are things you can do to help yourself stay dip-free for good.

Here are some tips that you can use every day to help you ditch the dip.

- Have a plan for getting through cravings. It's tempting to give in to dip when a craving hits. Having a plan can help. Your plan could include staying busy, talking with someone, or thinking about your reasons for quitting. Try to remind yourself that cravings generally last 5 to 10 minutes. Over time, these urges will fade if you stay away from dip.

- Focus on one day at a time. The first days and weeks without dipping can be really hard. Try to stay positive and take quitting one day at a time.

- Reward yourself for staying off dip. Quitting dip is a big deal— recognize that by rewarding yourself. Get tickets to see your favorite band or buy yourself something you really want with the money you're saving by quitting dip.

- Continue to get support. If you're having cravings, try reaching out to someone you trust, like a person who supported you when you first quit. If you find yourself craving dip, don't let it make you feel badly about yourself. These cravings are normal.

- Keep your guard up. It's important to know that your body has changed since you started using dip. Your brain learned to crave nicotine and tells your body to want it. No matter how long it's been since you quit, if you understand these changes, you can take steps not to give in and dip again.

Boost Your Mood after Quitting Dip

Feeling down? Mood changes are common after you quit dip. You might be irritable, restless, or feeling depressed.

If you have these feelings after quitting, there are things you can do to help lift your mood:

1. **Stay active.** Any kind of physical activity can help—for example, taking a walk, going to the gym, or joining a team sport. If you need to, start small and build-up overtime. This can be hard to do because feeling down can drain your energy. But making the effort will pay off. It will help you feel better.

2. **Come up with a daily routine.** Create a plan to stay busy. Find time to get out of the house whenever you can. Pick up an activity that you stopped doing when you started dipping, or take up something new.

3. **Talk and do things with other people.** Some people who feel down about themselves are cut off from others. Having daily contact with other people can help your mood.

4. **Build rewards into your life.** Some people who feel down don't have rewards or fun activities in their life. Find ways to reward yourself. Even small things, like reading a magazine or listening to music, add up and can help your mood.

5. **Talk with friends and loved ones.** Getting support from the important people in your life can make a big difference as you quit. They can be key to helping you feel better. Focus on spending time with people who make you feel good about yourself and want you to succeed in staying tobacco-free.

Section 4.3

Tobacco Brand Preferences

This section includes text excerpted from "Tobacco Brand Preferences," Centers for Disease Control and Prevention (CDC), August 1, 2018.

Cigarettes
Market Share Information

- According to 2017 sales data, Marlboro is the most popular cigarette brand in the United States, with sales greater than the next seven leading competitors combined.

- The three most heavily advertised brands—Marlboro, Newport, and Camel—continue to be the preferred brands of cigarettes smoked by young people.

Industry Marketing Practices

Tobacco industry marketing practices can influence the brands that certain groups prefer. For example:

- The packaging and design of certain cigarette brands appeal to adolescents and young adults.

84

Table 4.1. 2017 Market Shares for Leading Cigarette Brands

Brand	Market %
Marlboro	40%
Newport	14%
Camel (filter only)	8%
Pall Mall Box	7%
Maverick	2%
Santa Fe	2%
Winston	2%
Kool	2%

NOTE: Market share—or market percentage—is defined as the percentage of total sales in the United States.

- Historically, menthol cigarettes have been targeted heavily toward certain racial/ethnic groups, especially African Americans.

    - Among African American adult, adolescent, and young adult cigarette smokers, the most popular brands are all mentholated.

- Cigarettes with brand names containing words such as "thins" and "slims" have been manufactured to be longer and slimmer than traditional cigarettes to appeal directly to women—e.g., Virginia Slims and Capri brands.

Brand Characteristics

- Of all the cigarettes sold in the United States in 2016—

    - 99.7% were filtered

    - 35.0% were mentholated brands

- Use of mentholated brands varies widely by race/ethnicity. The percentage of individuals aged 12 years or older who reported using mentholated brands in 2010 was:

    - 19.1% Black

    - 3.6% Asian

    - 7.8% Hispanic

    - 6.5% White

- Before 2010, manufacturers were allowed to label cigarettes as "light" or "ultra light" if they delivered less than 15 mg of tar when measured by an automated smoking machine.

  - Such labeling allowed tobacco companies to deliberately misrepresent "light" cigarettes as being less harmful and an acceptable alternative to quitting smoking.

  - The 2009 Family Smoking Prevention and Tobacco Control Act (FSPTCA), however, prohibits use of terms like "light," "low," and "mild" on tobacco product labels.

Other Tobacco Products
Cigars

According to 2015 sales data, Swisher Little is the most popular brand of cigars in the United States, with sales substantially greater than any little cigar competitor and the leading large cigars and cigarillos competitors.

Table 4.2. 2015 Market Shares for Leading Cigar Brands

Brand	Category	Market %
Swisher Little	Little cigars	60%
Swisher Sweets	Large cigars and cigarillos	16%
Black & Mild	Large cigars and cigarillos	11%
Garcia y Vega	Large cigars and cigarillos	5%
White Owl	Large cigars and cigarillos	5%

NOTE: Market share—or market percentage—is defined as the percentage of total sales in the United States.

Smokeless Tobacco

The U.S. smokeless tobacco industry grew by 1.7 percent from 2010 to 2011, increasing its sales from 122.6 million pounds to 124.6 million pounds. The greatest growth occurred in the moist snuff category.

Table 4.3. 2011 Market Shares for Leading Smokeless Tobacco Brands

Brand	Category	Market %
Levi Garrett Plug	Moist plug tobacco	52%
Ds Work	Plug tobacco	45%

Table 4.3. Continued

Brand	Category	Market %
Red Man Plug	Moist plug tobacco	36%
Grizzly	Moist snuff and fine cut tobacco	26%
Copenhagen	Moist snuff and fine cut tobacco	25%
Garrett	Dry snuff	24%
Skoal	Moist snuff and fine cut tobacco	24%
Red Man	Loose leaf tobacco	18%

NOTE: Market share—or market percentage—is defined as the percentage of total sales in the United States.

Chapter 5

Smokeless Tobacco Products

Smokeless tobacco:

- Is not burned
- Includes tobacco that can be sucked or chewed
- Can be spit or swallowed, depending on the product
- Can be spitless, depending on the product
- Contains nicotine and is addictive
- May appeal to youth because it comes in flavors such as cinnamon, berry, vanilla, and apple

Types of smokeless tobacco:

- Chewing tobacco (loose leaf, plug, or twist and may come in flavors)
- Snuff (moist, dry, or in packets [U.S. snus])
- Dissolvables (lozenges, sticks, strips, orbs)

Chewing Tobacco

Chewing tobacco comes in the form of loose leaf, plug, or twist

---

This chapter includes text excerpted from "Smokeless Tobacco: Products and Marketing," Centers for Disease Control and Prevention (CDC), July 30, 2018.

Loose Chewing Tobacco   **Plug Chewing Tobacco**

Figure 5.1. *Chewing Tobacco*

Table 5.1. Market Share of Types of Chewing Tobacco

Form	Description	Use	Market Share (In 2011)*
Loose leaf	Cured (aged) tobacco, typically sweetened and packaged in foil pouches	Piece taken from pouch and placed between cheek and gums	17.50%
Plug	Cured tobacco leaves pressed together into a cake or "plug" form and wrapped in a tobacco leaf	Piece taken from pouch and placed between cheek and gums	0.50%
Twist or roll	Cured (aged) tobacco leaves twisted together like a rope	Piece cut off from twist and placed between cheek and gums	0.20%

* Market share is the percentage of the U.S. smokeless tobacco market for a specific product. For example, almost 2 of every 10 smokeless products (17.5%) sold in the United States in 2011 were loose-leaf smokeless tobacco products.

Snuff

Snuff is finely ground tobacco that can be dry, moist, or packaged in pouches or packets (dip, U.S. snus).

- Some types of snuff are sniffed or inhaled into the nose; other types are placed in the mouth.

- Snus is a newer form of moist snuff used in the United States.

Moist snuff   **Dry snuff**   **Snus**

Figure 5.2. *Snuff*

Table 5.2. Market Share of Types of Snuff

Form	Description	Use	Market Share (In 2011)*
Moist	Cured (aged) and fermented tobacco processed into fine particles and often packaged in round cans	Pinch or "dip" is placed between cheek or lip and gums; requires spitting	80.70%
Dry	Fire-cured tobacco in powder form	Pinch of powder is put in the mouth or inhaled through the nose; may require spitting	1.10%
U.S. snus	Moist snuff packaged in ready-to-use pouches that resemble small tea bags	Pouch is placed between cheek or teeth and gums; does not require spitting	Data unavailable

* Market share is the percentage of the U.S. smokeless tobacco market for a specific product. For example, more than 8 of every 10 snuff products sold in the United States in 2011 were moist snuff products.

Other Tobacco Products That Are Not Burned

Dissolvables are finely ground tobacco pressed into shapes such as tablets, sticks, or strips.

- Dissolvable tobacco products slowly dissolve in the mouth.

- These products may appeal to youth because they come in attractive packaging, look like candy or small mints, and can be easily hidden from view.

Table 5.3. Market Share of Other Types of Tobacco

Form	Description	Market Share (In 2011)
Lozenges	Resemble pellets or tablets	Data unavailable
Orbs	Resemble small mints	Data unavailable
Sticks	Have a toothpick-like appearance	Data unavailable
Strips	Thin sheets that work like dissolvable breath strips or medication strips	Data unavailable

Marketing Information

Some cigarette companies now make and sell smokeless tobacco products.

- In 2016, $759.3 million was spent on advertising and promotion of smokeless tobacco products, an increase from $684.9 million spent in 2015.

- Some cigarette companies advertise that smokeless tobacco can be used in places where tobacco smoking is not allowed.

- Additional research is needed to examine long-term effects of newer smokeless tobacco products, such as dissolvables and U.S. snus.

Chapter 6

Recent Trends in Tobacco Consumption

The consumption of cigarettes, small cigars, and chewing tobacco in the United States has declined over the past 15 years, according to an analysis from researchers at the Centers for Disease Control and Prevention (CDC).

These declines were partially offset, however, by a rise in consumption of large cigars, pipe tobacco, and snuff. But despite these changes, the researchers found that cigarettes still remain the most commonly consumed tobacco product by far.

"Notable shifts have occurred in the tobacco product landscape," the study authors wrote.

These shifts may be influenced by several factors, including changes in product pricing and marketing, as well as alterations in social norms, explained Michele Bloch, M.D., Ph.D., chief of National Cancer Institute's (NCI) Tobacco Control Research Branch (TCRB), who was not involved in the study. Analyses of tobacco consumption trends are crucial to identifying research priorities and informing public health strategies in the future, she said.

"This highlights the importance of continuing to monitor and reduce all forms of smoked and smokeless tobacco use in the United States," said the study's first author Teresa Wang, Ph.D.

This chapter includes text excerpted from "Shifts Seen in Tobacco Product Types Purchased in the U.S.," National Cancer Institute (NCI), January 23, 2017.

A Changing Landscape

Tobacco use is the leading preventable cause of death and disease in the United States, and public health policies have been implemented to help eliminate both tobacco use and exposure to secondhand smoke.

Tobacco smoking causes lung cancer and many other cancers, including those of the bladder, liver, and pancreas. In addition, tobacco smoking can cause or exacerbate other diseases such as heart disease, lung disease, and diabetes.

Smokeless tobacco products, such as chewing tobacco and snuff, can also cause serious detrimental health effects and lead to diseases, including cancer.

Tobacco use trends are studied to help develop public health policies and interventions. Because most tobacco products are taxed, trends in the type of tobacco products Americans purchase—that is, tobacco consumption—can be inferred from federal excise tax data.

To keep a pulse on the types of tobacco products Americans are using, the CDC regularly analyzes trends in tobacco product consumption over time. The research team estimated total and per person consumption of tobacco between 2000 and 2015. And for the first time, they analyzed consumption of smokeless tobacco products in addition to smoked tobacco products.

During this time period, total consumption of all smoked tobacco products combined dropped by 33.5 percent, the CDC investigators reported. Although total consumption of cigarettes fell by nearly 40 percent, total consumption of all other smoked tobacco products combined rose by approximately 117 percent. This increase was driven by a substantial rise in the consumption of large cigars and pipe tobacco, they found. In contrast, consumption of small cigars and roll-your-own tobacco decreased.

They also found that, between 2000 and 2015, a decline in chewing tobacco consumption was offset by a steady increase in snuff consumption.

"We think one reason for this trend is that tobacco companies spent much more on advertising snuff versus chewing tobacco," said Dr. Wang. For example, in 2013, tobacco companies spent almost $411 million on snuff advertisements, compared with just under $12 million for chewing tobacco advertisements, she explained.

Trends in per person consumption of smoked and smokeless tobacco between 2000 and 2015 were similar to those for total consumption.

When the research team focused on changes in tobacco consumption that occurred between 2014 and 2015, however, they found that some trends differed from those that occurred between 2000 and 2015. For example, from 2014 to 2015, total and per person cigarette consumption increased slightly, although both total and per person consumption of all other smoked tobacco products combined fell.

This was the first time since 1973 that more cigarettes were consumed than in the previous year, the authors noted.

Public Health Impact

The authors noted that their study was not without limitations: Consumption estimates based on tax data excludes illegal tobacco sales, sales on American Indian sovereign lands, and sales of certain tobacco products for which taxes are not reported. In addition, they wrote, "sales data might not reflect actual consumption, because all purchased products might not be used by the consumer because of loss, damage, or tobacco cessation."

And because per person estimates of tobacco consumption are not exact, the authors point out that an upward trend in tobacco consumption does not necessarily mean that more Americans are using tobacco products.

Overall, these results show that "cigarettes remain our biggest problem, but not our only problem," said Dr. Bloch. Although cigarette consumption has steadily decreased over the past 15 years, Americans continue to consume several million more cigarettes than equivalents of any other tobacco product, she explained.

"By using proven strategies to address the diversity of tobacco products consumed, we can continue to reduce tobacco-related disease and death," Dr. Wang said.

Chapter 7

Economic Costs of Smoking

Tobacco-Related Spending

In 2016, tobacco companies spent $9.5 billion on marketing cigarettes and smokeless tobacco in the United States. This amount translates to about $26 million each day, or more than $1 million every hour.

- Cigarette advertising and promotional expenses totaled approximately $8.7 billion in 2016—an increase from $8.3 billion during 2015. In 2016, about 66.7 percent (about $5.8 billion) of this was spent on price discounts paid to cigarette retailers to reduce the cost of cigarettes to consumers.

- Manufacturers spent a total of $759.3 million on smokeless tobacco advertising and promotion during 2016—an increase from $684.9 million in 2015.

- A number of electronic products, such as electronic cigarettes (e-cigarettes), electronic cigars (e-cigars), and electronic pipes (e-pipes), are being introduced in the marketplace.

- Sales of e-cigarettes grew considerably from 2011 to 2015. During 2014 to 2015 alone, unit sales for e-cigarettes grew 14.4 percent overall. During this period, unit sales of disposable e-cigarettes declined 42.7 percent, unit sales of rechargeable e-cigarettes increased 5.3 percent, unit sales of e-cigarette

This chapter includes text excerpted from "Economic Trends in Tobacco," Centers for Disease Control and Prevention (CDC), May 4, 2018.

liquid refills increased 307.7 percent, and unit sales of prefilled e-cigarette cartridges increased 31.6 percent.

Tobacco Production in the United States

Although U.S. tobacco production has decreased significantly since the 1980s (from nearly 180,000 tobacco-growing farms to about 10,000 in 2012), the United States continues to be a leading producer of tobacco leaves.

- The United States is the fourth largest tobacco-producing country in the world, following China, India, and Brazil.

- Farms in the United States harvested more than 700 million pounds of tobacco in 2015.

- In 2015, two states—North Carolina and Kentucky—accounted for more than 70 percent of total tobacco cultivation.

Tobacco Sales
Cigarette Sales

- During 2017, about 249 billion cigarettes were sold in the United States—a 3.5 percent decrease from the 258 billion sold in 2016.

  - Four companies—Philip Morris USA, Reynolds American Inc., ITG Brands, and Liggett—accounted for about 92 percent of U.S. cigarette sales.

  - Imports, primarily from Canada and South Korea, accounted for approximately 8.3 percent of U.S. cigarette inventories in 2016 and 7.9 percent in 2017.

- By state, the average retail price of a pack of 20 cigarettes (full-priced brands), including federal and state excise taxes, ranged from $5.12 in Missouri to a high of $10.66 in New York, as of November 2016.

- On average, federal and state excise taxes account for 44.3 percent of the retail price of cigarettes.

Other Tobacco Product Sales

- Nearly 12 billion cigars, including 11.9 billion large cigars and 0.6 billion little cigars, were sold in the United States in 2015.

- The total amount of smokeless tobacco sold by manufacturers to wholesalers and retailers in the United States was 129.36 million pounds in 2015, an increase from 127.81 million pounds sold in 2014.

  - During March 2017—March 2018, three companies—Altria Group Inc., British American Tobacco, and Swedish Match—accounted for 98 percent of U.S. dollar sales of smokeless tobacco.

Economic Costs Associated with Smoking
Cost of Smoking-Related Illness

Smoking-related illness in the United States costs more than $300 billion each year, including:

- Nearly $170 billion for direct medical care for adults

- More than $156 billion in lost productivity, including $5.6 billion in lost productivity due to secondhand smoke exposure

Effects of Increased Prices

Increasing the price of tobacco products is the single most effective way to reduce consumption.

- A 10 percent increase in price has been estimated to reduce overall cigarette consumption by 3 to 5 percent.

- Research on cigarette consumption suggests that both youth and young adults are two to three times more likely to respond to increases in price than adults.

Chapter 8

Marijuana: Are Hazards Similar to Those for Tobacco?

What Is Marijuana?

Marijuana refers to the dried leaves, flowers, stems, and seeds from the *Cannabis sativa* or *Cannabis indica* plant. The plant contains the mind-altering chemical tetrahydrocannabinol (THC) and other similar compounds. Extracts can also be made from the Cannabis plant.

Marijuana is the most commonly used illicit drug in the United States. Its use is widespread among young people. In 2015, more than 11 million young adults ages 18 to 25 used marijuana in the past year. According to the Monitoring the Future (MTF) survey, rates of marijuana use among middle and high school students have dropped or leveled off in the past few years after several years of increase. However, the number of young people who believe regular marijuana use is risky is decreasing.

Legalization of marijuana for medical use or adult recreational use in a growing number of states may affect these views.

This chapter contains text excerpted from the following sources: Text beginning with the heading "What Is Marijuana?" is excerpted from "Marijuana," National Institute on Drug Abuse (NIDA), June 2018; Text under the heading "Is Marijuana Medicine?" is excerpted from "Is Marijuana Medicine?" Centers for Disease Control and Prevention (CDC), March 7, 2018.

How Do People Use Marijuana?

People smoke marijuana in hand-rolled cigarettes (joints) or in pipes or waterpipes (bongs). They also smoke it in blunts—emptied cigars that have been partly or completely refilled with marijuana. To avoid inhaling smoke, some people are using vaporizers. These devices pull the active ingredients (including THC) from the marijuana and collect their vapor in a storage unit. A person then inhales the vapor, not the smoke. Some vaporizers use a liquid marijuana extract.

People can mix marijuana in food (edibles), such as brownies, cookies, or candy, or brew it as a tea. A newly popular method of use is smoking or eating different forms of THC-rich resins.

How Does Marijuana Affect the Brain?

Marijuana has both short-and long-term effects on the brain.

Short-Term Effects

When a person smokes marijuana, THC quickly passes from the lungs into the bloodstream. The blood carries the chemical to the brain and other organs throughout the body. The body absorbs THC more slowly when the person eats or drinks it. In that case, they generally feel the effects after 30 minutes to one hour.

THC acts on specific brain cell receptors that ordinarily react to natural THC-like chemicals. These natural chemicals play a role in normal brain development and function.

Marijuana overactivates parts of the brain that contain the highest number of these receptors. This causes the "high" that people feel. Other effects include:

- Altered senses (for example, seeing brighter colors)

- Altered sense of time

- Changes in mood

- Impaired body movement

- Difficulty with thinking and problem-solving

- Impaired memory

- Hallucinations (when taken in high doses)

- Delusions (when taken in high doses)

- Psychosis (when taken in high doses)

102

Long-Term Effects

Marijuana also affects brain development. When people begin using marijuana as teenagers, the drug may impair thinking, memory, and learning functions and affect how the brain builds connections between the areas necessary for these functions. Researchers are still studying how long marijuana's effects last and whether some changes may be permanent.

For example, a study from New Zealand conducted in part by researchers at Duke University showed that people who started smoking marijuana heavily in their teens and had an ongoing marijuana use disorder lost an average of 8 intelligence quotient (IQ) points between ages 13 and 38. The lost mental abilities didn't fully return in those who quit marijuana as adults. Those who started smoking marijuana as adults didn't show notable IQ declines.

In another study on twins, those who used marijuana showed a significant decline in general knowledge and in verbal ability (equivalent to 4 IQ points) between the preteen years and early adulthood, but no predictable difference was found between twins when one used marijuana and the other didn't. This suggests that the IQ decline in marijuana users may be caused by something other than marijuana, such as shared familial factors (e.g., genetics, family environment). The National Institute on Drug Abuse's (NIDA) Adolescent Brain Cognitive Development (ABCD) study, a major longitudinal study, is tracking a large sample of young Americans from late childhood to early adulthood to help clarify how and to what extent marijuana and other substances, alone and in combination, affect adolescent brain development.

What Are the Other Health Effects of Marijuana?

Marijuana use may have a wide range of effects, both physical and mental.

Physical Effects

Breathing problems. Marijuana smoke irritates the lungs, and people who smoke marijuana frequently can have the same breathing problems as those who smoke tobacco. These problems include daily cough and phlegm, more frequent lung illness, and a higher risk of lung infections. Researchers so far haven't found a higher risk for lung cancer in people who smoke marijuana.

Increased heart rate. Marijuana raises heart rate for up to three hours after smoking. This effect may increase the chance of heart attack. Older people and those with heart problems may be at higher risk.

Problems with child development during and after pregnancy. One study found that about 20 percent of pregnant women 24-years-old and younger screened positive for marijuana. However, this study also found that women were about twice as likely to screen positive for marijuana use via a drug test than they state in self-reported measures. This suggests that self-reported rates of marijuana use in pregnant females is not an accurate measure of marijuana use and may be underreporting their use. Additionally, in one study of dispensaries, nonmedical personnel at marijuana dispensaries were recommending marijuana to pregnant women for nausea, but medical experts warn against it. This concerns medical experts because marijuana use during pregnancy is linked to lower birth weight (LBW) and increased risk of both brain and behavioral problems in babies. If a pregnant woman uses marijuana, the drug may affect certain developing parts of the fetus's brain. Children exposed to marijuana in the womb have an increased risk of problems with attention, memory, and problem-solving compared to unexposed children. Some research also suggests that moderate amounts of THC are excreted into the breastmilk of nursing mothers. With regular use, THC can reach amounts in breast milk that could affect the baby's developing brain. More research is needed.

Intense nausea and vomiting. Regular, long-term marijuana use can lead to some people to develop cannabinoid hyperemesis syndrome (CHS). This causes users to experience regular cycles of severe nausea, vomiting, and dehydration, sometimes requiring emergency medical attention.

Mental Effects

Long-term marijuana use has been linked to mental illness in some people, such as:

- Temporary hallucinations

- Temporary paranoia

- Worsening symptoms in patients with schizophrenia—
  a severe mental disorder with symptoms such as hallucinations, paranoia, and disorganized thinking

Marijuana use has also been linked to other mental-health problems, such as depression, anxiety, and suicidal thoughts among teens. However, study findings have been mixed.

Are There Effects of Inhaling Secondhand Marijuana Smoke?

Failing a Drug Test

While it's possible to fail a drug test after inhaling secondhand marijuana smoke, it's unlikely. Studies show that very little THC is released in the air when a person exhales. Research findings suggest that, unless people are in an enclosed room, breathing in lots of smoke for hours at close range, they aren't likely to fail a drug test. Even if some THC was found in the blood, it wouldn't be enough to fail a test.

Getting High from Passive Exposure

Similarly, it's unlikely that secondhand marijuana smoke would give nonsmoking people in a confined space a high from passive exposure. Studies have shown that people who don't use marijuana report only mild effects of the drug from a nearby smoker, under extreme conditions (breathing in lots of marijuana smoke for hours in an enclosed room).

Other Health Effects

More research is needed to know if secondhand marijuana smoke has similar health risks as secondhand tobacco smoke. A study on rats suggests that secondhand marijuana smoke can do as much damage to the heart and blood vessels as secondhand tobacco smoke. But researchers haven't fully explored the effect of secondhand marijuana smoke on humans. What they do know is that the toxins and tar found in marijuana smoke could affect vulnerable people, such as children or people with asthma.

Is Marijuana a Gateway Drug?

Use of alcohol, tobacco, and marijuana are likely to come before use of other drugs. Animal studies have shown that early exposure to addictive substances, including THC, may change how the brain responds to other drugs. For example, when rodents are repeatedly exposed to THC when they're young, they later show an enhanced response to other addictive substances—such as morphine or nicotine—in the areas of the brain that control reward, and they're more likely to show addiction-like behaviors.

Although these findings support the idea of marijuana as a "gateway drug," the majority of people who use marijuana don't go on to use other "harder" drugs. It's also important to note that other factors besides biological mechanisms, such as a person's social environment, are also critical in a person's risk for drug use and addiction.

Can a Person Overdose on Marijuana?

An overdose occurs when a person uses enough of the drug to produce life-threatening symptoms or death. There are no reports of teens or adults dying from marijuana alone. However, some people who use marijuana can feel some very uncomfortable side effects, especially when using marijuana products with high THC levels. People have reported symptoms such as anxiety and paranoia, and in rare cases, an extreme psychotic reaction (which can include delusions and hallucinations) that can lead them to seek treatment in an emergency room.

While a psychotic reaction can occur following any method of use, emergency-room responders have seen an increasing number of cases involving marijuana edibles. Some people (especially preteens and teens) who know very little about edibles don't realize that it takes longer for the body to feel marijuana's effects when eaten rather than smoked. So they consume more of the edible, trying to get high faster or thinking they haven't taken enough. In addition, some babies and toddlers have been seriously ill after ingesting marijuana or marijuana edibles left around the house.

Is Marijuana Addictive?

Marijuana use can lead to the development of a substance-use disorder (SUD), a medical illness in which the person is unable to stop using even though it's causing health and social problems in their life. Severe SUDs are also known as addiction. Research suggests that between 9 and 30 percent of those who use marijuana may develop some degree of marijuana-use disorder. People who begin using marijuana before age 18 are four to seven times more likely than adults to develop a marijuana-use disorder.

Many people who use marijuana long-term and are trying to quit report mild withdrawal symptoms that make quitting difficult. These include:

- Grouchiness
- Sleeplessness
- Decreased appetite
- Anxiety
- Cravings

What Treatments Are Available for Marijuana-Use Disorder?

No medications are currently available to treat marijuana-use disorder, but behavioral support has been shown to be effective. Examples include therapy and motivational incentives (providing rewards to patients who remain drug-free). Continuing research may lead to new medications that help ease withdrawal symptoms, block the effects of marijuana, and prevent relapse.

Is Marijuana Medicine?

The marijuana plant has chemicals that may help symptoms for some health problems. More and more states are making it legal to use the plant as medicine for certain conditions. But there isn't enough research to show that the whole plant works to treat or cure these conditions. Also, the U.S. Food and Drug Administration (FDA) has not recognized or approved the marijuana plant as medicine.

Because marijuana is often smoked, it can damage your lungs and cardiovascular system (e.g., heart and blood vessels). These and other damaging effects on the brain and body could make marijuana more harmful than helpful as a medicine. Another problem with marijuana as a medicine is that the ingredients aren't exactly the same from plant to plant. There's no way to know what kind and how much of a chemical you're getting.

Two medicines have been made as pills from a chemical that's like THC, one of the chemicals found in the marijuana plant that makes people feel "high." These two medicines can treat nausea if you have cancer and make you hungry if you have acquired immunodeficiency syndrome (AIDS) and don't feel like eating. But the chemical used to make these medicines affects the brain also, so it can do things to your body other than just working as medicine.

Another marijuana chemical that scientists are studying, called cannabidiol (CBD), doesn't make you high because it acts on different parts of the nervous system than THC Scientists think this chemical might help children who have a lot of seizures (when your body starts twitching and jerking uncontrollably) that can't be controlled with other medicines. Some studies have started to see whether it can help.

Chapter 9

Tobacco-Use Statistics

Tobacco use is the leading cause of preventable disease, disability, and death in the United States. Nearly 40 million U.S. adults still smoke cigarettes, and about 4.7 million middle and high school students use at least one tobacco product, including e-cigarettes. Every day, more than 3,800 youth younger than 18 years smoke their first cigarette. Each year, nearly half a million Americans die prematurely of smoking or exposure to secondhand smoke. Another 16 million live with a serious illness caused by smoking. Each year, the United States spends nearly $170 billion on medical care to treat smoking-related disease in adults.

Fast Facts

- Smoking leads to disease and disability and harms nearly every organ of the body.

- Smoking is the leading cause of preventable death.

This chapter contains text excerpted from the following sources: Text in this chapter begins with excerpts from "Smoking and Tobacco Use—Data and Statistics," Centers for Disease Control and Prevention (CDC), January 17, 2018; Text beginning with the heading "Diseases and Death" is excerpted from "Smoking and Tobacco Use—Fast Facts," Centers for Disease Control and Prevention (CDC), February 20, 2018; Text under the heading "Smokeless Tobacco Use in the United States" is excerpted from "Smokeless Tobacco Use in the United States," Centers for Disease Control and Prevention (CDC), August 29, 2018.

109

- The tobacco industry spends billions of dollars each year on cigarette advertising and promotions.

- Smoking costs the United States billions of dollars each year.

- State spending on tobacco prevention and control does not meet Centers for Disease Control and Prevention (CDC)-recommended levels.

- 15.5 percent of all adults (37.8 million people): 17.5 percent of males, 13.5 percent of females were current cigarette smokers in 2016.

- Thousands of young people start smoking cigarettes every day.

- Many adult cigarette smokers want to quit smoking.

Diseases and Death

Smoking leads to disease and disability and harms nearly every organ of the body.

- More than 16 million Americans are living with a disease caused by smoking.

- For every person who dies because of smoking, at least 30 people live with a serious smoking-related illness.

- Smoking causes cancer, heart disease, stroke, lung diseases, diabetes, and chronic obstructive pulmonary disease (COPD), which includes emphysema and chronic bronchitis.

- Smoking also increases risk for tuberculosis, certain eye diseases, and problems of the immune system, including rheumatoid arthritis.

- Smoking is a known cause of erectile dysfunction in males.

Smoking is the leading cause of preventable death.

- Worldwide, tobacco use causes nearly 6 million deaths per year, and current trends show that tobacco use will cause more than 8 million deaths annually by 2030.

- Cigarette smoking is responsible for more than 480,000 deaths per year in the United States, including more than 41,000 deaths resulting from secondhand smoke exposure. This is about one in five deaths annually, or 1,300 deaths every day.

- On average, smokers die 10 years earlier than nonsmokers.

- If smoking continues at the current rate among U.S. youth, 5.6 million of today's Americans younger than 18 years of age are expected to die prematurely from a smoking-related illness. This represents about one in every 13 Americans aged 17 years or younger who are alive today.

Costs and Expenditures

The tobacco industry spends billions of dollars each year on cigarette and smokeless tobacco advertising and promotions.

- In 2016, $9.5 billion was spent on advertising and promotion of cigarettes and smokeless tobacco combined—about $26 million every day, and more than $1 million every hour.

- Price discounts to retailers account for 66.7 percent of all cigarette marketing (about $5.8 billion). These are discounts paid in order to reduce the price of cigarettes to consumers.

Smoking costs the United States billions of dollars each year.

- Total economic cost of smoking is more than $300 billion a year, including:

  - Nearly $170 billion in direct medical care for adults

  - More than $156 billion in lost productivity due to premature death and exposure to secondhand smoke

State spending on tobacco prevention and control does not meet CDC-recommended levels.

- States have billions of dollars from tobacco taxes and tobacco industry legal settlements to prevent and control tobacco use. However, states currently use a very small amount of these funds for tobacco control programs.

- In fiscal year 2018, states will collect a record $27.5 billion from tobacco taxes and legal settlements but will only spend $721.6 million—less than 3 percent—on prevention and cessation programs.

- Currently, not a single state funds tobacco control programs at the CDC's "recommended" level. Only two states (Alaska and California) provide more than 90 percent of recommended funding. Twenty-nine states and the District of Columbia are spending less than 20 percent of what the CDC recommends.

Two states (Connecticut and West Virginia) have allocated no state funds for tobacco use prevention.

- Spending 12 percent (i.e., $3.3 billion) of the $27.5 billion would fund every state tobacco control program at CDC-recommended levels.

Cigarette Smoking in the United States

Percentage of U.S. adults aged 18 years or older who were current cigarette smokers in 2016:

- 15.5 percent of all adults (37.8 million people): 17.5 percent of males, 13.5 percent of females

  - Nearly 32 of every 100 non-Hispanic American Indians/ Alaska Natives (31.8%)

  - About 25 of every 100 non-Hispanic multiple race individuals (25.2%)

  - Nearly 17 of every 100 non-Hispanic Blacks (16.5%)

  - Nearly 17 of every 100 non-Hispanic Whites (16.6%)

  - Nearly 11 of every 100 Hispanics (10.7%)

  - 9 of every 100 non-Hispanic Asians (9.0%)

Note: Current cigarette smokers are defined as persons who reported smoking at least 100 cigarettes during their lifetime and who, at the time they participated in a survey about this topic, reported smoking every day or some days.

Thousands of young people start smoking cigarettes every day.

- Each day, more than 3,200 people younger than 18 years of age smoke their first cigarette.

- Each day, an estimated 2,100 youth and young adults who have been occasional smokers become daily cigarette smokers.

Many adult cigarette smokers want to quit smoking.

- In 2015:

  - Nearly 7 in 10 (68.0%) adult cigarette smokers wanted to stop smoking.

  - More than 5 in 10 (55.4%) adult cigarette smokers had made a quit attempt in the past year.

- Since 2012, the *Tips From Former Smokers®* campaign has motivated at least 500,000 tobacco smokers to quit for good.

Smokeless Tobacco Use in the United States
Adult Smokeless Tobacco Use (National)

As shown in the graph below, smokeless tobacco use among females has remained low throughout the years. Among males, use decreased during 1986–2000 but has been increasing since then.

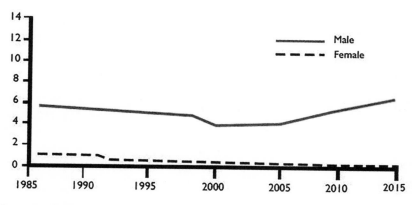

Figure 9.1. *Trends in Prevalence of Smokeless Tobacco Use among Adults 18 Years of Age and Older, by Gender and Selected Survey Years*

Percentage of Adults Who Were Current Smokeless Tobacco Users in 2016

- Adults aged 18 years and older: About 3 in every 100 (3.4%)

- Men: Nearly 7 in every 100 (6.6%)

- Women: Nearly 1 in every 100 (0.5%)

- Non-Hispanic African Americans: About 1 in every 100 (1.4%)

- Non-Hispanic American Indians/Alaska Natives: About 8 in every 100 (8.4%)

- Non-Hispanic Asians: About 1 in every 100 (0.4%)

- Hispanics: About 1 in every 100 (1.3%)

- Non-Hispanic Whites: Nearly 5 in every 100 (4.5%)

Adult Smokeless Tobacco Use (State-Specific)

- In 2016, current smokeless tobacco use was highest in:
  - Wyoming: Nearly 10 in every 100 people (9.8%)
  - West Virginia: Nearly 9 in every 100 people (8.5%)
  - Arkansas: Nearly 8 in every 100 people (7.8%)
  - Montana: Nearly 8 in every 100 people (7.7%)
- In 2016, current smokeless tobacco use was lowest in:
  - District of Columbia: About 1 in every 100 people (1.3%)
  - Rhode Island: Nearly 2 in every 100 people (1.5%)
  - Maryland: Nearly 2 in every 100 people (1.6%)
  - California: Nearly 2 in every 100 people (1.7%)

Youth Smokeless Tobacco Use

The table below shows the percentage of high school students who were current users of smokeless tobacco in 2017.

Table 9.1. Smokeless Tobacco Use among Youth

High School Students in 2017	Current Use of Smokeless Tobacco
Overall	5.50%
Males	7.70%
Females	3.00%
White non-Hispanic	7.20%
Black non-Hispanic	1.80%
Hispanic	3.70%

Multiple Product Use

According to the 2012 National Survey on Drug Use and Health (NSDUH):

- About 1 in every 100 youth aged 12 to 17 years (1.1%) and nearly 4 in every 100 young adults aged 18 to 25 years (3.9%) were current users of smokeless tobacco and at least one other tobacco product.
- About 1 in every 100 adults aged 26 years or older (1.2%) were current users of smokeless tobacco and at least one other tobacco product.

Chapter 10

Tobacco Use among Specific Populations

Chapter Contents

115

Section 10.1

Youth and Tobacco Use

This section includes text excerpted from "Youth and Tobacco Use,"
Centers for Disease Control and Prevention (CDC), February 5, 2019.

If smoking continues at the current rate among youth in this country, 5.6 million of today's Americans younger than 18 will die early
from a smoking-related illness. That's about 1 of every 13 Americans
aged 17 years or younger alive today.

Preventing Tobacco Use among Youth Is Critical to Ending the Tobacco Epidemic in the United States

- Tobacco product use is started and established primarily during
  adolescence.

  - Nearly 9 out of 10 cigarette smokers first try cigarette
    smoking by age 18, and 98% first try smoking by age 26

  - Each day in the United States, about 2,000 youth under 18
    years of age smoke their first cigarette, and more than 300
    youth under 18 years of age become daily cigarette smokers.

- Flavorings in tobacco products can make them more appealing to
  youth.

  - In 2014, 73 percent of high school students and 56 percent of
    middle school students who used tobacco products in the past
    30 days reported using a flavored tobacco product during that
    time.

Estimates of Tobacco Use among Youth
Electronic Cigarettes

- Current use of electronic cigarettes increased among middle-
  and high-school students from 2011 to 2018.

  - Nearly 1 of every 20 middle-school students (4.9%) reported in
    2018 that they used electronic cigarettes in the past 30 days—
    an increase from 0.6 percent in 2011.

  - Nearly 1 of every 5 high-school students (20.8%) reported in
    2018 that they used electronic cigarettes in the past 30 days—
    an increase from 1.5 percent in 2011.

Cigarettes

- From 2011 to 2018, current (past 30day) cigarette smoking went down among middle- and high-school students.

  - Nearly 1 of every 50 middle-school students (1.8%) reported in 2018 that they smoked cigarettes in the past 30 days—a decrease from 4.3 percent in 2011.

  - About 2 of every 25 high-school students (8.1%) reported in 2018 that they smoked cigarettes in the past 30 days—a decrease from 15.8 precent in 2011.

Cigars

- From 2011 to 2018, current use of cigars went down among middle- and high-school students.

  - Nearly 1 of every 50 middle-school students (1.6%) reported in 2018 that they had used cigars in the past 30 days—a decrease from 3.5 percent in 2011.

  - Nearly 4 of every 50 high-school students (7.6%) reported in 2018 that they had used cigars in the past 30 days—a decrease from 11.6 percent in 2011.

Smokeless Tobacco

- From 2011 to 2018, current use of smokeless tobacco went down among middle- and high-school students:

  - Nearly 2 of every 100 middle-school students (1.9%) reported current use of smokeless tobacco.

  - Nearly 6 of every 100 high-school students (5.9%) reported in 2018 that they had used smokeless tobacco in the past 30 days—a decrease from 7.9 percent in 2011.

Hookahs

- From 2011 to 2018, current use of hookahs did not change in a meaningful way among middle- and high-school students.

  - About 1 of every 100 middle-school students (1.2%) reported in 2018 that they had used hookah in the past 30 days. The rate was 1.0 percent in 2011.

- About 4 of every 100 high-school students (4.1%) reported in 2018 that they had used hookah in the past 30 days. The rate was 4.1 percent in 2011.

All Tobacco Product Use

- In 2018, about 7 of every 100 middle-school students (7.2%) and about 27 of every 100 high-school students (27.1%) reported current use of some type of tobacco product.

- In 2013, nearly 18 of every 100 middle-school students (17.7%) and nearly half (46.0%) of high-school students said they had ever tried a tobacco product.

Many Young People Use Two or More Tobacco Products

- In 2018, about 2 of every 100 middle-school students (2.4%) and about 11 of every 100 high-school students (11.3%) reported current use of two or more tobacco products in the past 30 days.

- In 2013, about 9 of every 100 middle-school students (9.4%) and about 31 of every 100 high-school students (31.4%) said they had ever tried two or more tobacco products.

Table 10.1. Tobacco Use* among High-School Students in 2018

Tobacco Product	Overall	Girls	Boys
Any tobacco product[†]	27.10%	24.90%	29.10%
Electronic cigarettes	20.80%	18.80%	22.60%
Cigarettes	8.10%	7.30%	8.80%
Cigars	7.60%	6.00%	9.00%
Smokeless tobacco	5.90%	3.30%	8.40%
Hookahs	4.10%	4.10%	4.00%
Pipe tobacco	1.10%	0.80%	1.40%

* *"Current use" is determined by respondents indicating that they have used a tobacco product on at least 1 day during the past 30 days.*
† *Any tobacco product includes cigarettes, cigars, smokeless tobacco (including chewing tobacco, snuff, dip, snus, and dissolvable tobacco), tobacco pipes, bidis, hookah, and electronic cigarettes.*

Table 10.2. Tobacco Use* among Middle-School Students in 2017

Tobacco Product	Overall	Girls	Boys
Any tobacco product†	7.20%	6.30%	8.00%
Electronic cigarettes	4.90%	4.80%	5.10%
Cigarettes	1.80%	1.50%	2.10%
Cigars	1.60%	1.60%	1.70%
Smokeless tobacco	1.80%	0.90%	2.70%
Hookahs	1.20%	1.00%	1.50%
Pipe tobacco	0.30%	0.40%	0.30%

* *"Current use" is determined by respondents indicating that they have used a tobacco product on at least 1 day during the past 30 days.*
† *Any tobacco product includes cigarettes, cigars, smokeless tobacco (including chewing tobacco, snuff, dip, snus, and dissolvable tobacco), tobacco pipes, bidis, hookah, and electronic cigarettes.*

Factors Associated with Youth Tobacco Use

Factors associated with youth tobacco product use include the following:

- Social and physical environments
  - The way mass media show tobacco product use as a normal activity can make young people want to try these products.
  - Youth are more likely to use tobacco products if they see people their age using these products.
  - High-school athletes are more likely to use smokeless tobacco than those of the same age who are not athletes.
  - Young people may be more likely to use tobacco products if a parent uses these products.
- Biological and genetic factors
  - There is evidence that youth may be sensitive to nicotine and that teens can feel dependent on nicotine sooner than adults.
  - Genetic factors may make quitting smoking harder for young people.

- A mother's smoking during pregnancy may increase the likelihood that her children will become regular smokers.

- Mental health: There is a strong relationship between youth smoking and depression, anxiety, and stress.

- Personal views: When young people expect positive things from smoking, such as coping with stress better or losing weight, they are more likely to smoke.

Other influences that affect youth tobacco use include:

- Lower socioeconomic status, including lower income or education

- Not knowing how to say "no" to tobacco product use

- Lack of support or involvement from parents

- Accessibility, availability, and price of tobacco products

- Doing poorly in school

- Low self-image or self-esteem

- Seeing tobacco product advertising in stores, on television, the Internet, in movies, or in magazines and newspapers

Reducing Youth Tobacco Use

National, state, and local program activities have been shown to reduce and prevent youth tobacco product use when implemented together. These activities include:

- Higher costs for tobacco products (for example, through increased taxes)

- Prohibiting smoking in indoor areas of workplaces and public places

- Raising the minimum age of sale for tobacco products to 21 years

- TV and radio commercials, posters, and other media messages aimed at kids and teens in order to counter tobacco product ads

- Community programs and school and college policies that encourage tobacco-free places and lifestyles

- Community programs that lower tobacco advertising, promotions, and help make tobacco products less easily available

Some social and environmental factors are related to lower smoking levels among youth. Among these are:

- Being part of a religious group or tradition

- Racial/ethnic pride and strong racial identity

- Higher academic achievement

It is important to keep working to prevent and reduce the use of all forms of tobacco product use among youth.

Section 10.2

Women and Tobacco Use

This section includes text excerpted from "Women and Smoking," Centers for Disease Control and Prevention (CDC), October 15, 2014. Reviewed February 2019.

In the last 50 years, a woman's risk of dying from smoking has more than tripled and is now equal to men's risk. The United States has more than 20 million women and girls who currently smoke cigarettes. Smoking puts them at risk for:

- Heart attacks

- Strokes

- Lung cancer

- Emphysema

- Other serious chronic illnesses such as diabetes.

More than 170,000 American women die of diseases caused by smoking each year, with additional deaths coming from the use of other tobacco products such as smokeless tobacco.

A Target Market

When the first Surgeon General's Report on smoking was released in 1964, it caused a rapid drop in smoking among men. Yet smoking rates among women continued to go up in the years immediately following the report as tobacco companies aggressively marketed to women. Documents from the tobacco industry show that cigarette companies created a line of slimmer cigarettes packaged in pastel colors to appeal to women, and implied that smoking could keep girls and women thin. They also used slogans, advertising, and sports sponsorships to tie their products to the women's rights movement throughout the 1960s and 1970s.

The women most likely to smoke today are among the most vulnerable—those disadvantaged by low income, less education, and mental health disorders. Women in these groups are also less likely to quit smoking when they become pregnant and are more likely to start smoking again after delivery. This worsens the dangerous health effects from smoking on mothers and their children.

Disease and Women Smokers

Many of the findings in the 2014 Surgeon General's Report are especially important for women who smoke. Between 1959 and 2010, lung cancer risk for smokers rose dramatically. While men's risk doubled, the risk among female smokers increased nearly ten-fold. Today, more women die from lung cancer than breast cancer.

Respiratory Diseases

Chronic obstructive pulmonary disease (COPD) includes emphysema, chronic bronchitis, and other conditions that damage airways. People with the disease suffer from shortness of breath and lack of oxygen that worsens over time. COPD has no cure. Nearly 9 out of 10 cases of COPD are caused by smoking. Women smokers in certain age groups are up to 38 times more likely to develop COPD than women who have never smoked. More women than men are now dying every year from COPD, and women appear more susceptible to developing severe COPD at younger ages.

Cardiovascular Disease

For more than half a century, the evidence that smoking causes cardiovascular disease has grown steadily. Today, women over age 35 who smoke have a slightly higher risk of dying from coronary heart disease than men who smoke. They are also slightly more likely to die from an abdominal aortic aneurysm—a weakened and bulging area of the artery that runs through the abdomen and carries blood to the major organs—than men who smoke.

Smoking and Pregnancy

Smoking during pregnancy causes premature birth, low birth weight, certain birth defects, and ectopic pregnancy in which the fertilized egg implants somewhere in the abdomen other than the womb. Smoking during pregnancy also causes complications with the placenta, the organ through which nutrients pass from mother to fetus. These complications include placenta previa and placental abruption, conditions that jeopardize the life and health of both mother and child. Women who are pregnant or who are planning a pregnancy should not smoke. It's important to encourage women to quit smoking before or early in pregnancy, when the most health benefits can be achieved, but cessation in all stages, even in late pregnancy, benefits maternal and fetal health.

Cessation

Nicotine addiction can be difficult to overcome, but over half of smokers in the United States have already quit. There are many support programs and cessation tools available to smokers who want to quit, including nicotine replacement products such as patches and gum, prescription medication, and free coaching. Benefits to women's health from quitting smoking are enormous and immediate. Heart attack risks drop dramatically in the first year and within five years, women who have quit smoking can see their stroke risk drop to that of a never smoker. In 10 years, a woman's risk of dying from lung cancer is cut in half. Women who want to quit smoking should ask their doctors for help, call 800-QUIT-NOW (800-7848-669), or visit women.smokefree.gov and cdc.gov/tips.

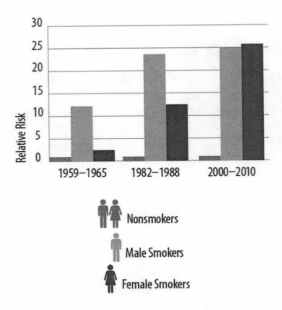

Figure 10.2. *Lung Cancer Risk for Smokers Compared to People Who Never Smoked*

Section 10.3

Adults and Tobacco Use

This section includes text excerpted from "Current Cigarette Smoking among Adults in the United States," Centers for Disease Control and Prevention (CDC), September 24, 2018.

Cigarette smoking is the leading cause of preventable disease and death in the United States, accounting for more than 480,000 deaths every year, or about 1 in 5 deaths.

In 2016, more than 15 of every 100 United States adults aged 18 years or older (15.5%) currently* smoked cigarettes. This means an estimated 37.8 million adults in the United States currently smoke cigarettes. More than 16 million Americans live with a smoking-related disease.

Current smoking has declined from 20.9 percent (nearly 21 of every 100 adults) in 2005 to 15.5 percent (more than 15 of every 100 adults) in 2016. The proportion of ever smokers who had quit increased; however, current smoking prevalence did not change significantly during 2015–2016.

Current smokers are defined as persons who reported smoking at least 100 cigarettes during their lifetime and who, at the time they participated in a survey about this topic, reported smoking every day or some days.

Smoking among Adults in 2016 (Nation)
By Gender

Men were more likely to be current cigarette smokers than women.

- Nearly 18 of every 100 adult men (17.5%)
- Nearly 14 of every 100 adult women (13.5%)

By Age

Current cigarette smoking was higher among persons aged 18 to 24 years, 25 to 44 years, and 45 to 64 years than among those aged 65 years and older.

- About 13 of every 100 adults aged 18 to 24 years (13.1%)
- Nearly 18 of every 100 adults aged 25 to 44 years (17.6%)
- 18 of every 100 adults aged 45 to 64 years (18.0%)
- Nearly 9 of every 100 adults aged 65 years and older (8.8%)

Figure 10.3. *Cigarette Smoking Is Down, but Almost 38 Million American Adults Still Smoke*

By Race / Ethnicity

Current cigarette smoking was highest among non-Hispanic American Indians/Alaska Natives and people of multiple races and lowest among Asians.

- Nearly 32 of every 100 non-Hispanic American Indians/Alaska Natives (31.8%)

- About 25 of every 100 non-Hispanic multiple race individuals (25.2%)

- Nearly 17 of every 100 non-Hispanic Blacks (16.5%)

- Nearly 17 of every 100 non-Hispanic Whites (16.6%)

- Nearly 11 of every 100 Hispanics (10.7%) of every 100 non-Hispanic Asians* (9.0%)

Non-Hispanic Asians does not include Native Hawaiians or Other Pacific Islanders.

By Education

Current cigarette smoking was highest among persons with a general education development (GED) certificate and lowest among those with a graduate degree.

- About 24 of every 100 adults with 12 or fewer years of education (no diploma) (24.1%)

- Nearly 41 of every 100 adults with a GED certificate (40.6%)

- Nearly 20 of every 100 adults with a high school diploma (19.7%)

- Nearly 19 of every 100 adults with some college (no degree) (18.9%)

- Nearly 17 of every 100 adults with an associate's degree (16.8%)

- Nearly 8 of every 100 adults with an undergraduate degree (7.7%)

- Nearly 5 of every 100 adults with a graduate degree (4.5%)

By Poverty Status

Current cigarette smoking was higher among persons living below the poverty* level than those living at or above this level.

- About 25 of every 100 adults who live below the poverty level (25.3%)

- About 14 of every 100 adults who live at or above the poverty level (14.3%)

Poverty thresholds are based on United States Census Bureau data.

By United States Census Region

Current cigarette smoking was highest in the Midwest and lowest in the West.

- Nearly 19 of every 100 adults who live in the Midwest (18.5%)

- Nearly 17 of every 100 adults who live in the South (16.9%)

- About 13 of every 100 adults who live in the Northeast (13.3%)

- About 12 of every 100 adults who live in the West (12.3%)

By Disability / Limitation

Current cigarette smoking was higher among persons with a disability/limitation than among those with no disability/limitation.

- About 21 of every 100 adults who reported having a disability/ limitation (21.2%)

- About 14 of every 100 adults who reported having no disability/ limitation (14.4%)

By Sexual Orientation

Lesbian/gay/bisexual adults were more likely to be current smokers than straight adults.

- Nearly 21 of every 100 lesbian/gay/bisexual adults (20.5%)

- About 15 of every 100 straight adults (15.3%)

By Serious Psychological Distress*

Adults that had experienced serious psychological distress were more likely to be current smokers than adults that did not report serious psychological distress.

- Nearly 36 of every 100 adults with serious psychological distress (35.8%)

- Nearly 15 of every 100 adults without serious psychological distress (14.7%)

* Measures of serious psychological distress are based on the Kessler psychological distress scale.

Smoking among Adults in 2016 (States)

- In 2016, current smoking ranged from nearly 9 of every 100 adults in Utah (8.8%) to nearly 25 of every 100 adults in West Virginia (24.8%).

Figure 10.4. *Percentage of Adults in Each State Who Were Current Smokers in 2016*

Section 10.4

Lesbian, Gay, Bisexual, and Transgender Persons and Tobacco Use

This section includes text excerpted from "Lesbian, Gay, Bisexual, and Transgender Persons and Tobacco Use," Centers for Disease Control and Prevention (CDC), August 29, 2018.

People who are lesbian, gay, bisexual, or transgender (LGBT) include all races and ethnicities, ages, and socioeconomic groups, and come from all parts of the United States It is estimated that lesbian, gay, and bisexual (LGB) persons make up approximately three percent of the total United States population.

Cigarette smoking among LGB individuals in the United States is higher than among heterosexual/straight individuals. About one in five LGB adults smoke cigarettes compared with about one in six heterosexual/straight adults.

Cigarette Smoking Prevalence

20.5 percent of LGB adults smoke cigarettes compared to 15.3 percent of straight adults.

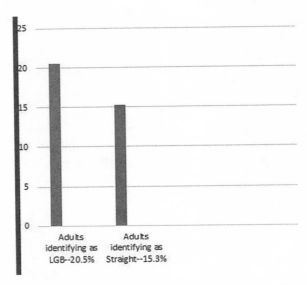

Figure 10.5. *Cigarette Smoking Prevalence among LGB and Straight Adults*

* Data taken from the 2016 National Health Interview Survey (NHIS) and refer to adults aged 18 years and older.

Transgender Individuals

Limited information exists on cigarette smoking prevalence among transgender people; however, cigarette smoking prevalence among transgender adults is reported to be higher than among the general population of adults.

The transgender population is considered especially vulnerable because of high rates of substance abuse, depression, human immunodeficiency virus (HIV) infection, and social and employment discrimination, all of which are associated with higher smoking prevalence.

Health Effects

- Gay men have high rates of human papillomavirus (HPV) infection which, when coupled with tobacco use, increases their risk for anal and other cancers.

- LGBT individuals often have risk factors for smoking that include daily stress related to prejudice and stigma that they may face.
- Bartenders and servers in LGBT nightclubs are exposed to high levels of secondhand smoke.
- Among women, secondhand smoke exposure is more common among nonsmoking lesbian women than among nonsmoking straight women.

Quitting Behavior

- LGB individuals are five times more likely than others to never intend to call a smoking cessation quitline.
- LGBT individuals are less likely to have health insurance than straight individuals, which may negatively affect health as well as access to cessation treatments, including counseling and medication.
- Gay, bisexual, and transgender men are 20 percent less likely than straight men to be aware of smoking quitlines despite LGBT individuals having exposure to tobacco cessation advertising similar to straight individuals' exposure.

Tobacco Industry Marketing and Influence

- High rates of tobacco use within the LGBT community are due in part to the aggressive marketing by tobacco companies that sponsor events, bar promotions, giveaways, and advertisements. Tobacco companies advertise at "gay pride" festivals and other LGBT community events and contribute to local and national LGBT and HIV/AIDS organizations.
- Tobacco advertisements in gay and lesbian publications often depict tobacco use as a "normal" part of LGBT life.
- The tobacco industry encourages menthol cigarette use among LGBT populations.
  - Approximately 36 percent of LGBT smokers report smoking menthol cigarettes compared to 29 percent of heterosexual/ straight smokers.
- The marketing campaign, Project SCUM (Subculture Urban Marketing), was created in the mid-1990s by a tobacco company to target LGBT and homeless populations.

Culturally appropriate antismoking health marketing strategies and mass media campaigns like Centers for Disease Control and Prevention's (CDC) Tips From Former Smokers national tobacco education campaign, as well as CDC-recommended tobacco prevention and control programs and policies, can help reduce the burden of disease among the LGBT population.

Section 10.5

Hispanics and Tobacco Use

This section includes text excerpted from "Hispanics/Latinos and Tobacco Use," Centers for Disease Control and Prevention (CDC), September 10, 2018.

Hispanic or Latinx is defined by the Office of Management and Budget (OMB) as "a person of Cuban, Mexican, Puerto Rican, South or Central American, or other Spanish culture or origin, regardless of race."

In the United States, there are more than 55 million people of Hispanic or Latinx ethnicity. This group makes up approximately 17 percent of the current estimated population, and is expected to comprise nearly 30 percent of the United States population by the year 2060.

Hispanic/Latinx adults generally have lower prevalence of cigarette smoking and other tobacco use than other racial/ethnic groups, with the exception of Asian Americans. However, prevalence varies among sub-groups within the Hispanic population.

Health Effects

Cancer, heart disease, and stroke—all of which can be caused by cigarette smoking—are among the five leading causes of death among Hispanics.

- Diabetes is the fifth leading cause of death among Hispanics. The risk of developing diabetes is 30 to 40 percent higher for cigarette smokers than nonsmokers.

Tobacco Use Prevalence

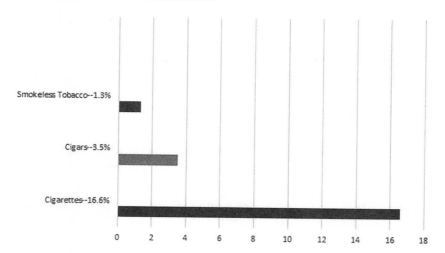

Figure 10.6. *Current Tobacco Use* among Hispanic/Latinx Adults†*
* "Current Use" is defined as self-reported consumption of cigarettes, cigars, or smokeless tobacco in the past month.
† Data taken from the National Survey on Drug Use and Health (NSDUH), 2016, and refer to Hispanic/Latino Americans aged 18 years and older.

- Prevalence of cigarette smoking is higher among Hispanic adults born in the United States than those who were foreign-born.

Cigarette smoking prevalence varies by Hispanic/Latinx subgroups:

Table 10.3. Prevalence of Cigarette Smoking among Hispanic/Latinx Sub-Groups

Hispanic/Latinx Sub-Group	Prevalence§
Puerto Rican	28.50%
Cuban	19.80%
Mexican	19.10%
Cent'al or South American	15.60%

§ Prevalence data for Hispanic/Latino sub-groups taken from National Survey on Drug Use and Health (NSDUH), 2010–2013, and include persons aged 18 years and older who reported smoking cigarettes during the past month.

Patterns of Tobacco Use

Variations in cigarette smoking exist among different Hispanic subgroups.

- Number of cigarettes smoked per day is highest among Cuban daily smokers than daily smokers within other Hispanic/Latinx groups.
  - 50 percent of Cuban men and more than 35 percent of Cuban women report smoking 20 or more cigarettes per day.
- Mexican men and women are less likely than other Hispanic/Latinx groups to report that they smoke 20 or more cigarettes per day.
- Intermittent current cigarette smoking (smoking only some days in the past month) is most common among Mexican men— 15.5 percent compared to 9.8 percent of Central American men, nine percent of Puerto Rican men, and 4.9 percent of Cuban men.
- Hispanic women generally have low prevalence of cigarette smoking during pregnancy.

Secondhand Smoke Exposure

During 2011–2012, nearly 58 million people were exposed to secondhand smoke in the United States, including 6.2 million Mexican American nonsmokers.

During 2011–2012:

- 29.9 percent of Mexican-American children aged 3 to 11 years were exposed to secondhand smoke.
- 16.9 percent of Mexican American adolescents aged 12 to 19 years and 23.8 percent of Mexican American adults aged 20 years and older were exposed to secondhand smoke.

Quitting Behavior

- Among Hispanic current daily cigarette smokers aged 18 years and older:
  - An estimated 67.4 percent report that they want to quit compared with 72.8 percent of African Americans, 67.5 percent of Whites, 69.6 percent of Asians, and 55.6 percent of American Indians/Alaska Natives.

- An estimated 56.2 percent report attempting to quit in the past year compared with 63.4 percent of African Americans, 53.3 percent of Whites, and 39.4 percent of Asians.

- Hispanics/Latinx have lower health insurance coverage and less healthcare access than Whites, making it less likely that they will be advised by a healthcare provider to quit smoking cigarettes or to have access to cessation treatments.

- Hispanics, however, still quit smoking at higher rates than Whites and the general population.

Tobacco Industry Marketing and Influence

Tobacco products are advertised and promoted disproportionately to racial/ethnic minority communities. Tobacco companies seek to appeal to the Hispanic population through branding, financial contributions, and targeted advertising.

- Historically, cigarette brand names such as "Rio" and "Dorado" have been heavily advertised and marketed to the Hispanic American community, including advertisements in many Hispanic publications.

- The tobacco industry has contributed to programs that enhance education of young people, such as funding universities and colleges and supporting scholarship programs targeting Hispanics.

- The tobacco industry has also provided significant support to Hispanic political organizations, cultural events, and the Hispanic art community.

Culturally appropriate antismoking health marketing strategies and mass media campaigns such as the CDC's Tips From Former Smokers national tobacco education campaign, as well as CDC-recommended tobacco prevention and control programs and policies, can help reduce the burden of disease among the Hispanic/Latinx population.

Section 10.6

Asian Americans and Pacific Islanders and Tobacco Use

This section includes text excerpted from "Asian Americans, Native
Hawaiians, or Pacific Islanders and Tobacco Use," Centers for
Disease Control and Prevention (CDC), August 28, 2018.

Asian American is defined by the Office of Management and Budget
(OMB) as "a person having origins in any of the original peoples of the
Far East, Southeast Asia, or the Indian subcontinent."

The six largest sub-groups of Asian Americans are from China, the
Philippines, India, Korea, Vietnam, and Japan.

Native Hawaiian or Other Pacific Islander is defined as "a person
having origins in any of the original peoples of Hawaii, Guam, Samoa,
or other Pacific Islands." There were over 16.6 million Asian Americans
in the United States in 2016—approximately 5.2 percent of the total
United States population. Native Hawaiians or Other Pacific Islanders
made up 0.2 percent of the population in 2016.

Although Asian Americans, Native Hawaiians, and Pacific Islanders (AA/NH/PI) are often combined together as one group in survey
data due to smaller numbers of the individual groups surveyed, they
are actually three distinct groups.

Cigarette smoking among Asian American/Pacific Islander adults
is lower than other racial/ethnic groups. However, prevalence among
Asian sub-groups varies and can be higher than that of the general
population.

Health Effects

The three leading causes of death among AA/NH/PI are cancer,
heart disease, and stroke, all of which can be caused by cigarette
smoking.

- Lung cancer is the leading cause of cancer death among AA/NH/
  PI. Within this population

  - Hawaiian men and women have the highest rates of lung
    cancer deaths.

  - Filipino men and women have the lowest rates of lung cancer
    deaths.

Tobacco Use Prevalence

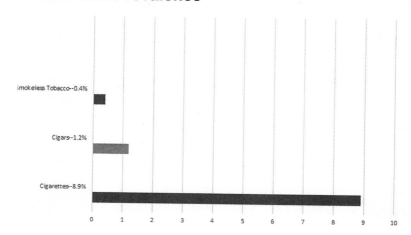

Figure 10.7. *Current Tobacco Use* among Asian American Adults†*

* *"Use" is defined as self-reported consumption of cigarettes, cigars, or smokeless tobacco in the past month.*
† *Data taken from the National Survey on Drug Use and Health (NSDUH), 2016, and refer to Asian American adults aged 18 years and older.*

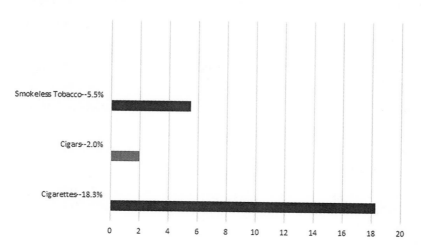

Figure 10.8. *Current Tobacco Use* among Native Hawaiian/Pacific Islander Adults†*

* *"Use" is defined as self-reported consumption of cigarettes, cigars, or smokeless tobacco in the past month.*
† *Data taken from the National Survey on Drug Use and Health (NSDUH), 2016, and refer to Asian American adults aged 18 years and older.*

By sub-groups, prevalence of cigarette smoking among Asian Americans varies considerably:

Table 10.4. Prevalence of Cigarette Smoking among Asian Americans

Asian Sub-Group	Cigarette Smoking Prevalence[§]
Chinese	7.60%
Asian Indian	7.60%
Japanese	10.20%
Filipino	12.60%
Vietnamese	16.30%
Korean	20.00%

§ *Data taken from National Survey on Drug Use and Health (NSDUH), 2010–2013, and include persons aged 18 years and older who reported smoking cigarettes during the past month.*

Patterns of Tobacco Use

Cigarette smoking varies by AA/NH/PI sub-groups as a result of a number of cultural, social, environmental, and individual factors.

- In general, Asian Americans and Pacific Islanders tend to be light, nondaily smokers.

- One study among Asian American and Pacific Islander college students in California showed that:

  - 48 percent of current smokers smoke cigarettes on fewer than two days per month.

  - 46 percent of current smokers smoke less than two cigarettes per day.

- Cigarette smoking during pregnancy is less common among Asian American/Pacific Islander women compared to other racial/ethnic groups.

Quitting Behavior

- Among Asian American current daily cigarette smokers aged 18 years and older:

  - 69.6 percent report that they want to quit compared to 72.8 percent of African Americans, 67.5 percent of Whites, 67.4

percent of Hispanics, and 55.6 percent of American Indians/
Alaska Natives.

- 69.4 percent report attempting to quit during the past year
  compared to 56.2 percent of Hispanics, 63.4 percent of African
  Americans, and 53.3 percent of whites.

Tobacco Industry Marketing and Influence

Tobacco companies have been creative in their efforts to reach different racial/ethnic groups. AA/NH/PI smokers are a key market for tobacco companies since cigarette smoking prevalence in most Asian countries is considerably higher than in the United States.

- The tobacco industry has sponsored Chinese New Year and
  Vietnamese New Year festivals and other activities related to
  Asian/Pacific American Heritage Month.

- Tobacco advertisements on billboards and in stores are more
  plentiful in predominantly urban Asian American communities
  than in other urban neighborhoods.

- Tobacco companies promote themselves among Asian American
  business owners by supporting their business associations and
  offering special retail sales materials.

Section 10.7

American Indians and Alaska Natives and Tobacco Use

This section includes text excerpted from "American Indians/ Alaska Natives and Tobacco Use," Centers for Disease Control and Prevention (CDC), August 28, 2018.

American Indian/Alaska Native is defined by the Office of Management and Budget (OMB) as "a person having origins in any of the original peoples of North and South America (including Central America), and who maintains tribal affiliation or community attachment." There are approximately 2.6 million American Indians/Alaska Natives in the United States—about one percent of the total population.

American Indians/Alaska Natives have the highest prevalence of cigarette smoking compared to all other racial/ethnic groups in the United States. Some American Indians use tobacco for ceremonial, religious, or medicinal purposes. For this reason, it is important to make the distinction between commercial and traditional tobacco use.

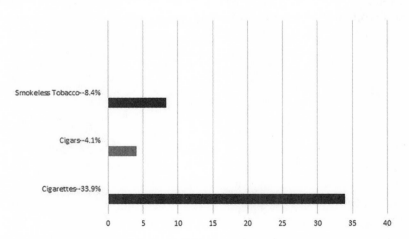

Figure 10.9. *Current Tobacco Use* among American Indian/Alaska Native Adults—2016†*

* "Current Use" is defined as self-reported consumption of cigarettes, cigars, or smokeless tobacco in the past month.
† Data taken from the National Survey on Drug Use and Health, 2016, and refer to American Indians/Alaska Natives aged 18 years and older.

140

Health Effects

American Indians/Alaska Natives have a higher risk of experiencing tobacco-related disease and death due to high prevalence of cigarette smoking and other commercial tobacco use.

- Cardiovascular disease is the leading cause of death among American Indians/Alaska Natives.

- Lung cancer is the leading cause of cancer deaths among American Indians/Alaska Natives.

- Diabetes is the fourth leading cause of death among American Indians/Alaska Natives. The risk of developing diabetes is 30 to 40 percent higher for smokers than nonsmokers.

Patterns of Tobacco Use

- American Indian/Alaska Native youth and adults have the highest prevalence of cigarette smoking among all racial/ethnic groups in the United States.

- Regional variations in cigarette smoking exist among American Indians/Alaska Natives, with lower prevalence in the Southwest and higher prevalence in the Northern Plains and Alaska.

- More American Indian/Alaska Native women smoke during their last 3 months of pregnancy—26.0 percent compared to 14.3 percent of whites, 8.9 percent of African Americans, 3.4 percent of Hispanics, and 2.1 percent of Asians/Pacific Islanders.

Quitting Behavior

- Quitting rates are relatively low among American Indians/Alaska Natives compared to other racial/ethnic groups.

- An estimated 55.6 percent of American Indians/Alaska Natives report that they want to quit compared to 72.8 percent of African Americans, 67.5 percent of Whites, 69.6 percent of Asians, and 67.4 percent of Hispanics.

Tobacco Industry Marketing and Influence

Tobacco companies target American Indian/Alaska Native communities through extensive promotions, sponsorships, and advertising campaigns.

- Historically, tobacco industry product promotions to American Indians/Alaska Natives featured symbols and names with special meanings to this group. For example, the American Spirit™ cigarettes were promoted as "natural" cigarettes, and their packaging featured an American Indian smoking a pipe.

Chapter 11

Minors' Access to Tobacco Products

Facts about Minors' Access to Tobacco Products

- It is illegal in all states to sell cigarettes to persons under age 18. Progress has been made in the past several years in reducing the percentage of retailers willing to sell tobacco to minors.

- In 1991, an estimated 225 million packs of cigarettes were sold illegally to minors, and in 1997 daily smokers aged 12 to 17 years smoked approximately 924 million packs of cigarettes.

- An estimated 20 to 70 percent of teenagers who smoke report purchasing their own tobacco; the proportion varies by age, social class, amount smoked, and factors related to availability.

- More than two-thirds of states restrict cigarette vending machines, but many of these restrictions are weak. Only two states (Idaho and Vermont) have total bans on vending machines.

This chapter contains text excerpted from the following sources: Text under the heading "Facts about Minors' Access to Tobacco" is excerpted from "Highlights: Minors' Access to Tobacco," Centers for Disease Control and Prevention (CDC), July 21, 2015. Reviewed February 2019; Text beginning with the heading "Report Potential Tobacco Product Violation" is excerpted from "Report Potential Tobacco Product Violation," U.S. Food and Drug Administration (FDA), October 31, 2018.

- Results from nine published studies found illegal vending machine sales to minors ranged from 82 to 100 percent between 1989 and 1992.

- More than 290 local jurisdictions, including New York City, successfully adopted and enforced outright bans on cigarette vending machines or restricted them to locations such as taverns and adult clubs where minors often are denied entry.

- Almost two-thirds of the states and many local jurisdictions require retailers to display signs that state the minimum age for purchase of tobacco products. Some regulations specify the size, wording, and location of these signs.

- All states have a specific restriction on the distribution of free tobacco samples to minors, and a few states or local jurisdictions prohibit free distribution altogether because of the difficulty of controlling who receives free samples.

- Several studies have found that single or loose cigarettes are sold in some locations. Such sales often are prohibited by state or local law, given single cigarettes do not display the required state tax stamp or federal health warning.

- Other regulations specify a minimum age for salespersons. These regulations recognize the difficulty young salespersons may have in refusing to sell cigarettes to their peers.

- Many state or local laws specify penalties only for the salesperson. However, applying penalties to business owners, who generally set hiring, training, supervising, and selling policies, is considered essential to preventing the sale of tobacco to minors.

- License suspensions or revocations imposed as penalties for repeated violation of youth access laws would communicate a clear message that illegal tobacco sales to minors should never be accepted or tolerated. Revenues from fines could be used for enforcement and retailer education programs.

- Numerous studies have shown that comprehensive merchant education and training programs help reduce illegal sales to minors.

- Growing number of states and local jurisdictions are imposing sanctions against minors who purchase, attempt to purchase, or possess tobacco products. Although these laws are a potential deterrent, some tobacco control advocates believe such laws deflect responsibility from retailers to underage youth.

- In 1992, the Synar Amendment (Public Law 102–321), was passed to curb the illegal sale of tobacco products to minors. An amended Synar Regulation, was issued by the Substance Abuse and Mental Health Services Administration in January 1996, and requires each state receiving federal grant money to conduct annual random, unannounced inspections of retail tobacco outlets to assess the extent of sales to minors. In 1999, seven states and the District of Columbia (DC) failed to attain their Synar Amendment targets. Failure to comply with the law puts states at risk of forfeiting federal block grant funds for substance abuse prevention and treatment services.

- On March 21, 2000, the United States Supreme Court ruled that the U.S. Food and Drug Administration (FDA) lacked jurisdiction to regulate tobacco products and to enforce rules to reduce the access and appeal of tobacco products for children and adolescents. The loss of the FDA's education and enforcement program eliminates vital federal support for state tobacco control programs.

Report Potential Tobacco Product Violation

The minimum legal age to purchase tobacco is 18, and yet, 87 percent of adult daily smokers begin smoking before age 18.

The Tobacco Control Act and related regulations give the FDA tools to help keep tobacco out of the hands of America's youth. But you also have an opportunity to play a key role. The public is crucial in helping the FDA enforce tobacco regulations to protect America's youth.

What Is a Potential Tobacco Product Violation?

Potential tobacco product violations include (but are not limited to):

- Sales to minors

- Flavored cigarette sales

- Illegal marketing and advertising—The Tobacco Control Act gives the FDA the ability to regulate certain marketing and advertising activities by the tobacco industry, including:

  - Describing tobacco products as "light," "mild," or "low"—or claiming a product is safer or less harmful without an FDA order

145

- Distributing T-shirts or other promotional or novelty items with brand names of cigarette or smokeless tobacco products

- Sponsoring events using the brand name of a tobacco product

- Distribution of free samples of tobacco products except in limited circumstances

- Placement of cigarette or smokeless tobacco product vending machines in prohibited areas (or providing access to self-service or direct access of tobacco products in prohibited areas)

- Sale of cigarettes in packages of less than 20

How Can I Report a Potential Tobacco Violation?

If you see what you believe to be a violation of the Tobacco Control Act or other related regulations, you can:

- Submit online: (www.accessdata.fda.gov/scripts/ptvr/index.cfm)

- Call the Tobacco Call Center using Center for Tobacco Product's (CTP) toll-free number: 877-287-1373

- Send an email: CTPCompliance@FDA.hhs.gov

- Print and mail: (www.fda.gov/downloads/AboutFDA/ReportsManualsForms/Forms/UCM343402.pdf) to

Potential Tobacco Products Violation Report

U.S. Food and Drug Administration (FDA)
Center for Tobacco Products
Office of Compliance and Enforcement
Document Control Center
Bldg. 71, Rm. G335
10903 New Hampshire Ave.
Silver Spring, MD 20993

What Happens When I Submit a Potential Violation?

The FDA will evaluate any report submitted to determine if the activity is a violation of the Tobacco Control Act or related regulations. Before deciding what follow-up action, if any, is necessary, they will check to see if the product named in the complaint is regulated by the FDA. If the product is regulated by a different federal or state agency, or different part of the FDA, they will forward the complaint to the applicable entity for review.

The FDA does not rely solely on what was submitted to take enforcement action. After reviewing a complaint, our investigation may include:

- Performing an inspection of a tobacco product manufacturer, distributor, or importer;

- Conducting a compliance check inspection of a tobacco retailer; or

- Initiating monitoring and surveillance of a tobacco product manufacturer's or retailer's website.

The FDA may determine that there is no evidence of a violation, or may find evidence of the reported violation or of other potential violations that requires additional surveillance, monitoring, and/or inspections.

Privacy and Anonymity

All reports to the FDA remain private to the extent allowed by law as explained in the FDA's Privacy Policy. Reports can be submitted anonymously; however, reports accompanied by names and contact information are helpful if the FDA regulators need to follow up for more information.

Children's Privacy: The FDA will not collect or store information from children under 13. If a child sends the FDA an e-mail inquiry or comment, they will answer it and then delete the e-mail from our files.

Chapter 12

Burden of Tobacco Use in the United States

Cigarette Smoking among U.S. Adults Aged 18 Years and Older

Tobacco use remains the single largest preventable cause of death and disease in the United States. Cigarette smoking kills more than 480,000 Americans each year, with more than 41,000 of these deaths from exposure to secondhand smoke. In addition, smoking-related illness in the United States costs more than $300 billion a year, including nearly $170 billion in direct medical care for adults and $156 billion in lost productivity.

In 2016, an estimated 15.5 percent (37.8 million) of United States adults were current* cigarette smokers. Of these, 76.1 percent smoked every day.

This chapter contains text excerpted from the following sources: Text beginning with the heading "Current Cigarette Smoking among U.S. Adults Aged 18 Years and Older" is excerpted from "Burden of Tobacco Use in the U.S.," Centers for Disease Control and Prevention (CDC), January 30, 2018; Text under the heading "Tobacco-Related Mortality" is excerpted from "Tobacco-Related Mortality," Centers for Disease Control and Prevention (CDC), January 17, 2018.

Table 12.1. By Race/Ethnicity

Race/Ethnicity	Prevalence
American Indian/Alaska Natives (non-Hispanic)	31.80%
Asians (non-Hispanic)	9.00%
Blacks (non-Hispanic)	16.50%
Hispanics	10.70%
Multiple Races (non-Hispanic)	25.20%
Whites (non-Hispanic)	**16.60%**

Table 12.2. By Sex

Sex	Prevalence
Men	17.50%
Women	13.50%

Table 12.3. By Age

Age	Prevalence
18–24 years	13.10%
25–44 years	17.60%
45–64 years	18.00%
65 years and older	8.80%

Table 12.4. By Education

Education Level	Prevalence
Less than high-school	24.10%
GED	40.60%
High-school graduate	19.70%
Some college	18.90%
Associate degree	16.80%
Undergraduate degree	7.70%
Graduate degree	4.50%

Table 12.5. By Poverty Status

Income Status	Prevalence
Below poverty level	25.30%
At or above poverty level	14.30%

* *Current cigarette smoking is defined as smoking 100 cigarettes or more during one's lifetime and currently smoking every day or some days.*

Cigarette Smoking among Specific Populations— United States

American Indians / Alaska Natives (Non-Hispanic)

American Indians/Alaska Native's (AI/AN) have a higher prevalence of current smoking than most other racial/ethnic groups in the United States. Factors that may affect smoking prevalence include sacred tobacco's ceremonial, religious, and medicinal roles in Native culture, which may affect attitudes, beliefs, and behaviors toward commercial tobacco use. Also, tobacco sold on tribal lands is typically not subject to state and national taxes, which reduces costs. Lower prices are connected with increased smoking rates.

- In 2016:

  - 31.8 percent of AI/AN adults in the United States smoked cigarettes, compared with 15.5 percent of U.S. adults overall.

  - The prevalence of cigarette smoking was nearly one in three (or 29.3%) among AI/AN men and about one in three (or 34.3%) among AI/AN women.

Asians (Non-Hispanic)

Asian Americans represent a wide variety of languages, dialects, and cultures. While non-Hispanic Asian adults have the lowest current cigarette smoking prevalence of any racial/ethnic group in the United States, there are significant differences in smoking prevalence among subgroups in this population. Many Asian Americans emigrate from countries where smoking rates are high and smoking among men is the social norm. However, research also shows an association between cigarette smoking and acculturation among Asian Americans, with those having higher English-language proficiency and those living in the United States longer being less likely to smoke.

- In 2016:

  - 9.0 percent of non-Hispanic Asian adults in the United States smoked cigarettes, compared with 15.5 percent of U.S. adults overall.

  - Cigarette smoking prevalence was significantly higher among non-Hispanic Asian men (14.0%) than among non-Hispanic Asian women (4.6%).

Surveys conducted during 2010–2013 by the National Survey on Drug Use and Health (NSDUH) reflect a more comprehensive representation of current smoking among Asian subpopulations:

- The percentage of respondents who reported smoking within the past 30 days by subpopulations surveyed were:

  - Chinese–7.6%

  - Asian Indian–7.6%

  - Japanese–10.2%

  - Filipino–12.6%

  - Vietnamese–16.3%

  - Korean–20.0%

- Among women, cigarette smoking prevalence ranged from 2.9 percent among Chinese women to 20.4 percent among Koreans.

- Among men, cigarette smoking prevalence ranged from 11.6 percent among Asian Indians to 24.4 percent among Vietnamese individuals.

Blacks (Non-Hispanic)

Although cigarette smoking rates are lower among Black high school students than among U.S. high school students overall (3.9% compared with 8.0% in 2016), this difference is not sustained in adulthood.

- In 2016:

  - 16.5 percent of non-Hispanic Black adults in the United States smoked cigarettes, compared with 15.5 percent of U.S. adults overall.

  - Smoking prevalence was significantly higher among non-Hispanic Black men (20.2%) than among non-Hispanic Black women (13.5%).

- From 2005 to 2016, current cigarette smoking prevalence decreased among non-Hispanic Blacks (from 21.5% in 2005 to 16.5% in 2016).

Hispanics

The prevalence of cigarette smoking among Hispanics is generally lower than the prevalence among other racial/ethnic groups in the United States, with the exception of non-Hispanic Asians. However, smoking prevalence among Hispanic men is significantly higher than among Hispanic women, and there are significant differences in smoking prevalence among subgroups in this population. Research also shows that acculturation plays a role and that smoking prevalence is higher among Hispanics who were born in the United States.

- In 2016:

  - 10.7 percent of Hispanic adults in the United States smoked cigarettes, compared with 15.5 percent among U.S. adults overall.

  - Cigarette smoking prevalence was significantly higher among Hispanic men (14.5%) than among Hispanic women (7.0%).

- From 2005 to 2016, a decline in current smoking prevalence among Hispanics was noted (from 16.2% in 2005 to 10.7% in 2016).

Surveys conducted during 2010–2013 by the National Survey on Drug Use and Health also reflect a more comprehensive representation of current smoking among Hispanic subpopulations:

- The percentage of respondents who reported smoking within the past 30 days by subpopulations surveyed were:

  - Central or South American–15.6%

  - Mexican–19.1%

  - Cuban–19.8%

  - Puerto Rican–28.5%

- Among women, smoking prevalence ranged from 11.4 percent among Central or South Americans to 25.1 percent among Puerto Ricans.

- Among men, smoking prevalence ranged from 19.8 percent among Central or South Americans to 32.1 percent among Puerto Ricans.

153

Lesbian, Gay, Bisexual, and Transgender

Smoking prevalence among lesbian, gay, bisexual, and transgender (LGBT) individuals in the United States is higher than among the total population. This may be in part due to the aggressive marketing of tobacco products to this community. LGBT individuals are also likely to have risk factors for smoking that include daily stress related to prejudice and stigma that they may face.

In 2016, the prevalence of current cigarette smoking among lesbian, gay, and bisexual individuals was 20.5 percent, compared with 15.3 percent among heterosexual/straight individuals.

Military Service Members and Veterans

In the United States, cigarette smoking prevalence is higher among people currently serving in the military than among the civilian population. Cigarette smoking prevalence is even higher among military personnel who have been deployed.

- In 2011, 24.0 percent of all active-duty military personnel reported currently smoking cigarettes, compared with 19.0 percent of civilians.

- During 2007–2010, male veterans aged 25–64 years were more likely to be current smokers than nonveterans (29.0% versus 24.0%).

Women Who Are Pregnant or Plan to Become Pregnant

In the United States, modest decreases in cigarette smoking during pregnancy and after delivery occurred between 2000 and 2010, according to a study of 39 states and New York City.

In 2010, data from 27 sites, representing 52 percent of live births, showed that among women with recent live births:

- About 23 percent reported smoking in the three months prior to pregnancy.

  - More than half of these smokers (54.3%) reported that they quit smoking by the last three months of pregnancy.

- Almost 11 percent reported smoking during the last three months of pregnancy.

- Almost 16 percent reported smoking after delivery.

Among racial and ethnic groups, smoking during pregnancy was highest among American Indians/Alaska Natives (26.0%) and lowest among Asians/Pacific Islanders (2.1%).

The highest prevalence of smoking after delivery was reported in women aged 20–24 years (25.5%), American Indians/Alaska Natives (40.1%), those who had less than 12 years of education (24.5%), and those who had Medicaid coverage during pregnancy or delivery (24.3%).

Table 12.6. Smoking Status and Women with Recent Live Births

Smoking Status—Women with Recent Live Births, 2010	Prevalence*
Smoked before pregnancy	23.2%[†]
Smoked during ᴾregnancy	10.70%
Smoked after delivery	15.90%

* *Pregnancy Risk Assessment Monitoring System 2010, 27 sites*
[†] *Among those who smoked before pregnancy, 54.3% quit smoking during pregnancy.*

People Living with Human Immunodeficiency Virus

Cigarette smoking prevalence is estimated to be at least two times higher among adults living with human immunodeficiency virus (HIV) than in the general population. Advances in science mean that HIV is now a chronic, manageable disease. Many people with HIV lead healthy, happy lives. However, smoking has serious health effects on people with HIV, including higher risks for cancer; chronic obstructive pulmonary disease (COPD); heart disease; stroke; and HIV-related infections, including bacterial pneumonia.

- In 2009, among adults with HIV, 42.4 percent were current cigarette smokers.

- Factors associated with higher smoking prevalence among adults with HIV include:

  - Age: Persons aged 40 to 49 years are more likely to smoke than those aged 18 to 29 years.

  - Race/ethnicity: Non-Hispanic Whites and Blacks are more likely to smoke than Hispanics and Latinos.

  - Education: Persons who only achieved a high school education or less are more likely to smoke than those with more than high school education.

  - Poverty level: Persons living below the poverty level are more likely to smoke than those living at or above the poverty level.

People with Mental-Health Conditions

Nationally, nearly one in five adults (or 45.7 million adults) have some form of a mental-health condition, and 36 percent of these people smoke cigarettes. In comparison, 21 percent of adults without mental-health conditions smoke cigarettes. (Mental-health conditions are defined here as diagnosable mental, behavioral, or emotional conditions, and does not include developmental and substance-use disorders (SUDs)).

Following are other key facts:

- 31 percent of all cigarettes are smoked by adults with a mental-health condition.

- 40 percent of men and 34 percent of women with a mental-health condition smoke.

- 48 percent of people with a mental-health condition who live below the poverty level smoke, compared with 33 percent of those with a mental-health condition who live above the poverty level.

Adults with Disabilities

Adults with disabilities are more likely to be cigarette smokers than those without disabilities. This might be because a smoker's disability is the result of smoking or because of possible higher stress associated with disabilities.

- In 2016, the prevalence of current cigarette smoking among adults with disabilities was 21.2 percent compared with 14.4 percent among adults with no disability.

Tobacco-Related Mortality

Overall mortality among both male and female smokers in the United States is about three times higher than that among similar people who never smoked.

The major causes of excess mortality among smokers are diseases that are related to smoking, including cancer and respiratory and vascular disease.

Smokeless tobacco is a known cause of cancer. In addition, the nicotine in smokeless tobacco may increase the risk of sudden death from a condition where the heart does not beat properly (ventricular arrhythmias).

Cigarettes and Death

Cigarette smoking causes about one of every five deaths in the United States each year. Cigarette smoking is estimated to cause the following:

- More than 480,000 deaths annually (including deaths from secondhand smoke)

- 278,544 deaths annually among men (including deaths from secondhand smoke)

- 201,773 deaths annually among women (including deaths from secondhand smoke)

Cigarette smoking causes premature death:

- Life expectancy for smokers is at least 10 years shorter than for nonsmokers.

- Quitting smoking before the age of 40 reduces the risk of dying from the smoking-related disease by about 90 percent.

Secondhand Smoke and Death

Exposure to secondhand smoke causes an estimated 41,000 deaths each year among adults in the United States:

- Secondhand smoke causes 7,333 annual deaths from lung cancer.

- Secondhand smoke causes 33,951 annual deaths from heart disease.

Increased Risk for Death among Men

- Men who smoke increase their risk of dying from bronchitis and emphysema by 17 times; from cancer of the trachea, lung, and bronchus by more than 23 times.

- Smoking increases the risk of dying from coronary heart disease among middle-aged men by almost four times.

Increased Risk for Death among Women

- Women who smoke increase their risk of dying from bronchitis and emphysema by 12 times; from cancer of the trachea, lung, and bronchus by more than 12 times.

157

- Between 1960 and 1990, deaths from lung cancer among women increased by more than 500 percent.

- In 1987, lung cancer surpassed breast cancer to become the leading cause of cancer death among U.S. women.

- In 2000, 67,600 women died from lung cancer.

- During 2010–2014, almost 282,000 women (56,359 women each year) will die from lung cancer.

- Smoking increases the risk of dying from coronary heart disease among middle-aged women by almost five times.

Death from Specific Diseases

The following table lists the estimated number of smokers aged 35 years and older who die each year from smoking-related diseases.

Table 12.7. Annual Cigarette Smoking-Related Mortality in the United States, 2005–2009

Disease	Male	Female	Total
Cancer			
Lung cancer	74,300	53,400	127,700
Other cancers[a]	26,000	10,000	36,000
Subtotal: Cancer	100,300	63,400	163,700
Cardiovascular Diseases and Metabolic Diseases			
Coronary heart disease	61,800	37,500	99,300
Other heart disease[b]	13,400	12,100	25,500
Cerebrovascular disease	8,200	7,100	15,300
Other vascular disease[c]	6,000	5,500	11,500
Diabetes mellitus	6,200	2,800	9,000
Subtotal: Cardiovascular and Metabolic	95,600	65,000	160,000
Respiratory Diseases			
Pneumonia, influenza, tuberculosis	7,800	4,700	12,500
COPD[d]	50,400	50,200	100,600
Subtotal: Respiratory	58,200	54,900	113,100
Total: Cancer, Cardiovascular, Metabolic, Respiratory	**254,100**	**183,300**	**437,400**

Table 12.7. Continued

Disease	Male	Female	Total
Perinatal Conditions			
Prenatal conditions	346	267	613
Sudden infant death syndrome	236	164	400
Total: Perinatal Conditions	**582**	**431**	**1,013**
Residential Fires	336	284	620
Secondhand Smoke			
Lung cancer	4,374	2,959	7,333
Coronary heart disease	19,152	14,799	33,951
Total: Secondhand smoke	**23,526**	**17,758**	**41,284**
TOTAL Attributable Deaths	**278,544**	**201,773**	**480,317**

[a] *Other cancers include cancers of the lip, pharynx and oral cavity, esophagus, stomach, pancreas, larynx, cervix uteri (women), kidney and renal pelvis, bladder, liver, colon, and rectum; also acute myeloid leukemia*

[b] *Other heart diseases include rheumatic heart disease, pulmonary heart disease, and other forms of heart disease.*

[c] *Other vascular diseases include atherosclerosis, aortic aneurysm, and other arterial diseases.*

[d] *COPD is chronic obstructive pulmonary disease and includes emphysema, bronchitis, and chronic airways obstruction.*

Part Two

Tobacco-Related Health Hazards

Chapter 13

The Health Consequences of Smoking

Chapter Contents

Section 13.1

Health Effects of Smoking

This section contains text excerpted from the following sources: Text in this section begins with excerpts from "Health Effects of Cigarette Smoking," Centers for Disease Control and Prevention (CDC), January 17, 2018; Text under the heading "Smoking Harms Reproduction" is excerpted from "What Smoking Means to You," Centers for Disease Control and Prevention (CDC), July 15, 2015. Reviewed February 2019.

Cigarette smoking harms nearly every organ of the body, causes many diseases, and reduces the health of smokers in general. Quitting smoking lowers your risk for smoking-related diseases and can add years to your life.

Smoking and Increased Health Risks

Smokers are more likely than nonsmokers to develop heart disease, stroke, and lung cancer:

- Estimates show smoking increases the risk:
  - For coronary heart disease by 2 to 4 times
  - For stroke by 2 to 4 times
  - Of men developing lung cancer by 25 times
  - Of women developing lung cancer by 25.7 times
- Smoking causes diminished overall health, increased absenteeism from work, and increased healthcare utilization and cost.

Smoking and Cardiovascular Disease

Smokers are at greater risk for diseases that affect the heart and blood vessels cardiovascular disease (CVD):

- Smoking causes stroke and coronary heart disease, which are among the leading causes of death in the United States.
- Even people who smoke fewer than five cigarettes a day can have early signs of cardiovascular disease.
- Smoking damages blood vessels and can make them thicken and grow narrower. This makes your heart beat faster and your blood pressure go up. Clots can also form.

- A stroke occurs when:
  - A clot blocks the blood flow to part of your brain;
  - A blood vessel in or around your brain bursts.
- Blockages caused by smoking can also reduce blood flow to your legs and skin.

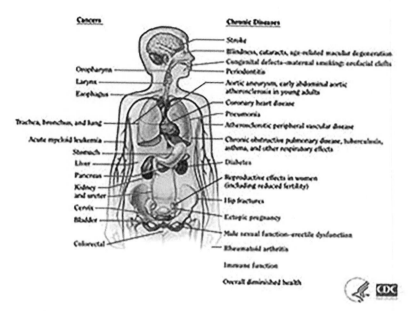

Figure 13.1. *Risk from Smoking*

Smoking and Respiratory Disease

Smoking can cause lung disease by damaging your airways and the small air sacs (alveoli) found in your lungs:

- Lung diseases caused by smoking include chronic obstructive pulmonary disease (COPD), which includes emphysema and chronic bronchitis.
- Cigarette smoking causes most cases of lung cancer.
- If you have asthma, tobacco smoke can trigger an attack or make an attack worse.
- Smokers are 12 to 13 times more likely to die from COPD than nonsmokers.

165

Smoking and Cancer

Smoking can cause cancer almost anywhere in your body:

- Bladder
- Blood (acute myeloid leukemia)
- Cervix
- Colon and rectum (colorectal)
- Esophagus
- Kidney and ureter
- Larynx
- Liver
- Oropharynx (includes parts of the throat, tongue, soft palate, and the tonsils)
- Pancreas
- Stomach
- Trachea, bronchus, and lung

Smoking also increases the risk of dying from cancer and other diseases in cancer patients and survivors.

Smoking and Death

Cigarette smoking is the leading preventable cause of death in the United States:

- Cigarette smoking causes more than 480,000 deaths each year in the United States. This is nearly one in five deaths. Smoking causes more deaths each year than the following causes combined:
  - Human immunodeficiency virus (HIV)
  - Illegal drug use
  - Alcohol use
  - Motor-vehicle injuries
  - Firearm-related incidents
- More than 10 times as many U.S. citizens have died prematurely from cigarette smoking than have died in all the wars fought by the United States.

- Smoking causes about 90 percent (or 9 out of 10) of all lung cancer deaths. More women die from lung cancer each year than from breast cancer.

- Smoking causes about 80 percent (or 8 out of 10) of all deaths from chronic obstructive pulmonary disease (COPD).

- Cigarette smoking increases risk for death from all causes in men and women.

- The risk of dying from cigarette smoking has increased over the last 50 years in the United States.

Smoking Harms Reproduction

Babies whose mothers smoked during pregnancy weigh less and have a greater risk of infant death and disease.

Smoking harms every phase of reproduction. Women who smoke have more difficulty becoming pregnant and have a higher risk of never becoming pregnant. Women who smoke during pregnancy have a greater chance of complications, premature birth, low birth weight infants, stillbirth, and infant mortality.

Low birth weight is a leading cause of infant deaths. Many of these deaths are linked to smoking. Even though everyone now know the danger of smoking during pregnancy, fewer than one out of four women quit smoking once they become pregnant.

High-risk pregnancy. Smoking makes it more difficult for women to become pregnant. Once they are pregnant, women who smoke have more complications. One complication is placenta previa, a condition where the placenta (the organ that nourishes the baby) grows too close to the opening of the womb. This condition frequently requires delivery by cesarean section. Pregnant women who smoke are also more likely to have placental abruption. In this condition, the placenta separates from the wall of the womb earlier than it should. This can lead to preterm delivery, stillbirth, and early infant death. If you smoke while you are pregnant, you are also at a 17 higher risk that your water will break before labor begins. All these conditions make it more likely that, if you smoke, your baby will be born too early.

Low birth weight babies. Babies of mothers who smoked during pregnancy have lower birth weights, often weighing less than 5.5 pounds. Low birth weight babies are at greater risk for childhood and adult illnesses and even death. Babies of smokers have less muscle mass and more fat than babies of nonsmokers. Nicotine causes the blood vessels to constrict in the umbilical cord and womb. This decreases the amount of oxygen to the unborn baby. This can lead to low birth weight. It also reduces the amount of blood in the baby's system. Pregnant smokers actually eat more than pregnant nonsmokers, yet their babies weigh less. If you quit smoking before your third trimester (the last three months), your baby is more likely to be close to normal weight.

Sudden infant death syndrome (SIDS). The death rate from sudden infant death syndrome has fallen by more than half since the "Back to Sleep" campaign began in the 1990s. This campaign reminds parents that babies should lie on their backs while sleeping. Yet more can be done. Babies exposed to secondhand smoke after birth have double the risk of SIDS. Babies whose mothers smoke before and after birth are three to four times more likely to die from SIDS.

Facts You Should Know

- Smoking causes lower fertility in women.

- Babies of women who smoke are more likely to be born too early.

- Smoking during pregnancy causes placenta previa and placental abruption. These conditions can cause a baby to be born too early and then be sick.

- The nicotine in cigarette smoke reduces the amount of oxygen reaching the fetus.

- Smoking causes reduced fetal growth and low birth weight.

- Smoking by the mother can cause SIDS.

Smoking Harms People of All Ages

- **Infants.** Effects of smoking on lung development can begin before birth. When mothers smoke during pregnancy, it hurts their babies' lungs.

- **Children.** Children and teens who smoke are less physically fit and have more breathing problems. Smoking at this age can slow lung growth. If you smoke as a teenager, your lung function begins to decline years earlier than nonsmokers. This hurts you when you want to be active.

- **All ages.** At any age, smoking damages your lungs. The more cigarettes you smoke, the faster this happens. Air pollution, being overweight, and not eating enough fresh fruit increase your risk of lung disease even more if you smoke. However, if you quit smoking, your lungs can gradually return to normal for your age.

Facts You Should Know

- Smoking causes injury to the airways and lungs, leading to a deadly lung condition.

- Smokers are more likely than nonsmokers to have upper and lower breathing tract infections.

- Mothers who smoke during pregnancy hurt the lungs of their babies.

- If you smoke during childhood and teenage years, it slows your lung growth and causes your lungs to decline at a younger age.

- Smoking is related to chronic coughing, wheezing, and asthma among children and teens.

- Smoking is related to chronic coughing and wheezing among adults.

- After stopping smoking, former smokers eventually return to normal age-related lung function.

Other Effects of Smoking

Overall health in smokers is poorer than in nonsmokers.

Smoking damages your health in many other ways. Smokers are less healthy overall than nonsmokers. Smoking harms your immune system and increases your risk of infections. The toxic ingredients in cigarette smoke travel throughout your body. For example, nicotine reaches your brain within 10 seconds after you inhale smoke. It has been found in every organ of the body, as well as in breast milk. If you smoke, your cells will not get the amount of oxygen needed to work properly. This is because carbon monoxide keeps red blood cells from carrying a full load of oxygen. Carcinogens, or cancer-causing poisons, in tobacco smoke, bind to cells in your airways and throughout your body.

Smoking harms your whole body. It increases your risk of fractures, dental diseases, sexual problems, eye diseases, and peptic ulcers. If you smoke, your illnesses last longer and you are more likely to be absent from work. In a study of U.S. military personnel, those who smoked were hospitalized 28 to 55 percent longer than nonsmokers. And the more cigarettes they smoked, the longer their hospitalization. Smokers also use more medical services than nonsmokers.

Section 13.2

Risks Associated with Short-Term and Low-Level Smoking

This section contains text excerpted from the following sources: Text under the heading "No Safe Level of Smoking" is excerpted from "No Safe Level of Smoking: Even Low-Intensity Smokers Are at Increased Risk of Earlier Death," National Cancer Institute (NCI), December 5, 2016; Text beginning with the heading "Dual Use of Tobacco Products" is excerpted from "Dual Use of Tobacco Products," Centers for Disease Control and Prevention (CDC), March 22, 2018.

No Safe Level of Smoking

People who consistently smoked an average of less than one cigarette per day over their lifetimes had a 64 percent higher risk of earlier death than never smokers, and those who smoked between one and 10 cigarettes a day had an 87 percent higher risk of earlier death than never smokers, according to a study from researchers at the National Cancer Institute (NCI). Risks were lower among former low-intensity smokers compared to those who were still smokers, and risk fell with earlier age at quitting.

When researchers looked at specific causes of death among study participants, a particularly strong association was observed for lung cancer mortality. Those who consistently averaged less than one cigarette per day over their lifetime had nine times the risk of dying from lung cancer than never smokers. Among people who smoked between one and 10 cigarettes per day, the risk of dying from lung cancer was nearly 12 times higher than that of never smokers.

The researchers looked at risk of death from respiratory disease, such as emphysema, as well as the risk of death from cardiovascular disease. People who smoked between one and 10 cigarettes a day had over six times the risk of dying from respiratory diseases than never smokers and about one and one-half times the risk of dying of cardiovascular disease than never smokers.

Smoking has many harmful effects on health, which have been detailed in numerous studies since the U.S. Surgeon General's 1964 report linking smoking to lung cancer. The health effects of consistent low-intensity smoking, however, have not been well studied and many smokers believe that low-intensity smoking does not affect their health.

To better understand the effects of low-intensity smoking on mortality from all causes and for specific causes of death, the scientists analyzed data on over 290,000 adults in the NIH-AARP Diet and Health Study. Low-intensity smoking was defined as 10 or fewer cigarettes per day. All participants were age 59 to age 82 at the start of the study.

Participants were asked about their smoking behaviors during nine periods across their lives, beginning with before they reached their 15th birthday until after they reached the age of 70 (for the older participants). Among current smokers, 159 reported smoking less than one cigarette per day consistently throughout the years that they smoked; nearly 1,500 reported smoking between one and 10 cigarettes per day.

The study relied on people recalling their smoking history over many decades, which introduced a degree of uncertainty into the findings. Also, despite the large number of people surveyed, the number of consistent low-intensity smokers was relatively small.

Another limitation of the study is that the participants were mostly white and in their 60s and 70s, so the smoking patterns collected in the study reflect only a particular set of age groups in the United States. Future studies among younger populations and other racial and ethnic groups are needed, particularly as low-intensity smoking has historically been more common among racial and ethnic minorities in the United States. The study also lacked detailed information about usage patterns among participants who reported smoking less than one cigarette per day. Hence, the researchers could not compare the effects of smoking every other day, every few days, or weekly, for example.

"The results of this study support health warnings that there is no safe level of exposure to tobacco smoke," said Maki Inoue-Choi,

Ph.D., NCI, Division of Cancer Epidemiology and Genetics (DCEG), and lead author of the study. "Together, these findings indicate that smoking even a small number of cigarettes per day has substantial negative health effects and provide further evidence that smoking cessation benefits all smokers, regardless of how few cigarettes they smoke."

Dual Use of Tobacco Products

When you cut down on cigarettes—by adding another tobacco product—you may feel that you're improving your health. This is called "dual use." It is not an effective way to fully safeguard your health, whether you're using electronic cigarettes (e-cigarettes), smokeless tobacco, or other tobacco products in addition to regular cigarettes. Quitting smoking completely is very important to protect your health. Smoking even a few cigarettes a day can be dangerous.

Risks of Smoking Less Compared to Quitting Completely

Even light smokers or social smokers can have serious health problems from smoking.

- Smoking just five cigarettes a day doubles your risk of dying from heart disease.

- Just cutting back on cigarettes may not protect you from an early death. Heavy smokers who reduce their cigarette use by half still have a very high risk for early death. And on average, smokers die 10 years earlier than nonsmokers.

- Social smokers—people who do not smoke cigarettes on a daily basis but who smoke in certain social situations on a regular basis—can harm their cardiovascular system. Social smokers have similar blood pressure and cholesterol levels of people who smoke regularly.

In contrast, completely quitting smoking improves your health. The health differences are dramatic for heart disease and illnesses that can cause an early death.

- When you quit smoking completely, you begin to reduce your heart disease risk right away. Your risk is cut in half 1 year after quitting and continues to drop over time.

172

- People who quit smoking completely live longer than those who keep smoking. The earlier you quit, the lower your risk for early death. Even quitting at age 50 cuts your risk in half for early death from a smoking-related disease.

In the United States, there are more former cigarette smokers than current cigarette smokers. It often takes several tries to quit smoking. Using proven quitting methods, such as medications and counseling, can double your chances for success.

Chapter 14

Tobacco and Cancer

Chapter Contents

Section 14.1

Understanding Cancer due to Tobacco Use

This section contains text excerpted from the following sources:
Text beginning with the heading "What Is Cancer?" is excerpted
from "Cancer," Centers for Disease Control and Prevention (CDC),
February 8, 2018; Text under the heading "Tobacco Use Causes
Many Cancers" is excerpted from "Cancer and Tobacco Use," Centers
for Disease Control and Prevention (CDC), November 2016.

What Is Cancer?

Cancer refers to diseases in which abnormal cells divide out of control and are able to invade other tissues. Cancer cells can spread to other parts of the body through the blood and lymph systems, which help the body get rid of toxins.

There are more than 100 different types of cancer. Most cancers are named for the organ or type of cell in which they start—for example, lung cancer begins in the lung and laryngeal cancer begins in the larynx.

Symptoms can include:

- A thickening or lump in any part of the body

- Weight loss or gain with no known reason

- A sore that does not heal

- Hoarseness or a cough that does not go away

- A hard time swallowing

- Discomfort after eating

- Changes in bowel or bladder habits

- Unusual bleeding or discharge

- Feeling weak or very tired

How Is Smoking Related to Cancer?

Smoking can cause cancer and then block your body from fighting it:

- Poisons in cigarette smoke can weaken the body's immune system, making it harder to kill cancer cells. When this happens, cancer cells keep growing without being stopped.

- Poisons in tobacco smoke can damage or change a cell's deoxyribonucleic acid (DNA). DNA is the cell's "instruction manual" that controls a cell's normal growth and function. When DNA is damaged, a cell can begin growing out of control and create a cancer tumor.

Doctors have known for years that smoking causes most lung cancers. It's still true today, when nearly 9 out of 10 lung cancers are caused by smoking cigarettes. In fact, smokers have a greater risk for lung cancer today than they did in 1964, even though they smoke fewer cigarettes. One reason may be changes in how cigarettes are made and what chemicals they contain.

Treatments are getting better for lung cancer, but it still kills more men and women than any other type of cancer. In the United States, more than 7,300 nonsmokers die each year from lung cancer caused by secondhand smoke. Secondhand smoke is the combination of smoke from the burning end of a cigarette and the smoke breathed out by smokers.

Smoking can cause cancer almost anywhere in your body, including the:

- Blood (acute myeloid leukemia)
- Bladder Cervix
- Colon and rectum
- Esophagus
- Kidney and renal pelvis
- Larynx
- Liver
- Lungs
- Mouth and throat
- Pancreas
- Stomach
- Trachea, lung, and bronchus

Men with prostate cancer who smoke may be more likely to die from these diseases than nonsmokers. Smokeless tobacco, such as chewing tobacco, also causes cancer, including cancers of the:

- Esophagus
- Mouth and throat
- Pancreas

How Can Smoking-Related Cancers Be Prevented?

- Quitting smoking lowers the risks for cancers of the lung, mouth, throat, esophagus, and larynx.

- Within five years of quitting, your chance of getting cancer of the mouth, throat, esophagus, and bladder is cut in half.

- Ten years after you quit smoking, your risk of dying from lung cancer drops by half.

- If nobody smoked, one of every three cancer deaths in the United States would not happen.

Cancer Screening
Screening for Cervical and Colorectal Cancers

Research shows that screening for cervical and colorectal cancers, as recommended, helps prevent these diseases. Screening for cervical and colorectal cancers helps find these diseases at an early, and often highly treatable, stage. Centers for Disease Control and Prevention (CDC) offers free or low-cost cervical cancer screening nationwide.

Healthcare reform through the Affordable Care Act (ACA) increases access to cervical and colorectal cancer screening through expanded insurance coverage and eliminating cost-sharing. In addition, CDC's Screen for Life: National Colorectal Cancer Action Campaign informs men and women aged 50 years and older about the importance of having regular colorectal cancer screening tests.

Screening for Lung Cancer

People who have smoked for many years may consider screening for lung cancer with low-dose computed tomography (LDCT). Talk to your doctor about lung cancer screening and the possible benefits and risks. Lung cancer screening is not a substitute for quitting smoking.

The U.S. Preventive Services Task Force (USPSTF) recommends annual screening for lung cancer with low-dose computed tomography in adults aged 55 to 80 years who have a 30 pack-year smoking history and currently smoke or have quit within the past 15 years. Screening should be discontinued once a person has not smoked for 15 years or develops a health problem that substantially limits life expectancy or the ability or willingness to have curative lung surgery.

How Is Cancer Treated?

The treatment for cancer depends on the type of cancer and the stage of the disease (how severe the cancer is and whether it has spread). Doctors may also consider the patient's age and general health. Often, the goal of treatment is to cure the cancer. In other cases, the goal is to control the disease or to reduce symptoms for as long as possible. The treatment plan for a person may change over time.

Most treatment plans include surgery, radiation therapy, or chemotherapy. Other plans involve biological therapy (a treatment that helps your immune system fight cancer).

Some cancers respond best to a single type of treatment. Other cancers may respond best to a combination of treatments.

For patients who get very high doses of chemotherapy or radiation therapy, a stem cell transplant, also known as a bone marrow transplant, may be recommended by their doctor. This is because high-dose therapies destroy both cancer cells and normal blood cells. A stem cell transplant can help the body to make healthy blood cells to replace the ones lost due to the cancer treatment. It's a complicated procedure with many side effects and risks.

Quitting smoking improves the outlook (the prognosis) for people with cancer. People who continue to smoke after diagnosis raise their risk for future cancers and death. They are more likely to die from cancer than nonsmokers and are more likely to develop a second (new) tobacco-related cancer.

Tobacco Use Causes Many Cancers

Tobacco use is the leading preventable cause of cancer and cancer deaths. It can cause not only lung cancer—but also cancers of the mouth and throat, voice box, esophagus, stomach, kidney, pancreas, liver, bladder, cervix, colon and rectum, and a type of leukemia. Each year, 660,000 people in the United States are diagnosed with and 343,000 people die from a cancer related to tobacco use. The progress is made: more than 1 million tobacco-related cancer deaths have been avoided since 1990 because of comprehensive cancer and tobacco control programs, early detection of cancer, and improvements in cancer treatment. However, not all states or all people have experienced the benefits of these efforts. When states make greater and longer investments in comprehensive cancer and tobacco control programs, fewer people use tobacco and get or die from tobacco-related cancers.

179

States and Communities Can

- Support comprehensive cancer control programs focusing on cancer prevention, education, screening, access to care, support for cancer survivors, and good health for all.

- Fund comprehensive tobacco prevention and control programs at CDC-recommended levels. Make tobacco cessation treatments more available to people who want to quit.

- Protect nonsmokers from secondhand smoke in all indoor public places and worksites, including all restaurants, bars, and casinos.

People Are Still Dying from Cancers Caused by Tobacco Use

People Who Use Tobacco or Are Exposed to Secondhand Smoke Are More Likely to Get and Die from Cancer

- Tobacco smoke has at least 70 chemicals that cause cancer, also known as carcinogens.

- Lung and colorectal cancers make up more than half of all cancers linked to tobacco use.

- Secondhand smoke exposure causes about 7,300 lung cancer deaths among nonsmoking adults each year.

Tobacco Use Is the Leading Preventable Cause of Cancer and Cancer Deaths

- About 3 in 10 cancer deaths are caused by cigarette smoking. Lung cancer is the leading cause of cancer death for both men and women.

- Quitting tobacco use at any age can reduce the risk of getting or dying from cancer.

- Getting screened for cancer can lead to fewer people getting or dying from some tobacco-related cancers (cervix, colorectal, and lung).

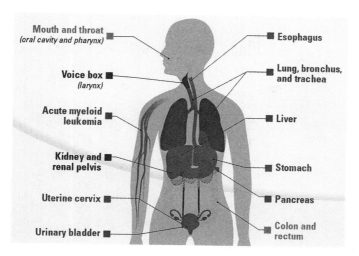

Figure 14.1. *Tobacco Use* Causes Cancer throughout the Body*

* Tobacco use includes smoked (cigarettes and cigars) and smokeless (snuff and chewing tobacco) tobacco products that, to date, have been shown to cause cancer.

Prevent Cancer Deaths from Tobacco Use

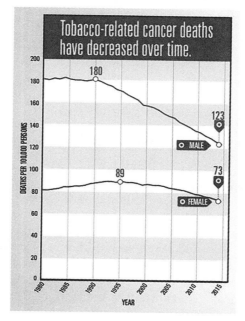

Figure 14.2. *Tobacco-Related Cancer Deaths Have Decreased over Time* (Source: National Vital Statistics System.)

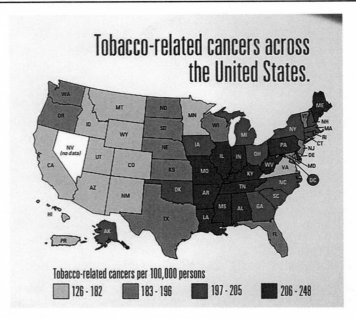

Figure 14.3. *Tobacco-Related Cancers across the United States*

Section 14.2

Bladder Cancer

This section includes text excerpted from "Bladder Cancer
Symptoms, Tests, Prognosis, and Stages (PDQ®)–Patient
Version," National Cancer Institute (NCI), December 27, 2018.

Bladder cancer is a disease in which malignant (cancer) cells form
in the tissues of the bladder. The bladder is a hollow organ in the
lower part of the abdomen. It is shaped like a small balloon and has
a muscular wall that allows it to get larger or smaller to store urine
made by the kidneys. There are two kidneys, one on each side of the
backbone, above the waist. Tiny tubules in the kidneys filter and clean
the blood. They take out waste products and make urine. The urine
passes from each kidney through a long tube called a ureter into the

bladder. The bladder holds the urine until it passes through the urethra and leaves the body.

Types of Bladder Cancer

There are three types of bladder cancer that begin in cells in the lining of the bladder. These cancers are named for the type of cells that become malignant (cancerous):

- **Transitional cell carcinoma:** Cancer that begins in cells in the innermost tissue layer of the bladder. These cells are able to stretch when the bladder is full and shrink when it is emptied. Most bladder cancers begin in the transitional cells. Transitional cell carcinoma can be low-grade or high-grade:

  - Low-grade transitional cell carcinoma often recurs (comes back) after treatment, but rarely spreads into the muscle layer of the bladder or to other parts of the body.

  - High-grade transitional cell carcinoma often recurs (comes back) after treatment and often spreads into the muscle layer of the bladder, to other parts of the body, and to lymph nodes. Almost all deaths from bladder cancer are due to high-grade disease.

- **Squamous cell carcinoma:** Cancer that begins in squamous cells (thin, flat cells lining the inside of the bladder). Cancer may form after long-term infection or irritation.

- **Adenocarcinoma:** Cancer that begins in glandular cells that are found in the lining of the bladder. Glandular cells in the bladder make substances such as mucus. This is a very rare type of bladder cancer.

Cancer that is in the lining of the bladder is called superficial bladder cancer. Cancer that has spread through the lining of the bladder and invades the muscle wall of the bladder or has spread to nearby organs and lymph nodes is called invasive bladder cancer.

Smoking Can Affect the Risk of Bladder Cancer

Anything that increases your chance of getting a disease is called a risk factor. Having a risk factor does not mean that you will get cancer; not having risk factors doesn't mean that you will not get cancer. Talk to your doctor if you think you may be at risk for bladder cancer.

Risk factors for bladder cancer include the following:

- Using tobacco, especially smoking cigarettes

- Having a family history of bladder cancer

- Having certain changes in the genes that are linked to bladder cancer

- Being exposed to paints, dyes, metals, or petroleum products in the workplace

- Past treatment with radiation therapy to the pelvis or with certain anticancer drugs, such as cyclophosphamide or ifosfamide

- Taking Aristolochia fangchi, a Chinese herb

- Drinking water from a well that has high levels of arsenic

- Drinking water that has been treated with chlorine

- Having a history of bladder infections, including bladder infections caused by Schistosoma haematobium

- Using urinary catheters for a long time

Older age is a risk factor for most cancers. The chance of getting cancer increases as you get older.

Signs and Symptoms of Bladder Cancer

These and other signs and symptoms may be caused by bladder cancer or by other conditions. Check with your doctor if you have any of the following:

- Blood in the urine (slightly rusty to bright red in color)

- Frequent urination

- Pain during urination

- Lower back pain

Diagnosing Bladder Cancer

The following tests and procedures may be used:

- Physical exam and history

- Internal exam

- Urinalysis
- Urine cytology
- Cystoscopy
- Intravenous pyelogram (IVP)
- Biopsy

Certain Factors Affect Prognosis (Chance of Recovery) and Treatment Options

The prognosis (chance of recovery) depends on the following:

- The stage of the cancer (whether it is superficial or invasive bladder cancer, and whether it has spread to other places in the body). Bladder cancer in the early stages can often be cured.
- The type of bladder cancer cells and how they look under a microscope.
- Whether there is carcinoma in situ in other parts of the bladder.
- The patient's age and general health.

If the cancer is superficial, prognosis also depends on the following:

- How many tumors there are.
- The size of the tumors. Whether the tumor has recurred (come back) after treatment.
- Treatment options depend on the stage of bladder cancer.

Section 14.3

Head and Neck Cancer

This section includes text excerpted from "Head and Neck Cancers," National Cancer Institute (NCI), March 28, 2017.

What Are Cancers of the Head and Neck?

Cancers that are known collectively as head and neck cancers usually begin in the squamous cells that line the moist, mucosal surfaces inside the head and neck (for example, inside the mouth, the nose, and the throat). These squamous cell cancers are often referred to as squamous cell carcinomas of the head and neck. Head and neck cancers can also begin in the salivary glands, but salivary gland cancers are relatively uncommon. Salivary glands contain many different types of cells that can become cancerous, so there are many different types of salivary gland cancer.

Cancers of the head and neck are further categorized by the area of the head or neck in which they begin. These areas are described below and labeled in the image of head and neck cancer regions.

Oral cavity: Includes the lips, the front two-thirds of the tongue, the gums, the lining inside the cheeks and lips, the floor (bottom) of the mouth under the tongue, the hard palate (bony top of the mouth), and the small area of the gum behind the wisdom teeth.

Pharynx: The pharynx (throat) is a hollow tube about 5 inches long that starts behind the nose and leads to the esophagus. It has three parts: the nasopharynx (the upper part of the pharynx, behind the nose); the oropharynx (the middle part of the pharynx, including the soft palate [the back of the mouth], the base of the tongue, and the tonsils); the hypopharynx (the lower part of the pharynx).

Larynx: The larynx, also called the voicebox, is a short passageway formed by cartilage just below the pharynx in the neck. The larynx contains the vocal cords. It also has a small piece of tissue, called the epiglottis, which moves to cover the larynx to prevent food from entering the air passages.

Paranasal sinuses and nasal cavity: The paranasal sinuses are small hollow spaces in the bones of the head surrounding the nose. The nasal cavity is the hollow space inside the nose.

Salivary glands: The major salivary glands are in the floor of the mouth and near the jawbone. The salivary glands produce saliva.

What Causes Cancers of the Head and Neck

Alcohol and tobacco use (including smokeless tobacco, sometimes called "chewing tobacco" or "snuff") are the two most important risk factors for head and neck cancers, especially cancers of the oral cavity, oropharynx, hypopharynx, and larynx. At least 75 percent of head and neck cancers are caused by tobacco and alcohol use. People who use both tobacco and alcohol are at greater risk of developing these cancers than people who use either tobacco or alcohol alone. Tobacco and alcohol use are not risk factors for salivary gland cancers.

Infection with cancer-causing types of human papillomavirus (HPV), especially HPV type 16, is a risk factor for some types of head and neck cancers, particularly oropharyngeal cancers that involve the tonsils or the base of the tongue. In the United States, the incidence of oropharyngeal cancers caused by HPV infection is increasing, while the incidence of oropharyngeal cancers related to other causes is falling. More information is available in the HPV and Cancer fact sheet.

Other risk factors for cancers of the head and neck include the following:

- **Paan (betel quid).** Immigrants from Southeast Asia who use paan (betel quid) in the mouth should be aware that this habit has been strongly associated with an increased risk of oral cancer.

- **Preserved or salted foods.** Consumption of certain preserved or salted foods during childhood is a risk factor for nasopharyngeal cancer.

- **Oral health.** Poor oral hygiene and missing teeth may be weak risk factors for cancers of the oral cavity. Use of mouthwash that has a high alcohol content is a possible, but not proven, risk factor for cancers of the oral cavity.

- **Occupational exposure.** Occupational exposure to wood dust is a risk factor for nasopharyngeal cancer. Certain industrial exposures, including exposures to asbestos and synthetic fibers, have been associated with cancer of the larynx, but the increase in risk remains controversial. People working in certain jobs in the construction, metal, textile, ceramic, logging, and food industries may have an increased risk of cancer of the larynx.

Industrial exposure to wood or nickel dust or formaldehyde is a risk factor for cancers of the paranasal sinuses and nasal cavity.

- **Radiation exposure.** Radiation to the head and neck, for noncancerous conditions or cancer, is a risk factor for cancer of the salivary glands.

- **Epstein-Barr virus infection.** Infection with the Epstein-Barr virus is a risk factor for nasopharyngeal cancer and cancer of the salivary glands.

- **Ancestry.** Asian ancestry, particularly Chinese ancestry, is a risk factor for nasopharyngeal cancer.

What Are the Symptoms of Head and Neck Cancers?

The symptoms of head and neck cancers may include a lump or a sore that does not heal, a sore throat that does not go away, difficulty in swallowing, and a change or hoarseness in the voice. These symptoms may also be caused by other, less serious conditions. It is important to check with a doctor or dentist about any of these symptoms. Symptoms that may affect specific areas of the head and neck include the following:

- **Oral cavity.** A white or red patch on the gums, the tongue, or the lining of the mouth; a swelling of the jaw that causes dentures to fit poorly or become uncomfortable; and unusual bleeding or pain in the mouth.

- **Pharynx.** Trouble breathing or speaking; pain when swallowing; pain in the neck or the throat that does not go away; frequent headaches, pain, or ringing in the ears; or trouble hearing.

- **Larynx.** Pain when swallowing or ear pain.

- **Paranasal sinuses and nasal cavity.** Sinuses that are blocked and do not clear; chronic sinus infections that do not respond to treatment with antibiotics; bleeding through the nose; frequent headaches, swelling or other trouble with the eyes; pain in the upper teeth; or problems with dentures.

- **Salivary glands.** Swelling under the chin or around the jawbone, numbness or paralysis of the muscles in the face, or pain in the face, the chin, or the neck that does not go away.

How Common Are Head and Neck Cancers?

Head and neck cancers account for approximately four percent of all cancers in the United States. These cancers are more than twice as common among men as they are among women. Head and neck cancers are also diagnosed more often among people over age 50 than they are among younger people.

Researchers estimated that more than 65,000 men and women in this country would be diagnosed with head and neck cancers in 2017.

How Can I Reduce My Risk of Developing Head and Neck Cancers?

People who are at risk of head and neck cancers—particularly those who use tobacco—should talk with their doctor about ways that they may be able to reduce their risk. They should also discuss with their doctor how often to have checkups. In addition, ongoing clinical trials are testing the effectiveness of various medications in preventing head and neck cancers in people who have a high risk of developing these diseases. Descriptions of these clinical trials can be accessed by searching National Cancer Institute's (NCI) list of cancer clinical trials. NCI's list of cancer clinical trials includes all NCI-supported clinical trials that are taking place across the United States and Canada, including the NIH Clinical Center in Bethesda, MD.

Information specialists from NCI's Cancer Information Service (CIS) can also help people find clinical trials for the prevention of head and neck cancers. The CIS can be reached at 800-4-CANCER (800-422-6237) or by chatting with a cancer information specialist online through LiveHelp.

Avoiding oral HPV infection may reduce the risk of HPV-associated head and neck cancers. However, it is not yet known whether the Food and Drug Administration-approved HPV vaccines Gardasil®, Gardasil 9®, and Cervarix® prevent HPV infection of the oral cavity, and none of these vaccines has yet been approved for the prevention of oropharyngeal cancer. More information about these vaccines is in the NCI fact sheet Human Papillomavirus (HPV) Vaccines.

How Are Head and Neck Cancers Diagnosed?

To find the cause of the signs or symptoms of a problem in the head and neck area, a doctor evaluates a person's medical history, performs a physical examination, and orders diagnostic tests. The exams and tests may vary depending on the symptoms. Examination of a sample of tissue under a microscope is always necessary to confirm a diagnosis of cancer.

If the diagnosis is cancer, the doctor will want to learn the stage (or extent) of disease. Staging is a careful attempt to find out whether the cancer has spread and, if so, to which parts of the body. Staging may involve an examination under anesthesia (in an operating room), X-rays and other imaging procedures, and laboratory tests. Knowing the stage of the disease helps the doctor plan treatment.

How Are Head and Neck Cancers Treated?

The treatment plan for an individual patient depends on a number of factors, including the exact location of the tumor, the stage of the cancer, and the person's age and general health. Treatment for head and neck cancer can include surgery, radiation therapy, chemotherapy, targeted therapy, or a combination of treatments.

People who are diagnosed with HPV-positive oropharyngeal cancer may be treated differently than people with oropharyngeal cancers that are HPV-negative. Some research has shown that patients with HPV-positive oropharyngeal tumors have a better prognosis and may do just as well on less intense treatment. An ongoing clinical trial is investigating this question.

What Are the Side Effects of Treatment?

Surgery for head and neck cancers often changes the patient's ability to chew, swallow, or talk. The patient may look different after surgery, and the face and neck may be swollen. The swelling usually goes away within a few weeks. However, if lymph nodes are removed, the flow of lymph in the area where they were removed may be slower and lymph could collect in the tissues, causing additional swelling; this swelling may last for a long time.

After a laryngectomy (surgery to remove the larynx) or other surgery in the neck, parts of the neck and throat may feel numb because nerves have been cut. If lymph nodes in the neck were removed, the shoulder and neck may become weak and stiff.

Patients who receive radiation to the head and neck may experience redness, irritation, and sores in the mouth; a dry mouth or thickened saliva; difficulty in swallowing; changes in taste; or nausea. Other problems that may occur during treatment are loss of taste, which may decrease appetite and affect nutrition, and earaches (caused by the hardening of ear wax). Patients may also notice some swelling or drooping of the skin under the chin and changes in the texture of the skin. The jaw may feel stiff, and patients may not be able to open their mouth as wide as before treatment.

Patients should report any side effects to their doctor or nurse, and discuss how to deal with them.

Section 14.4

Lung Cancer

This section includes text excerpted from "Lung Cancer and Smoking," U.S. Food and Drug Administration (FDA), January 15, 2019.

Lung cancer is the leading cancer killer of American men and women. Not all people who get lung cancer are smokers, but many people who smoke do get lung cancer. In fact, smoking is directly responsible for more than 80 percent of lung cancer deaths. U.S. Food and Drug Administration's (FDA) Center for Tobacco Products (CTP), established in 2009, has joined in the fight against lung cancer by working to:

- Reduce the number of people who start to use tobacco products

- Encourage more people to stop using these products

- Reduce the adverse health impact for those who continue to use these products

You may be familiar with some of the statistics, but if you, a coworker, friend, or loved one is a smoker, it's worth taking another look at what cigarettes can do to our bodies.

Cigarette smoke contains over 7,000 chemicals, including more than 70 that can cause cancer. Smoking can cause cancer almost anywhere in your body, including the esophagus, larynx, mouth, nose, throat, trachea, kidney, bladder, pancreas, stomach, cervix, bone marrow, and blood.

And, despite major progress over the past half-century, tobacco use continues to be the leading cause of preventable death and disease in the United States.

- More than 160,000 people die each year from cancers caused by cigarette smoking.
- Secondhand smoke causes more than 7,000 lung cancer deaths each year.

The statistics around young people are even more sobering. Each day in the United States:

- About 2,000 youth smoked their first cigarette prior to turning 18.
- More than 300 youth under 18 years of age become daily cigarette smokers.

The good news is that you can do something about it now.

Ways to Reduce Your Risk of Lung Cancer

You, your friends, and family can get engaged in the following ways:

- Find a quitting method that works for you.
- Educate yourself with the latest information on the health effects of tobacco use.

Section 14.5

Oral Cancer

This section includes text excerpted from "Oral Cancer," National Institute of Dental and Craniofacial Research (NIDCR), July 2018.

Oral cancer includes cancers of the mouth and the back of the throat. Oral cancers develop on the tongue, the tissue lining the mouth and gums, under the tongue, at the base of the tongue, and the area of the throat at the back of the mouth.

Oral cancer accounts for roughly three percent of all cancers diagnosed annually in the United States, or about 49,700 new cases each year.

Oral cancer most often occurs in people over the age of 40 and affects more than twice as many men as women. Most oral cancers are related to tobacco use, alcohol use (or both), or infection by the human papillomavirus (HPV).

Causes

- **Tobacco and alcohol use.** Tobacco use of any kind, including cigarette smoking, puts you at risk for developing oral cancers. Heavy alcohol use also increases the risk. Using both tobacco and alcohol increases the risk even further.

- **HPV.** Infection with the sexually transmitted human papillomavirus (specifically the HPV 16 type) has been linked to oral cancers.

- **Age.** Risk increases with age. Oral cancers most often occur in people over the age of 40.

- **Sun Exposure.** Cancer of the lip can be caused by sun exposure.

- **Diet.** A diet low in fruits and vegetables may play a role in oral cancer development.

Symptoms

If you have any of these symptoms for more than two weeks, see a dentist or a doctor.

- A sore, irritation, lump, or thick patch in your mouth, lip, or throat
- A white or red patch in your mouth
- A feeling that something is caught in your throat
- Difficulty chewing or swallowing
- Difficulty moving your jaw or tongue
- Swelling in your jaw
- Numbness in your tongue or other areas of your mouth
- Pain in one ear without hearing loss

193

Diagnosis

Because oral cancer can spread quickly, early detection is important. An oral cancer examination can detect early signs of cancer. The exam is painless and takes only a few minutes. Many dentists will perform the test during your regular dental check-up.

During the exam, your dentist or dental hygienist will check your face, neck, lips, and entire mouth for possible signs of cancer.

Treatment

When oral cancer is detected early, it is treated with surgery or radiation therapy. Oral cancer that is further along when it is diagnosed may use a combination of treatments.

For example, radiation therapy and chemotherapy are often given at the same time. Another treatment option is targeted therapy, which is a newer type of cancer treatment that uses drugs or other substances to precisely identify and attack cancer cells. The choice of treatment depends on your general health, where in your mouth or throat the cancer began, the size and type of the tumor, and whether the cancer has spread.

Your doctor may refer you to a specialist. Specialists who treat oral cancer include:

• Head and neck surgeons

• Dentists who specialize in surgery of the mouth, face, and jaw (oral and maxillofacial surgeons)

• Ear, nose, and throat doctors (otolaryngologists)

• Doctors who specifically treat cancer (medical and radiation oncologists)

Other healthcare professionals who may be part of a treatment team include dentists, plastic surgeons, reconstructive surgeons, speech pathologists, oncology nurses, registered dietitians, and mental health counselors.

Helpful Tips

Oral cancer and its treatment can cause dental problems. It's important that your mouth is in good health before cancer treatment begins.

- See a dentist for a thorough exam one month, if possible, before starting cancer treatment to give your mouth time to heal after any dental work you might need.

- Before, during, and after cancer treatment, ask your healthcare provider for ways to control pain and other symptoms, and to relieve the side effects of therapy.

- Talk to your healthcare team about financial aid, transportation, home care, emotional, and social support for yourself and your family.

Section 14.6

Smokeless Tobacco and Cancer

This section includes text excerpted from "Smokeless Tobacco and Cancer," National Cancer Institute (NCI), October 25, 2010. Reviewed February 2019.

What Is Smokeless Tobacco?

Smokeless tobacco is tobacco that is not burned. It is also known as chewing tobacco, oral tobacco, spit or spitting tobacco, dip, chew, and snuff. Most people chew or suck (dip) the tobacco in their mouth and spit out the tobacco juices that build up, although "spitless" smokeless tobacco has also been developed. Nicotine in the tobacco is absorbed through the lining of the mouth.

People in many regions and countries, including North America, northern Europe, India and other Asian countries, and parts of Africa, have a long history of using smokeless tobacco products.

Types of Smokeless Tobacco

There are two main types of smokeless tobacco:

- **Chewing tobacco**, which is available as loose leaves, plugs (bricks), or twists of rope. A piece of tobacco is placed between

195

the cheek and lower lip, typically toward the back of the mouth. It is either chewed or held in place. Saliva is spit or swallowed.

- **Snuff**, which is finely cut or powdered tobacco. It may be sold in different scents and flavors. It is packaged moist or dry; most American snuff is moist. It is available loose, in dissolvable lozenges or strips, or in small pouches similar to tea bags. The user places a pinch or pouch of moist snuff between the cheek and gums or behind the upper or lower lip. Another name for moist snuff is snus. Some people inhale dry snuff into the nose.

Are There Harmful Chemicals in Smokeless Tobacco?

Yes. There is no safe form of tobacco. At least 28 chemicals in smokeless tobacco have been found to cause cancer. The most harmful chemicals in smokeless tobacco are tobacco-specific nitrosamines, which are formed during the growing, curing, fermenting, and aging of tobacco. The level of tobacco-specific nitrosamines varies by product. Scientists have found that the nitrosamine level is directly related to the risk of cancer.

In addition to a variety of nitrosamines, other cancer-causing substances in smokeless tobacco include polonium-210 (a radioactive element found in tobacco fertilizer) and polynuclear aromatic hydrocarbons (also known as polycyclic aromatic hydrocarbons).

Does Smokeless Tobacco Cause Cancer?

Yes. Smokeless tobacco causes oral cancer, esophageal cancer, and pancreatic cancer.

Does Smokeless Tobacco Cause Other Diseases?

Yes. Using smokeless tobacco may also cause heart disease, gum disease, and oral lesions other than cancer, such as leukoplakia (precancerous white patches in the mouth).

Can a User Get Addicted to Smokeless Tobacco?

Yes. All tobacco products, including smokeless tobacco, contain nicotine, which is addictive. Users of smokeless tobacco and users of cigarettes have comparable levels of nicotine in the blood. In users of smokeless tobacco, nicotine is absorbed through the mouth tissues directly into the blood, where it goes to the brain. Even after the tobacco is removed from the mouth, nicotine continues to be absorbed into the bloodstream. Also, the nicotine stays in the blood longer for users of smokeless tobacco than for smokers.

The level of nicotine in the blood depends on the amount of nicotine in the smokeless tobacco product, the tobacco cut size, the product's pH (a measure of its acidity or basicity), and other factors.

A Centers for Disease Control and Prevention (CDC) study of the 40 most widely used popular brands of moist snuff showed that the amount of nicotine per gram of tobacco ranged from 4.4 milligrams to 25.0 milligrams. Other studies have shown that moist snuff had between 4.7 and 24.3 milligrams per gram of tobacco, dry snuff had between 10.5 and 24.8 milligrams per gram of tobacco, and chewing tobacco had between 3.4 and 39.7 milligrams per gram of tobacco.

Is Using Smokeless Tobacco Less Hazardous Than Smoking Cigarettes?

Because all tobacco products are harmful and cause cancer, the use of all of these products should be strongly discouraged. There is no safe level of tobacco use. People who use any type of tobacco product should be urged to quit.

As long ago as 1986, the advisory committee to the Surgeon General concluded that the use of smokeless tobacco "is not a safe substitute for smoking cigarettes. It can cause cancer and a number of noncancerous oral conditions and can lead to nicotine addiction and dependence." Furthermore, a panel of experts convened by the National Institutes of Health (NIH) in 2006 stated that the "range of risks, including nicotine addiction, from smokeless tobacco products may vary extensively because of differing levels of nicotine, carcinogens, and other toxins in different products."

Should Smokeless Tobacco Be Used to Help a Person Quit Smoking?

No. There is no scientific evidence that using smokeless tobacco can help a person quit smoking. Because all tobacco products are harmful and cause cancer, the use of all tobacco products is strongly discouraged. There is no safe level of tobacco use. People who use any type of tobacco product should be urged to quit. For help with quitting, ask your doctor about individual or group counseling, telephone quitlines, or other methods.

How Can I Get Help Quitting Smokeless Tobacco?

National Cancer Institute (NCI) offers free information about quitting smokeless tobacco:

- Call NCI's Smoking Quitline at 877-44U-QUIT (877-448-7848). Talk with a smoking cessation counselor about quitting smokeless tobacco. You can call the quitline, within the United States, Monday through Friday, 9:00 a.m. to 9:00 p.m., Eastern time.

- Use LiveHelp online chat. You can have a confidential online text chat with an NCI smoking cessation counselor Monday through Friday, 9:00 a.m. to 9:00 p.m., Eastern time.

Chapter 15

Smoking and Respiratory Diseases

Chapter Contents

Section 15.1

Chronic Obstructive Pulmonary Disease

This section includes text excerpted from "Chronic Obstructive Pulmonary Disease (COPD)," Centers for Disease Control and Prevention (CDC), February 8, 2018.

What Is Chronic Obstructive Pulmonary Disease?

Chronic obstructive pulmonary disease (COPD) refers to a group of diseases that cause airflow blockage and breathing-related problems. COPD includes emphysema; chronic bronchitis; and in some cases, asthma.

With COPD, less air flows through the airways—the tubes that carry air in and out of your lungs—because of one or more of the following:

- The airways and tiny air sacs in the lungs lose their ability to stretch and shrink back.

- The walls between many of the air sacs are destroyed.

- The walls of the airways become thick and inflamed (irritated and swollen).

- The airways make more mucus than usual, which can clog them and block air flow.

In the early stages of COPD, there may be no symptoms, or you may only have mild symptoms, such as:

- A nagging cough (often called "smoker's cough")

- Shortness of breath, especially with physical activity

- Wheezing (a whistling sound when you breathe)

- Tightness in the chest

As the disease gets worse, symptoms may include:

- Having trouble catching your breath or talking

- Blue or gray lips and/or fingernails (a sign of low oxygen levels in your blood)

- Trouble with mental alertness

- A very fast heartbeat

- Swelling in the feet and ankles
- Weight loss

How severe your COPD symptoms depend on how damaged your lungs are. If you keep smoking, the damage will get worse faster than if you stop smoking. Among 15 million U.S. adults with COPD, 39 percent continue to smoke.

How Is Smoking Related to Chronic Obstructive Pulmonary Disease?

COPD is usually caused by smoking. Smoking accounts for as many as 8 out of 10 COPD-related deaths. However, as many as 1 out of 4 Americans with COPD never smoked cigarettes. Smoking during childhood and teenage years can slow how lungs grow and develop. This can increase the risk of developing COPD in adulthood.

How Can Chronic Obstructive Pulmonary Disease Be Prevented?

The best way to prevent COPD is to never start smoking, and if you smoke, quit. Talk with your doctor about programs and products that can help you quit. Also, stay away from secondhand smoke, which is smoke from burning tobacco products, such as cigarettes, cigars, or pipes. Secondhand smoke also is smoke that has been exhaled, or breathed out, by a person smoking.

How Is Chronic Obstructive Pulmonary Disease Treated?

Treatment of COPD requires a careful and thorough exam by a doctor. Quitting smoking is the most important first step you can take to treat COPD. Avoiding secondhand smoke is also critical. Other lifestyle changes and treatments include one or more of the following:

- For people with COPD who have trouble eating because of shortness of breath or being tired:

  - Following a special meal plan with smaller, more frequent meals

  - Resting before eating

  - Taking vitamins and nutritional supplements

201

- A broad program that helps improve the well-being of people who have chronic (ongoing) breathing problems and includes the following:
  - Exercise training
  - Nutritional counseling
  - Education on your lung disease or condition and how to manage it
  - Energy-conserving techniques
  - Breathing strategies
  - Psychological counseling and/or group support
- Medicines such as:
  - A bronchodilator to relax the muscles around the airways. This helps open airways and makes breathing easier. Most bronchodilators are taken with a device called an inhaler.
  - A steroid drug you inhale to reduce swelling in the airways
  - Antibiotics to treat respiratory infections, if appropriate
  - A vaccination during the flu season
- Oxygen therapy, which can help people who have severe COPD and low levels of oxygen in their blood to breathe better
- Surgery for people who have severe symptoms that have not improved with other treatments
  - Lung volume reduction surgery (LVRS): Surgery to remove diseased parts of the lung so healthier lung tissue can work better. LVRS is not a cure for COPD.
  - A lung transplant: Surgery in which one or two healthy lungs from a donor are put in the patient's body to replace diseased lungs. A lung transplant is a last resort.

Even though there is no cure for COPD, these lifestyle changes and treatments can help you breathe easier, stay more active, and slow the progress of the disease.

Section 15.2

Asthma and Smoking

This section includes text excerpted from "Asthma and Secondhand Smoke," Centers for Disease Control and Prevention (CDC), March 21, 2018.

What Is Asthma?

Asthma is a chronic disease that affects the airways of the lungs. During an asthma attack, airways (tubes that carry air to your lungs) become swollen, making it hard to breathe. As the walls of the airways swell, they narrow, and less air gets in and out of the lungs. Cells in the airways can make more mucus (a sticky, thick liquid) than usual, which can make breathing even harder.

Symptoms of an asthma attack include:

- Coughing

- Shortness of breath or trouble breathing

- Wheezing

- Tightness or pain in the chest

Asthma attacks can be mild, moderate, or serious—and even life-threatening.

How Is Smoking Related to Asthma?

If you have asthma, an asthma attack can occur when something irritates your airways and "triggers" an attack. Your triggers might be different from other people's triggers.

Tobacco smoke is one of the most common asthma triggers. Tobacco smoke—including secondhand smoke—is unhealthy for everyone, especially people with asthma. Secondhand smoke is a mixture of gases and fine particles that includes:

- Smoke from a burning cigarette, cigar, or pipe tip

- Smoke that has been exhaled (breathed out) by someone who smokes

Secondhand smoke contains more than 7,000 chemicals, including hundreds that are toxic and about 70 that can cause cancer.

If you have asthma, it's important that you avoid exposure to secondhand smoke.

If you are among the 21 percent of U.S. adults who have asthma and smoke, quit smoking.

How Can Asthma Attacks Be Prevented?

If you or a family member has asthma, you can manage it with the help of your healthcare provider (for example, by taking your medicines exactly as your doctor tells you) and by avoiding triggers. Staying far away from tobacco smoke is one important way to avoid asthma attacks. Some other helpful tips are:

- Do not smoke or allow others to smoke in your home or car. Opening a window does not protect you from the smoke.

- If your state still allows smoking in public areas, look for restaurants and other places that do not allow smoking. "No-smoking sections" in the same restaurant with "smoking sections" do not protect adequately from secondhand smoke— even if there is a filter or ventilation system.

- Make sure your children's daycare centers and schools are tobacco-free. For schools, a tobacco-free campus policy means no tobacco use or advertising on school property is allowed by anyone at any time. This includes off-campus school events.

- Teach children to stay away from secondhand smoke. Be a good role model by not smoking.

How Is Asthma Treated?

There is no cure for asthma. However, to help control your asthma and avoid attacks:

- Take your medicine exactly as your doctor tells you.

- Stay away from things that can trigger an attack.

Everyone with asthma does not take the same medicine. Some medicines can be breathed in, and some can be taken as a pill. There are two kinds of asthma medicines:

- Quick-relief (can help control symptoms of an asthma attack)

- Long-term control (can help you have fewer and milder attacks, but they don't help you while you are having an asthma attack)

Chapter 16

Smoking and Cardiovascular Diseases

Chapter Contents

Section 16.1

Smoking and Heart Health

This section includes text excerpted from documents published by two public domain sources. Text under headings marked 1 are excerpted from "How Smoking Affects Heart Health," U.S. Food and Drug Administration (FDA), February 1, 2019; Text under headings marked 2 are excerpted from "Smoking and Your Heart," National Heart, Lung, and Blood Institute (NHLBI), March 12, 2013. Reviewed February 2019.

How Does Smoking Affect Your Cardiovascular Health?[1]

Cigarette smoking is the chief cause of preventable disease and death in the United States and can harm nearly any part of the body. Cigarette smoke is a toxic mix of more than 7,000 chemicals and, when inhaled, can interfere with important processes in the body that keep it functioning normally. One of these processes is the delivery of oxygen-rich blood to your heart and the rest of your body.

When you breathe in air from the atmosphere, the lungs take in oxygen and deliver it to the heart, which pumps this oxygen-rich blood to the rest of the body through the blood vessels. But when you breathe in cigarette smoke, the blood that is distributed to the rest of the body becomes contaminated with the smoke's chemicals. These chemicals can damage your heart and blood vessels, which can lead to cardiovascular disease (CVD)—the leading cause of all deaths in the United States.

In addition to permanently damaging your heart and blood vessels, cigarette smoke can also cause CVD by changing your blood chemistry and causing plaque—a waxy substance comprises of cholesterol, scar tissue, calcium, fat, and other material—to build up in the arteries, the major blood vessels that carry blood from your heart to your body. This plaque buildup can lead to a disease called atherosclerosis. When the chemicals in cigarette smoke cause atherosclerosis and thickened blood in the arteries, it becomes more difficult for blood cells to move through arteries and other blood vessels to get to vital organs like the heart and brain. This can create blood clots and ultimately lead to a heart attack or stroke, even death. Other rare but serious cardiovascular conditions that can be caused by smoking include:

- **Peripheral artery disease (PAD) and peripheral vascular disease (PVD):** A condition in which the narrowing of blood vessels results in insufficient blood flow to arms, legs, hands, and feet. Smoking is the leading preventable cause of this condition, which can result in amputation.

- **Abdominal aortic aneurysm (AAA):** A bulge that is formed in an area of the aorta—the main artery that distributes blood through the body—that sits in the abdomen. When an abdominal aortic aneurysm bursts, it can result in sudden death. More women than men die from aortic aneurysms, and nearly all deaths from this condition are caused by smoking.

Impact of Cardiovascular Disease Caused by Smoking[1]

According to the American Heart Association (AHA), CVD accounts for about 800,000 U.S. deaths every year, making it the leading cause of all deaths in the United States. Of those, nearly 20 percent are due to cigarette smoking.

While smoking is a direct cause of cardiovascular disease and death, you don't have to be a smoker to be at risk. Nonsmokers who are regularly exposed to secondhand smoke have a 25 to 30 percent increased risk of coronary heart disease than those not exposed. In fact, 30,000 United States coronary heart disease deaths per year are caused by secondhand smoke. Secondhand smoke exposure also increases your risk of having a heart attack or stroke.

How Does Smoking Affect the Heart and Blood Vessels?[2]

Cigarette smoking causes about one in every five deaths in the United States each year. It's the main preventable cause of death and illness in the United States. Smoking harms nearly every organ in the body, including the heart, blood vessels, lungs, eyes, mouth, reproductive organs, bones, bladder, and digestive organs. The chemicals in tobacco smoke harm your blood cells. They also can damage the function of your heart and the structure and function of your blood vessels. This damage increases your risk of atherosclerosis.

- Atherosclerosis is a disease in which a waxy substance called plaque builds up in the arteries. Over time, plaque hardens and narrows your arteries. This limits the flow of oxygen-rich blood to your organs and other parts of your body.

- Ischemic heart disease occurs if plaque builds up in the arteries that supply blood to the heart, called coronary arteries. Over time, heart disease can lead to chest pain, heart attack, heart failure, arrhythmias, or even death.

- Smoking is a major risk factor for heart disease. When combined with other risk factors—such as unhealthy blood cholesterol levels, high blood pressure, and overweight or obesity—smoking further raises the risk of heart disease.

- Smoking also is a major risk factor for PAD. PAD is a condition in which plaque builds up in the arteries that carry blood to the head, organs, and limbs. People who have PAD are at increased risk for heart disease, heart attack, and stroke.

Smoking and Atherosclerosis

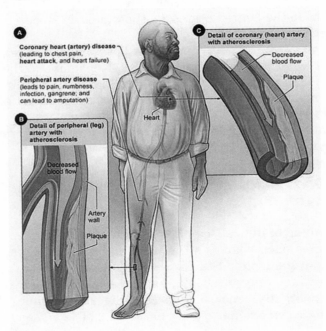

Figure 16.1. *Smoking and Atherosclerosis*

The figure shows how smoking can affect arteries in the heart and legs. Figure A shows the location of coronary heart disease and peripheral artery disease. Figure B shows a detailed view of a leg artery with atherosclerosis—plaque buildup that is partially blocking blood flow. Figure C shows a detailed view of a coronary (heart) artery with atherosclerosis.

The image shows how smoking can affect arteries in the heart and legs. Figure A shows the location of coronary heart disease and peripheral artery disease. Figure B shows a detailed view of a leg artery with atherosclerosis—plaque buildup that partially blocks blood flow. Figure C shows a detailed view of a coronary artery with atherosclerosis.

Smoking and Cardiovascular Risks

Any amount of smoking, even light smoking or occasional smoking, damages the heart and blood vessels. For some people, such as women who use birth control pills and people who have diabetes, smoking poses an even greater risk to the heart and blood vessels.

The chemicals in tobacco smoke harm your heart and blood vessels in many ways. For example, they:

- Contribute to inflammation, which may trigger plaque buildup in your arteries

- Damage blood vessel walls, making them stiff and less elastic (stretchy). This damage narrows the blood vessels and contributes to the damage caused by unhealthy cholesterol levels.

- Disturb normal heart rhythms

- Increase your blood pressure and heart rate, making your heart work harder than normal

- Lower your HDL ("good") cholesterol and raise your LDL ("bad") cholesterol. Smoking also increases your triglyceride level. Triglycerides are a type of fat found in the blood.

- Thicken your blood and make it harder for your blood to carry oxygen

Smoking and Ischemic Heart Disease Risk

Smoking is a major risk factor for ischemic heart disease, a condition in which plaque builds up inside the coronary arteries. These arteries supply your heart muscle with oxygen-rich blood.

When plaque builds up in the arteries, the condition is called atherosclerosis.

Plaque narrows the arteries and reduces blood flow to your heart muscle. The buildup of plaque also makes it more likely that blood clots will form in your arteries. Blood clots can partially or completely block blood flow.

Over time, smoking contributes to atherosclerosis and increases your risk of having and dying from heart disease, heart failure, or a heart attack.

Compared with nonsmokers, people who smoke are more likely to have heart disease and suffer from a heart attack. The risk of having or dying from a heart attack is even higher among people who smoke and already have heart disease.

For some people, such as women who use birth control pills and people who have diabetes, smoking poses an even greater risk to the heart and blood vessels.

Smoking is a major risk factor for heart disease. When combined with other risk factors—such as unhealthy blood cholesterol levels, high blood pressure, and overweight or obesity—smoking further raises the risk of heart disease.

Smoking and the Risk of Peripheral Artery Disease

Peripheral artery disease (PAD) is a disease in which plaque builds up in the arteries that carry blood to your head, organs, and limbs. Smoking is a major risk factor for PAD.

PAD usually affects the arteries that carry blood to your legs. Blocked blood flow in the leg arteries can cause cramping, pain, weakness, and numbness in your hips, thighs, and calf muscles.

Blocked blood flow also can raise your risk of getting an infection in the affected limb. Your body might have a hard time fighting the infection.

If severe enough, blocked blood flow can cause gangrene (tissue death). In very serious cases, this can lead to leg amputation.

If you have PAD, your risk of heart disease and heart attack is higher than the risk for people who don't have PAD.

Smoking even one or two cigarettes a day can interfere with PAD treatments. People who smoke and people who have diabetes are at highest risk for PAD complications, including gangrene in the leg from decreased blood flow.

Secondhand Smoke Risks[2]

Secondhand smoke is the smoke that comes from the burning end of a cigarette, cigar, or pipe. Secondhand smoke also refers to smoke that's breathed out by a person who is smoking.

Secondhand smoke contains many of the same harmful chemicals that people inhale when they smoke. It can damage the heart and blood vessels of people who don't smoke in the same way that active smoking harms people who do smoke. Secondhand smoke greatly increases adults' risk of heart attack and death.

Secondhand smoke also raises the risk of future coronary heart disease in children and teens because it:

- Damages heart tissues
- Lowers HDL cholesterol
- Raises blood pressure

The risks of secondhand smoke are especially high for premature babies who have respiratory distress syndrome and children who have conditions such as asthma.

Cigar and Pipe Smoke Risks[2]

Researchers know less about how cigar and pipe smoke affects the heart and blood vessels than they do about cigarette smoke.

However, the smoke from cigars and pipes contains the same harmful chemicals as the smoke from cigarettes. Also, studies have shown that people who smoke cigars are at increased risk of heart disease.

How Can You Protect Your Heart?[1]

The best way to safeguard your heart from smoking-related disease and death is to never start using cigarettes, but if you are a smoker, the earlier you are able to quit, the better. Quitting smoking benefits your heart and cardiovascular system now and in the future:

- Twenty minutes after you quit smoking, your heart rate drops.
- Just 12 hours after quitting smoking, the carbon monoxide level in the blood drops to normal, allowing more oxygen to vital organs like your heart.
- Within four years of quitting, your risk of stroke drops to that of lifetime nonsmokers.

Although quitting smoking is difficult, it is achievable, and medicinal cessation therapies such as nicotine replacement therapy (NRT) may be able to help you on your quit journey. Many addicted smokers find that U.S. Food and Drug Administration (FDA)-approved NRT helps them get through the hardest parts of quitting by lessening cravings and symptoms of withdrawal. NRTs are proven safe and effective to help you quit smoking by delivering measured amounts of nicotine without the toxic chemicals found in cigarette smoke.

If you are a smoker and you are concerned about your cardiovascular health, consulting with your doctor about NRTs or other cessation options and seeking help with quitting may help you protect your heart long term.

What Are the Benefits of Quitting Smoking?[2]

One of the best ways to reduce your risk of coronary heart disease is to avoid tobacco smoke. Don't ever start smoking. If you already smoke, quit. No matter how much or how long you've smoked, quitting will benefit you. Also, try to avoid secondhand smoke. Don't go to places where smoking is allowed. Ask friends and family members to not smoke in the house and car. Quitting smoking will benefit your heart and blood vessels. For example:

- Among persons diagnosed with coronary heart disease, quitting smoking greatly reduces the risk of recurrent heart attack and cardiovascular death. In many studies, this reduction in risk has been 50 percent or more.

- Heart disease risk associated with smoking begins to decrease soon after you quit, and for many people it continues to decrease over time.

- Your risk of atherosclerosis and blood clots related to smoking declines over time after you quit smoking.

- Quitting smoking can lower your risk of heart disease as much as, or more than, common medicines used to lower heart disease risk, including aspirin, statins, beta-blockers, and ACE inhibitors.

Benefits of Quitting Smoking and Avoiding Secondhand Smoke[2]

One of the best ways to reduce your risk of heart disease is to avoid tobacco smoke. Don't ever start smoking. If you already smoke, quit. No matter how much or how long you've smoked, quitting will benefit you.

Also, try to avoid secondhand smoke. Don't go to places where smoking is allowed. Ask friends and family members who smoke not to do it in the house and car.

Quitting smoking will reduce your risk of developing and dying from heart disease. Over time, quitting also will lower your risk of atherosclerosis and blood clots.

If you smoke and already have heart disease, quitting smoking will reduce your risk of sudden cardiac death, a second heart attack, and death from other chronic diseases.

Researchers have studied communities that have banned smoking at worksites and in public places. The number of heart attacks in these communities dropped quite a bit. Researchers think these results are due to a decrease in active smoking and reduced exposure to secondhand smoke.

Section 16.2

Smoking and Coronary Heart Disease

This section includes text excerpted from "At a Glance: Coronary Heart Disease," National Heart, Lung, and Blood Institute (NHLBI), August 15, 2009. Reviewed February 2019.

Coronary heart disease (CHD), also called coronary artery disease, is the leading cause of death in the United States for both men and women. CHD occurs when plaque builds up inside the coronary arteries. These arteries supply your heart muscle with oxygen-rich blood.

Plaque is made up of fat, cholesterol, calcium, and other substances found in the blood. Over time, plaque hardens and narrows the arteries, reducing blood flow to your heart muscle.

Eventually, an area of plaque can rupture, causing a blood clot to form on the surface of the plaque. If the clot becomes large enough, it can mostly or completely block the flow of oxygen-rich blood to the part of the heart muscle fed by the artery. This can lead to angina or a heart attack.

Angina is chest pain or discomfort that occurs when not enough oxygen-rich blood is flowing

to an area of your heart muscle. Angina may feel like pressure or squeezing in your chest. The pain also may occur in your shoulders, arms, neck, jaw, or back.

A heart attack occurs when blood flow to an area of your heart muscle is completely blocked. This prevents oxygen-rich blood from reaching that area of heart muscle, causing it to die. Without quick treatment, a heart attack can lead to serious problems or death.

Over time, CHD can weaken the heart muscle and lead to heart failure and arrhythmias. Heart failure is a condition in which your heart can't pump enough blood throughout your body. Arrhythmias are problems with the rate or rhythm of your heartbeat.

Causes and Risk Factors

Research suggests that CHD starts when certain factors damage the inner layers of the coronary arteries. These factors include smoking, high amounts of certain fats and cholesterol in the blood, high blood pressure, and high amounts of sugar in the blood due to insulin resistance or diabetes.

When damage occurs, your body starts a healing process. This process causes plaque to build up where the arteries are damaged. The buildup of plaque in the coronary arteries may start in Childhood.

Certain traits, conditions, or habits raise your risk for CHD. These conditions are known as risk factors. The major risk factors for CHD include:

- Smoking

- Unhealthy blood cholesterol levels

- High blood pressure

- Insulin resistance

- Diabetes

- Overweight or obesity

- Metabolic syndrome

- Lack of physical activity

- Age (as you get older, your risk for CHD increases)

- Family history of early heart disease

Lifestyle changes, medicines, and/or medical procedures can prevent or treat CHD in most people.

Treatment and Prevention

Taking action to control your risk factors can help prevent or delay CHD. Your chance of developing CHD goes up with the number of risk factors you have.

For some people, lifestyle changes may be the only treatment needed. Lifestyle changes include following a heart-healthy diet, doing physical activity regularly, maintaining a healthy weight, quitting smoking, and reducing stress.

You may need medicines to treat CHD if lifestyle changes aren't enough. Medicines can help control CHD risk factors and relieve CHD symptoms. Some people who have CHD also need a medical procedure to treat the disease. Angioplasty and coronary artery bypass grafting are two procedures used to treat CHD.

You may need medicines to treat CHD if lifestyle changes aren't enough. Medicines can help control CHD risk factors and relieve CHD symptoms. Some people who have CHD also need a medical procedure to treat the disease. Angioplasty and coronary artery bypass grafting are two procedures used to treat CHD. If you've been diagnosed with CHD, see your doctor for ongoing care. Follow your treatment plan and take all medicines as your doctor prescribes. Call your doctor if you have new or worsening symptoms.

Section 16.3

Smoking and Aneurysms

This section includes text excerpted from "When Blood Vessels Bulge—All about Aneurysms," *NIH News in Health*, National Institutes of Health (NIH), March 2010. Reviewed February 2019.

An aneurysm—a balloon-like bulge in an artery—can develop and grow for years without causing any symptoms. But an aneurysm is a silent threat to your health.

If an aneurysm grows too large, it can burst open, or rupture, and lead to dangerous bleeding inside the body. Aneurysms can also cause a split within the layers of an artery wall. This split, called a dissection, can lead to bleeding within the artery's layers. Aneurysms that rupture or dissect can cause sudden death.

Many aneurysms occur in deep inside the chest, abdomen or brain. Most appear in the aorta—the main artery that carries blood from the heart down through the center of the body. About 14,000 Americans die each year from aortic aneurysms. Three-fourths of these aneurysms arise in the lower part of the aorta. These are called abdominal aortic aneurysms (AAA). Aneurysms that occur in the chest are called thoracic aortic aneurysms (TAA).

About 1 in 50 people has some type of aneurysm in the brain. Many of these are small and cause no real problems. But each year about 27,000 people in the United States have a ruptured brain aneurysm, which is a type of stroke.

Age and gender also contribute to risk. Aortic aneurysms are most common in men after age 65. Brain aneurysms appear more often in women between 30 and 60 years of age.

The symptoms of an aneurysm can vary widely and depend on its location and size. Large AAAs might cause a throbbing in the abdomen. Large TAAs may lead to pain in the back, jaw, neck, or chest. Brain aneurysms can cause pain around the eye or numbness on one side of the face.

If an aneurysm ruptures or dissects, get immediate medical attention. Sudden, severe pain in the lower abdomen and back can indicate a ruptured AAA. Dissected or ruptured TAAs may cause sharp pain that travels from the upper back to the abdomen. Ruptured brain aneurysms can cause a sudden, intense headache.

However, because most aneurysms have no symptoms, they're often found by chance during a doctor visit. "Many aneurysms are found when a patient is getting images—such as magnetic resonance imaging (MRI) or computed tomography (CT) scans—done for another reason," says Tolunay. "Sometimes, if an abdominal aortic aneurysm is big enough, the doctor might be able to feel it during a routine physical."

If you have an aneurysm, medications can help lower your blood pressure and reduce the risk of rupture. If the aneurysm is small, your doctor may recommend regular checkups to monitor its size. Large or quickly growing aneurysms may be treated with surgery, although surgery for brain aneurysms carries many risks. Options for aortic aneurysms include open surgery, which removes the aneurysm, or endovascular repair, which strengthens the aorta by inserting a tube, or stent.

Talk with your doctor if you have a family history of aneurysms or other risk factors. People at high risk may need routine screening to find and monitor an aneurysm. "If you're concerned, get checked," says Tolunay. "And certainly lifestyle changes—like stopping smoking—can help."

Reduce Your Risk

You can't control your genes or certain other risk factors, but taking these steps might lower your risk of aneurysm and its complications:

• Quit smoking

• Keep your blood pressure and cholesterol levels in check

• Follow a healthy diet and exercise regularly

• If your family members have had aneurysms, talk to your doctor about being screened

Section 16.4

Smoking and Peripheral Arterial Disease

This section contains text excerpted from the following sources: Text in this section begins with excerpts from "Peripheral Artery Disease," National Heart, Lung, and Blood Institute (NHLBI), January 23, 2019; Text under the heading "Smoking Linked to Peripheral Artery Disease in African Americans" is excerpted from "Smoking Linked to Higher Risk of Peripheral Artery Disease in African Americans," National Institutes of Health (NIH), January 23, 2019.

Peripheral artery disease (PAD) is a disease in which plaque builds up in the arteries that carry blood to your head, organs, and limbs. Plaque is made up of fat, cholesterol, calcium, fibrous tissue, and other substances in the blood.

When plaque builds up in the body's arteries, the condition is called atherosclerosis. Over time, plaque can harden and narrow the arteries. This limits the flow of oxygen-rich blood to your organs and other parts of your body.

PAD usually affects the arteries in the legs, but it also can affect the arteries that carry blood from your heart to your head, arms, kidneys, and stomach. This focuses on PAD that affects blood flow to the legs.

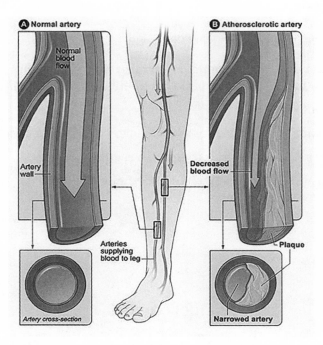

Figure 16.2. *Normal Artery and Artery with Plaque Buildup*

The illustration shows how PAD can affect arteries in the legs. Figure A shows a normal artery with normal blood flow. The inset image shows a cross-section of the normal artery. Figure B shows an artery with plaque buildup that's partially blocking blood flow. The inset image shows a cross-section of the narrowed artery.

Blocked blood flow to your legs can cause pain and numbness. It also can raise your risk of getting an infection in the affected limbs. Your body may have a hard time fighting the infection.

If severe enough, blocked blood flow can cause gangrene (tissue death). In very serious cases, this can lead to leg amputation.

If you have leg pain when you walk or climb stairs, talk with your doctor. Sometimes older people think that leg pain is just a symptom of aging. However, the cause of the pain could be PAD. Tell your doctor if you're feeling pain in your legs and discuss whether you should be tested for PAD.

Smoking is the main risk factor for PAD If you smoke or have a history of smoking, your risk of PAD increases. Other factors, such as age and having certain diseases or conditions, also increase your risk of PAD.

PAD increases your risk of ischemic heart disease, heart attack, stroke, and transient ischemic attack ("mini-stroke"). Although PAD is serious, it's treatable. If you have the disease, see your doctor regularly and treat the underlying atherosclerosis. PAD treatment may slow or stop disease progress and reduce the risk of complications. Treatments include lifestyle changes, medicines, and surgery or procedures. Researchers continue to explore new therapies for PAD.

Causes

The most common cause of PAD is atherosclerosis. Atherosclerosis is a disease in which plaque builds up in your arteries. The exact cause of atherosclerosis isn't known.

The disease may start if certain factors damage the inner layers of the arteries. These factors include:

- Smoking
- High amounts of certain fats and cholesterol in the blood
- High blood pressure
- High amounts of sugar in the blood due to insulin resistance, or
- Diabetes

When damage occurs, your body starts a healing process. The healing may cause plaque to build up where the arteries are damaged.

Eventually, a section of plaque can rupture (break open), causing a blood clot to form at the site. The buildup of plaque or blood clots can severely narrow or block the arteries and limit the flow of oxygen-rich blood to your body.

Risk Factors

Peripheral artery disease (PAD) affects millions of people in the United States. The disease is more common in blacks than any other racial or ethnic group. The major risk factors for PAD are smoking, older age, and having certain diseases or conditions.

- **Smoking**—is the main risk factor for PAD and your risk increases if you smoke or have a history of smoking. Quitting smoking slows the progress of PAD People who smoke and people who have diabetes are at highest risk for PAD complications, such as gangrene (tissue death) in the leg from decreased blood flow.

- **Older age**—also is a risk factor for PAD Plaque builds up in your arteries as you age. Older age combined with other risk factors, such as smoking or diabetes, also puts you at higher risk for PAD

- **Diseases and conditions**—Many diseases and conditions can raise your risk of PAD, including:

  - Diabetes

  - High blood pressure

  - High blood cholesterol

  - Ischemic heart disease

  - Stroke

  - Metabolic syndrome

Smoking Linked to Higher Risk of Peripheral Artery Disease in African Americans

African Americans who smoke appear to be at greater risk for peripheral artery disease, or PAD, a research has found. Additionally, the findings suggest that smoking intensity how many cigarettes a day and for how many years also affects the likelihood of getting the disease.

PAD affects 8 to 12 million people in the United States and 202 million worldwide, especially those age 50 and older. It develops when arteries in the legs become clogged with plaque, fatty deposits that limit blood flow to the legs. Clogged arteries in the legs can cause symptoms such as claudication, pain due to too little blood flow, and increased risk for heart attack and stroke.

The impact of cigarette smoking on PAD has been understudied in African Americans, even though PAD is nearly three times more prevalent in African Americans than in whites. The current study looked at the relationship between smoking and PAD in participants in the Jackson Heart Study, the largest single site cohort study investigating cardiovascular disease in African Americans.

The research, as well as the Jackson Heart Study, are funded by the National Heart, Lung, and Blood Institute (NHLBI), and the National Institute of Minority Health and Health Disparities

(NIMHD), both part of the National Institutes of Health. The findings appear in the January issue of the Journal of the American Heart Association.

"These findings demonstrate that smoking is associated with PAD in a dose-dependent manner," said lead researcher Donald Clark, III, M.D., an assistant professor of medicine at the University of Mississippi Medical Center (UMMC), Jackson. "This is particularly important in the African American community and supports the evaluation of smoking-cessation efforts to reduce the impact of PAD in this population."

Even though PAD is more prevalent in African Americans than in whites, prior studies about the disease did not include significant numbers of African Americans. This limited the researchers' ability to single out the specific effects of smoking in this population from other risk factors, such as hypertension, diabetes mellitus, and obesity.

For the study, researchers divided the 5,258 participants into smokers, past smokers and never smokers. After taking into account other risk factors, they assessed smoking intensity and found a dose-dependent link between cigarette smoking and PAD. Those smoking more than a pack a day had significantly higher risk than those smoking fewer than 19 cigarettes daily. Similarly, those with a longer history of smoking had an increased likelihood of the disease.

"Current and past smokers had higher odds of peripheral artery disease than never smokers; although the odds were lower among past smokers," Clark said. "Our findings add to the mountain of evidence of the negative effects of smoking and highlight the importance of smoking cessation, as well as prevention of smoking initiation."

Section 16.5

Smoking and Buerger Disease

This section includes text excerpted from
"Buerger's Disease," Centers for Disease Control and
Prevention (CDC), March 22, 2018.

What Is Buerger Disease?

Buerger disease affects blood vessels in the arms and legs. Blood vessels swell, which can prevent blood flow, causing clots to form. This can lead to pain, tissue damage, and even gangrene (the death or decay of body tissues). In some cases, amputation may be required.

The most common symptoms of Buerger disease are:

- Pale, red, or bluish hands or feet

- Cold hands or feet

- Pain in the hands and feet, which may be severe

- Pain in the legs, ankles, or feet when walking—often located in the arch of the foot

- Skin changes, painful sores, or ulcers on the hands or feet

How Is Smoking Related to Buerger Disease?

Almost everyone with Buerger disease smokes cigarettes. However, Buerger disease can occur in people who use other forms of tobacco, like chewing tobacco. People who smoke 1½ packs a day or more are most likely to develop Buerger disease.

Researchers are working to understand how tobacco increases the risk for Buerger disease. One idea is that chemicals in tobacco irritate the lining of the blood vessels and cause them to swell.

How Can Buerger Disease Be Prevented?

If you want to prevent getting Buerger disease, don't smoke or use any other tobacco products.

How Is Buerger Disease Treated?

There is no cure for Buerger disease. The only way to keep Buerger disease from getting worse is to stop using all tobacco products.

Medicines don't usually work well to treat the disease. The best they can do is to control the symptoms.

Surgery may help restore blood flow to some areas. It may be necessary to amputate the hand or foot if infection or widespread tissue death occurs.

Section 16.6

Cardiovascular Diseases and Stroke

This section includes text excerpted from "Heart Disease and Stroke," Centers for Disease Control and Prevention (CDC), April 19, 2018.

Heart disease and stroke are cardiovascular (heart and blood vessel) diseases (CVDs). Heart disease includes several types of heart conditions. The most common type in the United States is coronary heart disease (also known as coronary artery disease), which is narrowing of the blood vessels that carry blood to the heart. This can cause:

- Chest pain

- Heart attack (when blood flow to the heart becomes blocked and a section of the heart muscle is damaged or dies)

- Heart failure (when the heart cannot pump enough blood and oxygen to support other organs)

- Arrhythmia (when the heart beats too fast, too slow, or irregularly)

A stroke occurs when the blood supply to the brain is blocked or when a blood vessel in the brain bursts, causing brain tissue to die. Stroke can cause disability (such as paralysis, muscle weakness, trouble speaking, memory loss) or death.

How Is Smoking Related to Heart Disease and Stroke?

Smoking is a major cause of CVD and causes one of every three deaths from CVD. Smoking can:

- Raise triglycerides (a type of fat in your blood)

- Lower "good" cholesterol (high-density lipoproteins (HDL))

- Make blood sticky and more likely to clot, which can block blood flow to the heart and brain

- Damage cells that line the blood vessels

- Increase the buildup of plaque (fat, cholesterol, calcium, and other substances) in blood vessels

- Cause thickening and narrowing of blood vessels

How Is Breathing Secondhand Smoke Related to Heart Disease and Stroke?

Breathing secondhand smoke also harms your health. Secondhand smoke is the smoke from burning tobacco products. Secondhand smoke also is smoke breathed out by a smoker.

Breathing secondhand smoke can cause coronary heart disease, including heart attack and stroke.

- Secondhand smoke causes nearly 34,000 early deaths from coronary heart disease each year in the United States among nonsmokers.

- Nonsmokers who breathe secondhand smoke at home or at work increase their risk of developing heart disease by 25 to 30 percent. Secondhand smoke increases the risk for stroke by 20 to 30 percent.

- Each year, secondhand smoke exposure causes more than 8,000 deaths from stroke.

- Breathing secondhand smoke interferes with the normal functioning of the heart, blood, and vascular systems in ways that increase your risk of having a heart attack.

- Even briefly breathing secondhand smoke can damage the lining of blood vessels and cause your blood to become stickier. These changes can cause a deadly heart attack.

How Can Heart Disease and Stroke Be Prevented?

Heart disease and stroke are major causes of death and disability in the United States. Many people are at high risk for these diseases and don't know it. The good news is that many risk factors for heart disease and stroke can be prevented or controlled.

The federal government's Million Hearts® initiative aims to prevent 1 million heart attacks and strokes by 2017. It's important to know your risk for heart disease and stroke and to take action to reduce that risk. A good place to start is with the ABCS of heart health:

- **A**spirin: Aspirin may help reduce your risk for heart disease and stroke. But do not take aspirin if you think you are having a stroke. It can make some types of stroke worse. Before taking aspirin, talk to your doctor about whether aspirin is right for you.

- **B**lood pressure: Control your blood pressure.

- **C**holesterol: Manage your cholesterol.

- **S**moking: Quit smoking, or don't start.

In addition to your ABCS, several lifestyle choices can help protect your heart and brain health. These include the following:

- Avoid breathing secondhand smoke

- Eat low-fat, low-salt foods most of the time and fresh fruits and vegetables

- Maintain a healthy weight

- Exercise regularly

- Limit alcohol use

- Get other health conditions (such as diabetes) under control

Chapter 17

Smoking and Mental Health

Chapter Contents

Section 17.1

Tobacco Use among Adults with Mental Illness and Substance-Use Disorders

This section includes text excerpted from "Tobacco Use among Adults with Mental Illness and Substance Use Disorders," Centers for Disease Control and Prevention (CDC), January 14, 2019.

Tobacco Use Prevalence among Persons with Mental Illness

32.0 percent of adults with any mental illness reported current use* of tobacco in 2016 compared to 23.3 percent of adults with no mental illness.

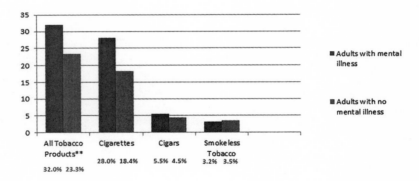

Figure 17.1. *Current Use* of Specific Tobacco Products among Adults with Mental Illness Compared with Adults with No Mental Illness†.*

* "Current Use" is defined as self-reported consumption of cigarettes, cigars, and smokeless tobacco in the past month (at the time of survey).
** All Tobacco Products includes cigarettes, smokeless tobacco (i.e., snuff, dip, chewing tobacco, or "snus"), cigars, and pipe tobacco.
† Data taken from the National Survey on Drug Use and Health, 2016, and refer to adults aged 18 years and older self-reporting any mental illness in the past year, excluding serious mental illness.

Tobacco Use Prevalence among Persons with Substance-Use Disorders

63.5 percent of adult cigarette smokers reported co-use of alcohol in 2016 compared to 52.8 percent of adult nonsmokers.

Table 17.1. Current Illicit Drug and Alcohol Use among Adult Cigarette Smokers Compared with Nonsmokers[‡]

Usage	Smokers	Nonsmokers
Current illicit drug (in past month)	25.30%	7.10%
Marijuana	21.80%	5.90%
Cocaine	2.50%	0.30%
Heroin	0.80%	0.00%
Hallucinogens	1.50%	0.30%
Inhalants	0.40%	0.10%
Nonmedical use of prescription drugs	5.90%	1.50%
Current alcohol use (in past month)	63.50%	52.80%
Binge drinking[§]	43.50%	21.70%
Heavy drinking[¶]	14.60%	4.50%

[‡] *Data taken from the National Survey on Drug Use and Health, 2016, and refer to persons aged 18 years and older reporting smoking, drug, and/or alcohol use in the past 30 days.*
[§] *Binge alcohol use is defined as drinking five or more drinks on the same occasion (i.e., at the same time or within a couple of hours of each other) on at least 1 day in the past 30 days.*
[¶] *Heavy alcohol use is defined as drinking five or more drinks on the same occasion on each of 5 or more days in the past 30 days; all heavy alcohol users are also binge alcohol users.*

Health Effects

- People with mental illness or substance-use disorders die about five years earlier than those without these disorders; many of these deaths are caused by smoking cigarettes.

- The most common causes of death among people with mental illness are heart disease, cancer, and lung disease, which can all be caused by smoking.

- Drug users who smoke cigarettes are four times more likely to die prematurely than those who do not smoke.

- Nicotine has mood-altering effects that can temporarily mask the negative symptoms of mental illness, putting people with mental illness at higher risk for cigarette use and nicotine addiction.

- Tobacco smoke can interact with and inhibit the effectiveness of certain medications taken by mental health and substance-abuse patients.

Secondhand Smoke Exposure

- Approximately half of mental health and one-third of substance-treatment facilities report having smoke-free campuses.

- Tobacco-free campus policies that prohibit all forms of tobacco product use can support tobacco cessation, reinforce tobacco-free norms, and eliminate exposure to secondhand tobacco product emissions.

Quitting Behavior

- People with mental illness are less likely to stop smoking than those without mental illness; however, many smokers with mental illness want to quit.

- People with mental illness are more likely to have stressful living conditions, have low annual household income, and lack access to health insurance, healthcare, and help quitting. All of these factors make it more challenging to quit.

- Fewer than half of mental health and substance-abuse treatment facilities in the United States (including Puerto Rico) offer evidence-based tobacco-cessation treatments.

Tobacco Industry Marketing and Influence

The tobacco industry has used multiple strategies to market cigarettes to populations with mental illnesses, including:

- Developing relationships with and making financial contributions to organizations that work with mentally ill patients

- Funding research to foster the myth that cessation would be too stressful because persons with mental illness use nicotine to alleviate negative mood (i.e., self-medicate)

- Providing free or cheap cigarettes to psychiatric facilities

- Supporting efforts to block smoke-free psychiatric hospital policies

- Creating marketing plans that target marginalized populations, including mentally ill, homeless, and Lesbian, Gay, Bisexual, and Transgender (LGBT) individuals. One example is "Project SCUM" (Sub Culture Urban Marketing), which was implemented in San Francisco in the mid-1990s.

Myths

Myths about tobacco use among people with mental illness and substance-abuse problems persist, including that:

- They are not interested in quitting

- They cannot quit

- Quitting interferes with recovery from mental illness or addictions

- Tobacco is not as harmful as other substances

- Tobacco is necessary for self-medication, and tobacco cessation would be too stressful

- Tobacco cessation efforts might prevent treatment of other addictions

However, with careful monitoring, delivering smoking-cessation interventions does not interfere with treatments for mental illness and can actually be part of the treatment.

Policy Strategies

A number of policy strategies have been suggested in the effort to encourage smoking cessation among people with mental illness and substance-use disorders:

- Mental-health facilities should consider going smokefree, including prohibiting tobacco use among employees or encouraging staff to quit.

- Counselors should ask clients who smoke cigarettes or use other tobacco products about their interest in quitting while in substance-abuse treatment.

- The practice of providing mental-health patients with cigarettes as an incentive or reward should be discontinued. Additionally, staff should not be allowed to smoke cigarettes with patients.

- Extra help to succeed in quitting should be offered to patients who smoke cigarettes. This can include more counseling, combining stop-smoking medicines or using them longer, and followup to care (e.g., telephone calls by a counselor, extended counseling, or continued medications).

Section 17.2

Common Comorbidities with Substance-Use Disorders

This section includes text excerpted from "Common Comorbidities with Substance Use Disorders," National Institute on Drug Abuse (NIDA), February 2018.

The high prevalence of comorbidity between Substance-use disorders and other mental illnesses does not necessarily mean that one caused the other, even if one appeared first. Establishing causality or directionality is difficult for several reasons. For example, behavioral or emotional problems may not be severe enough for a diagnosis (called subclinical symptoms), but subclinical mental-health issues may prompt drug use. Also, people's recollections of when drug use or addiction started may be imperfect, making it difficult to determine whether the substance-use or mental-health issues came first.

Three main pathways can contribute to the comorbidity between substance-use disorders and mental illnesses:

1. Common Risk Factors Can Contribute to Both Mental Illness and Substance-Use and Addiction

Both substance-use disorders and other mental illnesses are caused by overlapping factors such as genetic and epigenetic vulnerabilities, issues with similar areas of the brain, and environmental influences such as early exposure to stress or trauma.

Genetic Vulnerabilities

It is estimated that 40 to 60 percent of an individual's vulnerability to substance-use disorders is attributable to genetics. An active area of comorbidity research involves the search for that might predispose individuals to develop both a substance-use disorder and other mental illnesses, or to have a greater risk of a second disorder occurring after the first appears. Most of this vulnerability arises from complex interactions among multiple genes and genetic interactions with environmental influences. For example, frequent marijuana use during adolescence is associated with increased risk of psychosis in adulthood, specifically among individuals who carry a particular gene variant.

In some instances, a gene product may act directly, as when a protein influences how a person responds to a drug (e.g., whether the drug experience is pleasurable or not) or how long a drug remains in the body. Specific genetic factors have been identified that predispose an individual to alcohol dependence and cigarette smoking, and research is starting to uncover the link between genetic sequences and a higher risk of cocaine dependence, heavy opioid use, and cannabis craving and withdrawal. But genes can also act indirectly by altering how an individual responds to stress or by increasing the likelihood of risk-taking and novelty-seeking behaviors, which could influence the initiation of substance use as well as the development of substance use disorders and other mental illnesses. Research suggests that there are many genes that may contribute to the risk for both mental disorders and addiction, including those that influence the action of neurotransmitters—chemicals that carry messages from one neuron to another—that are affected by drugs and commonly dysregulated in mental illness, such as dopamine and serotonin.

Epigenetic Influences

Scientists are also beginning to understand the very powerful ways that genetic and environmental factors interact at the molecular level. Epigenetics refers to the study of changes in the regulation of gene activity and expression that are not dependent on gene sequence; that is, changes that affect how genetic information is read and acted on by cells in the body. Environmental factors such as chronic stress, trauma, or drug exposure can induce stable changes in gene expression, which can alter functioning in neural circuits and ultimately impact behavior.

Through epigenetic mechanisms, the environment can cause long-term genetic adaptations—influencing the pattern of genes that are active or silent in encoding proteins—without altering the deoxyribonucleic acid (DNA) sequence. These modifications can sometimes even be passed down to the next generation. There is also evidence that they can be reversed with interventions or environmental alteration.

The epigenetic impact of environment is highly dependent on developmental stage. Studies suggest that environmental factors interact with genetic vulnerability during particular developmental periods to increase the risk for mental illnesses and addiction. For example, animal studies indicate that a maternal diet high in fat during pregnancy can influence levels of key proteins involved in neurotransmission in the brain's reward pathway. Other animal research has shown that poor quality maternal care diminished the ability of offspring to respond to stress through epigenetic mechanisms. Researchers are using animal models to explore the epigenetic changes induced by chronic stress or drug administration, and how these changes contribute to depression- and addiction-related behaviors. A better understanding of the biological mechanisms that underlie the genetic and biological interactions that contribute to the development of these disorders will inform the design of improved treatment strategies.

Brain-Region Involvement

Many areas of the brain are affected by both substance-use disorders and other mental illnesses. For example, the circuits in the brain that mediate reward, decision making, impulse control, and emotions may be affected by addictive substances and disrupted in substance-use disorders, depression, schizophrenia, and other psychiatric disorders. In addition, multiple neurotransmitter systems have been implicated in both substance use disorders and other mental disorders including, but not limited to, dopamine, serotonin, glutamate, GABA, and norepinephrine.

Environmental Influences

Many environmental factors are associated with an increased risk for both substance-use disorders and mental illness including chronic stress, trauma, and adverse childhood experiences, among others. Many of these factors are modifiable and; thus, prevention interventions will often result in reductions in both substance-use disorders and mental illness, as discussed in the Surgeon General's report on alcohol, drugs, and health.

Stress

Stress is a known risk factor for a range of mental disorders and, therefore, provides one likely common neurobiological link between the disease processes of substance-use disorders and mental disorders. Exposure to stressors is also a major risk factor for relapse to drug use after periods of recovery. Stress responses are mediated through the hypothalamic-pituitary-adrenal (HPA) axis, which in turn can influence brain circuits that control motivation. Higher levels of stress have been shown to reduce activity in the prefrontal cortex and increase responsivity in the striatum, which leads to decreased behavioral control and increased impulsivity. Early life stress and chronic stress can cause long-term alterations in the HPA axis, which affects limbic brain circuits that are involved in motivation, learning, and adaptation, and are impaired in individuals with substance-use disorders and other mental illnesses.

Importantly, dopamine pathways have been implicated in the way in which stress can increase vulnerability to substance-use disorders. HPA axis hyperactivity has been shown to alter dopamine signaling, which may enhance the reinforcing properties of drugs. In turn, substance use causes changes to many neurotransmitter systems that are involved in responses to stress. These neurobiological changes are thought to underlie the link between stress and escalation of drug use as well as relapse. Treatments that target stress, such as mindfulness-based stress reduction, have been shown to be beneficial for reducing depression, anxiety, and substance use.

Trauma and Adverse Childhood Experiences

Physically or emotionally traumatized people are at much higher risk for drug use and SUDs. and the co-occurrence of these disorders is associated with inferior treatment outcomes. People with post-traumatic stress disorder (PTSD) may use substances in an attempt to reduce their anxiety and to avoid dealing with trauma and its consequences.

The link between substance-use disorder and PTSD is of particular concern for service members returning from tours of duty in Iraq and Afghanistan. Between 2004 and 2010, approximately 16 percent of veterans had an untreated substance-use disorder, and 8 percent needed treatment for serious psychological distress (SPD). Data from a survey that used a contemporary, national sample of veterans estimated that the rate of lifetime PTSD was eight percent, while approximately five percent reported current PTSD. Approximately one in five veterans with PTSD also has a co-occurring substance use disorder.

2. Mental Illnesses Can Contribute to Drug Use and Addiction

Substance use can lead to changes in some of the same brain areas that are disrupted in other mental disorders, such as schizophrenia, anxiety, mood, or impulse-control disorders. Drug use that precedes the first symptoms of a mental illness may produce changes in brain structure and function that kindle an underlying predisposition to develop that mental illness.

3. Substance Use and Addiction Can Contribute to the Development of Mental Illness

Certain mental disorders are established risk factors for developing a substance-use disorder. It is commonly hypothesized that individuals with severe, mild, or even subclinical mental disorders may use drugs as a form of self-medication. Although some drugs may temporarily reduce symptoms of a mental illness, they can also exacerbate symptoms, both acutely and in the long run. For example, evidence suggests that periods of cocaine use may worsen the symptoms of bipolar disorder and contribute to progression of this illness.

When an individual develops a mental illness, associated changes in brain activity may increase the vulnerability for problematic use of substances by enhancing their rewarding effects, reducing awareness of their negative effects, or alleviating the unpleasant symptoms of the mental disorder or the side effects of the medication used to treat it. For example, neuroimaging suggests that attention deficit hyperactivity disorder (ADHD) is associated with neurobiological changes in brain circuits that are also associated with drug cravings, perhaps partially explaining why patients substance-use use disorders report greater cravings when they have comorbid ADHD.

Section 17.3

The Relationship between Tobacco and Mental Health

This section includes text excerpted from documents published by three public domain sources. Text under headings marked 1 are excerpted from "Smoking Cessation For Persons with Mental Illness," Substance Abuse and Mental Health Services Administration (SAMHSA), November 30, 2016; Text under headings marked 2 are excerpted from "Tobacco, Nicotine, and E-Cigarettes— Do People with Mental Illness and Substance Use Disorders Use Tobacco More Often," National Institute on Drug Abuse (NIDA), January 2018; Text under heading marked 3 is excerpted from "Smoking and Mental Illness Among Adults in the United States," National Center for Biotechnology Information (NCBI), March 30, 2017.

Increased Tobacco Use among Persons with Mental Illnesses[1]

Researchers believe that a combination of biological, psychological, and social factors contribute to increased tobacco use among persons with mental illnesses.

Biological predisposition—Persons with mental illnesses have unique neurobiological features that may increase their tendency to use nicotine, make it more difficult to quit, and complicate withdrawal symptoms. Nicotine affects the actions of neurotransmitters (e.g., dopamine). For example, people with schizophrenia who use tobacco may experience less negative symptoms (lack of motivation, drive, and energy). Nicotine enhances concentration, information processing, and learning. (This is especially important for persons with psychotic disorders for whom cognitive dysfunction may be a part of their illness or a side effect of antipsychotic medications.) Other biological factors include nicotine's positive effects on mood, and feelings of pleasure and enjoyment. Some evidence also suggests that smoking is associated with a reduced risk of antipsychotic-induced Parkinsonism.

Specific Psychiatric and Co-Occurring Disorders[1]

- **Depression**—Among patients seeking smoking cessation treatment, 25 to 40 percent have a history of major depression and many have minor dysthymic symptoms. Depression has been shown to predict poorer smoking cessation rates. Consider starting or restarting psychotherapy or pharmacotherapy for depression in patients who state that depression intensified with cessation or that cessation caused depression. Cognitive behavioral therapy (CBT) for depression and antidepressants has been found to improve smoking cessation rates in those with a history of depression or symptoms of depression. For a smoker with a history of depression currently taking antidepressant medication, it is important to note that some antidepressant levels will increase with smoking cessation.

- **Schizophrenia**—Persons with schizophrenia who smoke may be less interested in tobacco cessation, making strategies to enhance motivation to quit especially important. When patients with schizophrenia do try to stop, many are unsuccessful; thus, intensive treatments are appropriate even with early attempts. The high prevalence of alcohol and illicit-drug abuse in patients with schizophrenia can interfere with smoking cessation. The blood levels of some antipsychotics can increase dramatically with cessation. Nicotine withdrawal can mimic the akathisia, depression, difficulty concentrating, and insomnia seen in patients with schizophrenia. Researchers are working to identify the brain circuits that contribute to the high prevalence of smoking among people with schizophrenia. Schizophrenia is associated with widespread reductions in functional connectivity between the dorsal anterior cingulate cortex and diverse parts of the limbic system. One report identified 15 circuits for which the reduction of functional connectivity correlated with severity of nicotine addiction.

Co-Occurring Substance Abuse and Dependence[1]

Tobacco use is strongly correlated with the development of other substance-use disorders and with more severe substance-use disorders. Tobacco appears to affect the same neural pathway—the mesolimbic dopamine system—as alcohol, opioids, cocaine, and marijuana. Tobacco use impedes recovery of brain function among clients whose brains have been damaged by chronic alcohol use. At the same time, concurrent use of alcohol and/or other drugs is a negative predictor of smoking cessation outcomes during smoking cessation treatment.

Do People with Mental Illness and Substance-Use Disorders Use Tobacco More Often?[2]

There is significant comorbidity between tobacco use and mental disorders. People with mental illness smoke at two to four times the rate of the general population. Among people with a mental illness, 36.1 percent smoked from 2009 to 2011, compared with 21.4 percent among adults with no mental illness. Smoking rates are particularly high among patients with serious mental illness (i.e., those who demonstrate greater functional impairment). While estimates vary, as many as 70 to 85 percent of people with schizophrenia and as many as 50 to 70 percent of people with bipolar disorder smoke.

Rates of smoking among people with mental illness were highest for those younger than 45, those with low levels of education, and those living below the poverty level. Longitudinal data from the National Survey on Drug Use and Health (NSDUH) (2005–2013) indicate that smoking among adults without chronic conditions has declined significantly, but remains particularly high among those reporting anxiety, depression, and substance-use disorders. Smoking is believed to be more prevalent among people with depression and schizophrenia because nicotine may temporarily lessen the symptoms of these illnesses, such as poor concentration, low mood, and stress. But it is important to note that smoking cessation has been linked with improved mental health—including reduced depression, anxiety, and stress, and enhanced mood and quality of life (QOL).

Additionally, smokers with a mental-health disorder tend to smoke more cigarettes than those in the general population. The average number of cigarettes smoked during the past month was higher among those with a mental illness compared with those without one—331 versus 310 cigarettes. High cigarette consumption is a particular problem for people with serious mental illness. Although adults with serious mental illness comprised only 6.9 percent of past month smokers, they consumed 8.7 percent of all cigarettes sold, according to data from the 2008 to 2012 National Survey on Drug Use and Health (NSDUH).

Smokers with Mental Illness are More Vulnerable to Relapse[2]

People with mental and substance-use disorders do not quit smoking at the same rate as those in the general population. Survey responses from people who have smoked at some point during their lives indicated that fewer smokers with mental illness had quit compared to those without psychiatric disorders: 47.4 percent of lifetime smokers without mental illness smoked during the past month, compared with 66 percent of those with mental illness. Having a mental disorder at the time of cessation is a risk factor for relapse to smoking, even for those who have sustained abstinence for more than a year. Many smokers with mental illness want to quit for the same reasons cited by others (such as health and family), but they may be more vulnerable to relapse related to stress and negative feelings.

The disparity in smoking prevalence is costing lives. A study found that tobacco-related diseases accounted for approximately 53 percent of deaths among people with schizophrenia, 48 percent among those with bipolar disorder, and 50 percent among those with depression.

Since the 1980s, many providers have believed that people with schizophrenia smoke to obtain relief from symptoms such as poor concentration, low mood, and stress. But research is now showing that smoking is associated with worse behavioral and physical-health outcomes in people with mental illness, and quitting smoking is showing clear benefits for this population. Comprehensive tobacco-control programs and enhanced efforts to prevent and treat nicotine addiction among those with mental illness would reduce illness and deaths. Integrated treatment—concurrent therapy for mental illness and nicotine addiction—will likely have the best outcomes.

Smokers who receive mental-health treatment have higher quit rates than those who do not. Moreover, evidence-based treatments that work in the general population are also effective for patients with mental illness. For example, people with schizophrenia showed better quit rates with the medication bupropion, compared with placebo, and showed no worsening of psychiatric symptoms. A combination of the medication varenicline and behavioral support has shown promise for helping people with bipolar and major depressive disorders quit, with no worsening of psychiatric symptoms. A clinical trial found that a combination of varenicline and CBT was more effective than CBT alone for helping people with serious mental illness stop smoking for a prolonged period—after one year of treatment and at six months after treatment ended.

Smoking Cessation, by Mental-Illness Status[3]

Tobacco use continues to be the leading cause of preventable death in the United States. Despite further evidence provided in the 2010 Surgeon General's Report that cigarette smoking causes disease and that no level of cigarette use is safe, rates of cigarette use among certain groups of Americans remain high. Cigarette use among people with mental illness has garnered attention and concern over the past decade. For example, previous research has shown that people with mental illness are not only more likely to smoke, but they also smoke more frequently than people with no mental illness. For example, people with mental illness or a substance use disorder account for 25 percent of the adult population, but they consumed 40 percent of cigarettes sold in the United States. People who are receiving treatment for mental illness also have lower cessation rates of smoking than those people who are not seeking mental health treatment.

In general, when persons with mental illnesses make an attempt at smoking cessation, they should be followed closely to monitor for more severe nicotine withdrawal, exacerbation of their psychiatric disorder and possible side effects due to cessation-induced increases in medication levels.

Current cigarette use among those who had smoked daily at some time in their lives (hereafter referred to as "lifetime daily smokers") is an indicator of smoking cessation. Cessation refers to temporarily or completely stopping cigarette smoking. High proportions of current cigarette use among lifetime daily smokers in specific population groups indicate that few people in those groups have quit smoking.

Quick Facts[1]

• Persons with mental illnesses and substance abuse disorders report over 44 percent of the U.S. tobacco market.

• Persons with mental illnesses or substance abuse disorders are nicotine dependent at rates that are two to three times higher than the general population.

• About 200,000 of the 435,000 annual deaths from smoking occur among patients with mental illnesses and/or substance use disorders.

• Smoking cessation is a key component of consumer-driven, individualized treatment planning.

• Persons with mental illnesses want to quit smoking and want information on cessation services and resources.

• Persons with mental illnesses can successfully quit using tobacco.

- Although smoking cessation rates for individuals with psychiatric illness are lower than those in the general population, these quit rates are still substantial.

- Because persons with mental illnesses use tobacco at greater rates, they suffer greater smoking-related medical illnesses and mortality.

Chapter 18

Women's Health and Smoking

Chapter Contents

Section 18.1

Impact on Women's Health

This section includes text excerpted from "Smoking:
A Women's Health Issue," U.S. Food and Drug
Administration (FDA), January 9, 2019.

Smoking for anyone, at any age, is dangerous and can lead to preventable disease, and even death. But, for women, smoking carries certain additional risks. While many Americans may know that smoking can cause cardiovascular disease and certain cancers, like lung cancer, they may not be aware that smoking can also negatively impact a woman's reproductive health, as well as lead to cervical cancer.

How Smoking Affects Reproductive Health

Because cigarette smoke comprises of a mix of over 7,000 chemicals, breathing these chemicals in can damage nearly any part of the body. For women, smoking cigarettes can lead to reproductive damage, reduced fertility, and difficulty conceiving. Research shows smoking may affect hormone production, making it difficult to become pregnant. Further, certain chemicals found in cigarettes, like 1,3-Butadiene and benzene, have been shown to harm the reproductive system and may reduce fertility.

If a woman is able to conceive but smokes during pregnancy, she may experience complications—such as ectopic pregnancy—as a result of the chemicals in cigarette smoke. An ectopic pregnancy occurs when a fertilized egg fails to reach the womb but instead begins to grow outside of the womb. This serious condition almost always results in the death of the fetus, and in some cases, maternal death as well. Additionally, there is some evidence that smoking during pregnancy may result in a miscarriage of the fetus.

Smoking and Pregnancy

Smoking during pregnancy can also result in negative outcomes for a woman's unborn baby. Approximately 400,000 U.S. infants per year are exposed to cigarette smoke and its chemicals in the womb. These babies are at risk of a number of complications including:

- Low birth weight
- Lungs that fail to develop properly

- Birth defects such as cleft lip and/or cleft palate
- Sudden infant death syndrome (SIDS)

While cigarettes are the tobacco product most harmful to public health, no tobacco product is safe for pregnant women to use. That's because nearly all tobacco products—including most e-cigarettes—contain nicotine, which can cross the placenta and interfere with fetal and postnatal development.

Smoking and Cancer in Women

Smoking cigarettes can cause cervical cancer, cancer that only affects women. And nearly all lung cancers, the number one killer of both men and women, is caused by cigarette smoking.

Quitting: The Healthiest Option for All Women

The best way for a woman to safeguard her health is to never start smoking. But for women who smoke, quitting is the best option, and it is never too late to do so. Within just a few years of quitting, a woman's cervical cancer risk is reduced, and lung cancer risk can drop by as much as half within 10 years after quitting.

Pregnant women who are looking to quit smoking should consult with a doctor about how to do so with the most success for themselves and their babies. Smokefree.gov also offers resources for quitting while pregnant, including a texting program to offer support for pregnant women trying to quit.

Because nicotine in any form may be unsafe for expectant mothers, medical guidance is necessary. Women who are not pregnant may find that nicotine replacement therapy (NRT) can help with quitting. NRT is designed to help addicted smokers through the hardest parts of quitting. U.S. Food and Drug Administration (FDA)-approved NRTs help addicted smokers withdraw from smoking by delivering measured amounts of nicotine without the toxic chemicals from cigarette smoke. When used properly, NRTs are safe and effective cessation methods and can double a smoker's chances of quitting cigarettes successfully.

How the U.S. Food and Drug Administration Plans to Further Reduce Smoking

In an effort to prevent a new generation of smokers and to help addicted smokers quit cigarettes, in July 2017, the FDA Commissioner

Scott Gottlieb, M.D., announced the agency's intention to lower nicotine levels in cigarettes to minimally addictive or nonaddictive levels. While this proposed action would not remove cigarettes from the market, it would reduce the addiction potential of cigarettes, offering addicted adult smokers the chance to attempt to quit with the help of NRTs, or to switch to other potentially less harmful tobacco products. Through this action, the FDA hopes to protect the health of not just women and their children, but of all Americans.

Section 18.2

Smoking Risks during Pregnancy

This section contains text excerpted from the following sources: Text beginning with the heading "How Does Smoking during Pregnancy Harm Your Health and Your Baby?" is excerpted from "Tobacco Use and Pregnancy," Centers for Disease Control and Prevention (CDC), September 29, 2017; Text beginning with the heading "An Outlook for Mother and Baby" is excerpted from "Smoking, Pregnancy, and Babies," Centers for Disease Control and Prevention (CDC), March 22, 2018.

How Does Smoking during Pregnancy Harm Your Health and Your Baby?

Most people know that smoking causes cancer, heart disease, and other major health problems. Smoking during pregnancy causes additional health problems, including premature birth (being born too early), certain birth defects, and infant death.

- Smoking makes it harder for a woman to get pregnant.

- Women who smoke during pregnancy are more likely than other women to have a miscarriage.

- Smoking can cause problems with the placenta—the source of the baby's food and oxygen during pregnancy. For example, the placenta can separate from the womb too early, causing bleeding, which is dangerous to the mother and baby.

- Smoking during pregnancy can cause a baby to be born too early or to have low birth weight—making it more likely the baby will be sick and have to stay in the hospital longer. A few babies may even die.

- Smoking during and after pregnancy is a risk factor of sudden infant death syndrome (SIDS). SIDS is an infant death for which a cause of the death cannot be found.

- Babies born to women who smoke are more likely to have certain birth defects, like a cleft lip or cleft palate.

How Many Women Smoke during Pregnancy?

According to the 2011 Pregnancy Risk Assessment and Monitoring System (PRAMS) data from 24 states.

- Approximately 10 percent of women reported smoking during the last three months of pregnancy.

- Of women who smoked three months before pregnancy, 55 percent quit during pregnancy. Among women who quit smoking during pregnancy, 40 percent started smoking again within six months after delivery.

How Does Other People's Smoke (Secondhand Smoke) Harm My Health and My Child's Health?

Breathing other people's smoke make children and adults who do not smoke sick. There is no safe level of breathing others people's smoke.

- Pregnant women who breathe other people's cigarette smoke are more likely to have a baby who weighs less.

- Babies who breathe in other people's cigarette smoke are more likely to have ear infections and more frequent asthma attacks.

- Babies who breathe in other people's cigarette smoke are more likely to die from SIDS. SIDS is an infant death for which a cause of the death cannot be found.

In the United States, 58 million children and adults who do not smoke are exposed to other people's smoke. Almost 25 million children and adolescents aged 3 to 19 years, or about 4 out of 10 children in this age group, are exposed to other people's cigarette smoke. Home and vehicles are the places where children are most exposed to cigarette smoke, and a major location of smoke exposure for adults too. Also, people can be exposed to cigarette smoke in public places, restaurants, and at work.

An Outlook for Mother and Baby

Most people know that smoking causes cancer and other major health problems. And smoking while you're pregnant can cause serious problems, too. Your baby could be born too early, have a birth defect, or die from SIDS. Even being around cigarette smoke can cause health problems for you and your baby.

It's best to quit smoking before you get pregnant. But if you're already pregnant, quitting can still help protect you and your baby from health problems. It's never too late to quit smoking.

If you smoked and had a healthy pregnancy in the past, there's no guarantee that your next pregnancy will be healthy. When you smoke during pregnancy, you put your health and your baby's health at risk.

How Does Smoking Affect Fertility?

Smoking can cause fertility problems for you or your partner. Women who smoke have more trouble getting pregnant than women who don't smoke. In men, smoking can damage sperm and contribute to impotence (erectile dysfunction, or ED). Both problems can make it harder for a man to father a baby when he and his partner are ready.

How Can Smoking Harm You and Your Baby?

- Your baby may be born too small, even after a full-term pregnancy. Smoking slows your baby's growth before birth.

- Your baby may be born too early (premature birth). Premature babies often have health problems.

- Smoking can damage your baby's developing lungs and brain. The damage can last through childhood and into the teen years.

- Smoking doubles your risk of abnormal bleeding during pregnancy and delivery. This can put both you and your baby in danger.

- Smoking raises your baby's risk for birth defects, including cleft lip, cleft palate, or both. A cleft is an opening in your baby's lip or in the roof of her mouth (palate). She or he can have trouble eating properly and is likely to need surgery.

- Babies of moms who smoke during pregnancy—and babies exposed to cigarette smoke after birth—have a higher risk for SIDS.

How Can a Premature Birth Harm Your Baby?

If you smoke during pregnancy, you are more likely to give birth too early. A baby born three weeks or more before your due date is premature. Babies born too early miss important growth that happens in the womb during the final weeks and months of pregnancy.

- Low birth weight
- Feeding difficulties

The earlier a baby is born, the greater the chances for serious health problems or death. Premature babies can have:

- Breathing problems right away
- Breathing problems that last into childhood
- Cerebral palsy (brain damage that causes trouble with movement and muscle tone)
- Developmental delays (when a baby or child is behind in language, thinking, or movement skills)
- Problems with hearing or eyesight

Premature babies may need to stay at the hospital for days, weeks, or even months.

How Can Quitting Help You and Your Baby?

The best time to quit smoking is before you get pregnant, but quitting at any time during pregnancy can help your baby get a better start on life. Talk to your doctor about the best ways to quit while you're pregnant or trying to get pregnant.

When you stop smoking:

- Your baby gets more oxygen, even after just one day
- Your baby will grow better
- Your baby is less likely to be born too early

- You'll have more energy and breathe more easily

- You will be less likely to develop heart disease, stroke, lung cancer, lung disease, and other smoking-related diseases

Section 18.3

E-Cigarettes and Pregnancy

This section includes text excerpted from
"E-Cigarettes and Pregnancy," Centers for Disease
Control and Prevention (CDC), August 3, 2016.

You may have heard that e-cigarettes are safer than regular cigarettes or that they can help you to quit smoking. Quitting can be hard—but if you're pregnant, quitting all forms of tobacco products, including e-cigarettes, is best for you and your baby.

What Are E-Cigarettes?

Electronic cigarettes (also called electronic nicotine delivery systems or e-cigarettes) come in different sizes and shapes, including pens, mods, (i.e., these types are modified by the user) and tanks. Most e-cigarettes contain a battery, a heating device, and a cartridge to hold liquid. The liquid typically contains nicotine, flavorings, and other chemicals. The battery-powered device heats the liquid in the cartridge into an aerosol that the user inhales.

Are E-Cigarettes Safer Than Regular Cigarettes in Pregnancy?

Although the aerosol of e-cigarettes generally has fewer harmful substances than cigarette smoke, e-cigarettes and other products containing nicotine are not safe to use during pregnancy. Nicotine is a health danger for pregnant women and developing babies and can damage a developing baby's brain and lungs. Also, some of the flavorings used in e-cigarettes may be harmful to a developing baby.

Can E-Cigarettes Help Me Quit Smoking during Pregnancy or Stay Smoke-Free after My Baby Is Born?

There is not enough evidence to know whether or not e-cigarettes help people to quit smoking. Individual, group, and telephone counseling have been found to be effective in helping people to quit. If you are pregnant and haven't been able to quit smoking on your own or with counseling, you can discuss the risks and benefits of U.S. Food and Drug Administration (FDA)-approved medications with your health-care provider. After your baby is born, you will have more options for FDA-approved quit-smoking medications. E-cigarettes are not approved by the FDA to help people quit smoking. It's important to know that nicotine is poisonous, and parents and caregivers should keep all products containing nicotine, including e-cigarettes and stop-smoking medications, out of the reach of children.

Chapter 19

Men's Health and Smoking

Smoking continues to have a profound impact on the health and well-being of men and their families in the United States.

- Almost 16 percent (15.8%) of men smoke cigarettes.

- Each day in the United States, about 1,050 boys under 18 years of age smoke their first cigarette.

- 7.6 percent of high school aged boys smoke cigarettes.

Impacts of Men Smoking

There is abundant research about the many harms of smoking—whether it's the dangerous chemicals, the addictive properties, or the damage smoking causes to the lungs, the heart, and nearly every organ in the body. For men who smoke, these effects can have a profound impact on your body and your life, including diminished overall health, increased absenteeism from work, and increased healthcare needs and costs. Smoking also exposes your family to the harmful effects of secondhand smoke. Here are some facts about smoking's effects to you and those around you.

This chapter includes text excerpted from "Men's Health and Smoking," U.S. Food and Drug Administration (FDA), January 15, 2019.

For Men

- Smoking causes heart disease, cancer, and stroke—the first, second, and fifth leading causes of death among men in the United States.

- Smoking cigarettes cause chronic obstructive pulmonary disease (COPD). People with COPD have trouble breathing and slowly start to die from lack of air. Approximately 80 percent of COPD deaths are caused by smoking. Smokers are 12 to 13 times more likely to die from COPD than nonsmokers.

- Smokers are up to 20 times more likely to develop lung cancer than nonsmokers.

- Life expectancy for smokers is at least a decade less than for nonsmokers.

- Smokers with prostate cancer may be more likely to die from the disease than nonsmokers.

- Smoking can damage DNA (deoxyribonucleic acid) in men's sperm, which can cause an increased risk of infertility.

For Families

Secondhand smoke causes disease and premature death in non-smoking adults and children.

The U.S. Surgeon General estimates that living with a smoker increases a nonsmoker's risk of developing lung cancer by 20 to 30 percent.

Exposure to secondhand smoke increases school children's risk for ear infections, lower respiratory illnesses, more frequent and more severe asthma attacks, and slowed lung growth, and can cause coughing, wheezing, phlegm, and breathlessness.

Teens are more likely to smoke if they have friends or family who smoke.

Next Steps

The good news is that you can do something about it now—smoking truly is what the Centers for Disease Control and Prevention (CDC) terms a "modifiable" risk factor.

Encourage the men in your life—the fathers, sons, brothers, and friends—to take a moment to care for themselves and put their own health first by finding a quit method that works for them.

Chapter 20

Smoking and Other Specific Health Consequences

Chapter Contents

Section 20.1

Smoking and Eye Disease

This section includes text excerpted from "Vision Loss and
Blindness," Centers for Disease Control and
Prevention (CDC), March 22, 2018.

Overview of Smoking and Eyesight

Smoking is as bad for your eyes as it is for the rest of your body. If
you smoke, you can develop serious eye conditions that can cause vision
loss or blindness. Two of the greatest threats to your eyesight are:

- Macular degeneration

- Cataracts

Macular degeneration, also called age-related macular degeneration
(AMD), is an eye disease that affects central vision. You need central
vision to see objects clearly and for common tasks such as reading,
recognizing faces, and driving.

There are two forms of AMD: dry AMD and wet AMD. Macular
degeneration always begins in the dry form, and sometimes progresses
to the more advanced wet form, where vision loss can be very rapid
if untreated.

Cataracts cause blurry vision that worsens over time. Without surgery, cataracts can lead to serious vision loss. The best way to protect
your sight from damage linked to smoking is to quit or never start
smoking.

Symptoms of Eye Diseases Related to Smoking

You may think your eyes are fine, but the only way to know for
sure is by getting a full eye exam. AMD often has no early symptoms,
so an eye exam is the best way to spot this eye disease early. An eye
specialist will place special drops in your eyes to widen your pupils.
This offers a better view of the back of your eye, where a thin layer of
tissue (the retina) changes light into signals that go to the brain. The
macula is a small part of the retina that you need for sharp, central
vision.

When symptoms of AMD do occur, they can include:

- Blurred vision or a blurry spot in your central vision

- The need for more light to read or do other tasks
- Straight lines that look wavy
- Trouble recognizing faces

Eye injections are often the preferred treatment for wet AMD. Your doctor can inject a drug to stop the growth of these blood vessels and stop further damage to your eyes. You may need injections on a regular basis to save your vision.

How Does Smoking Affect Your Eyes?

Smoking causes changes in the eyes that can lead to vision loss. If you smoke:

- You are twice as likely to develop AMD compared with a nonsmoker
- You are two to three times more likely to develop cataracts compared with a nonsmoker

How Can You Prevent Vision Loss Related to Smoking?

If you smoke, stop. Quitting may lower your risk for both AMD and cataracts. If you already have AMD, quitting smoking may slow the disease. AMD tends to get worse over time. Quitting smoking is something within your control that may help save your sight. Other healthy habits may also help protect your eyes from cataracts and AMD:

- Exercise regularly
- Maintain normal blood pressure and cholesterol levels
- Eat a healthy diet rich in green, leafy vegetables and fish
- Wear sunglasses and a hat with a brim to protect your eyes from sunlight

How Is a Cataract Treated?

The symptoms of an early cataract may improve with new eyeglasses, brighter lighting, antiglare sunglasses, or magnifying lenses.

When glasses and brighter lighting don't help, you may need surgery. A doctor will remove the cloudy lens and replace it with an artificial lens. This clear, plastic lens becomes a permanent part of your eye.

Help for Vision Loss

Coping with vision loss can be frightening, but there is help to make the most of the vision you have left and to continue enjoying your friends, family, and special interests. If you've already lost some sight, ask your healthcare professional about low-vision counseling and devices such as high-powered lenses, magnifiers, and talking computers.

Section 20.2

Smoking and Diabetes

This section contains text excerpted from the following sources: Text in this section begins with excerpts from "Diabetes and Tobacco Use," U.S. Food and Drug Administration (FDA), November 20, 2018; Text beginning with the heading "What You Need to Know about Smoking and Diabetes" is excerpted from "Smoking and Diabetes," Centers for Disease Control and Prevention (CDC), October 15, 2014. Reviewed February 2019.

It is a known fact that smoking poses serious health risks, negatively impacting vital organs like the lungs and the heart. But did you know that smoking can increase your risk for type 2 diabetes?

What Is Diabetes?

Diabetes is a disease in which the body's blood sugar levels are abnormally elevated. When digested, carbohydrates from food are turned into a natural sugar called "glucose," which is used by the body's cells for energy. Glucose is ushered into the cells by a hormone called insulin. People with diabetes are unable to make or efficiently use insulin, causing glucose to build up in the blood without making its way to the cells.

There are two types of diabetes: type 1 and type 2. The most common of these is type 2, or adult-onset diabetes, which accounts for more than 90 percent of all diabetes cases and is the seventh leading cause of death in the United States. Research has found that smoking is a direct cause of type 2 diabetes. In fact, smokers are 30 to 40 percent more likely to develop type 2 diabetes than nonsmokers, and smoking is responsible for about 9,000 diabetes deaths in the United States per year.

How Does Smoking Affect Diabetes Management?

Diabetes is a serious yet manageable health condition, but smoking can worsen the disease, causing additional problems. Smokers with diabetes are more likely to have trouble regulating their blood sugar levels with treatment than diabetic nonsmokers. Diabetics who smoke are also at a higher risk for disease complications than nonsmokers with diabetes, including:

- Poor blood circulation in the legs and feet that can lead to infections, ulcers, and even amputation of toes and feet

- Heart disease

- Kidney disease

- Retinopathy (an eye disease which can lead to blindness)

- Nerve damage in the arms and legs that can cause numbness, pain, weakness, and poor mobility

How Can You Lower Your Risk for Type 2 Diabetes?

- Don't start smoking. Smoking increases your chance of having type 2 diabetes.

- If you smoke, lower your risk for type 2 diabetes by quitting. Find a quitting method that works for you.

What You Need to Know about Smoking and Diabetes

The Surgeon General's Report has found that smoking is a cause of type 2 diabetes, which is also known as adult-onset diabetes. Smokers have a greater risk of developing type 2 diabetes than do nonsmokers.

The risk of developing diabetes increases with the number of cigarettes smoked per day.

Diabetes is a disease that causes blood sugar levels in the body to be too high and puts the body at risk for many serious health conditions. More than 25 million adults suffer from diabetes in the United States, where the disease is the seventh leading cause of death. It is also a growing health crisis around the world.

How Smoking Causes Type 2 Diabetes

Smoking increases inflammation in the body. Inflammation occurs when chemicals in cigarette smoke injure cells, causing swelling and interfering with proper cell function. Smoking also causes oxidative stress, a condition that occurs as chemicals from cigarette smoke combine with oxygen in the body. This causes damage to cells. Evidence strongly suggests that both inflammation and oxidative stress may be related to an increased risk of diabetes.

The evidence also shows that smoking is associated with a higher risk of abdominal obesity, or belly fat. Abdominal obesity is a known risk factor for diabetes because it encourages the production of cortisol, a hormone that increases blood sugar. Smokers tend to have higher concentrations of cortisol than nonsmokers.

What Smoking Means to People with Diabetes

Studies have confirmed that when people with type 2 diabetes are exposed to high levels of nicotine, insulin (the hormone that lowers blood sugar levels) is less effective. People with diabetes who smoke need larger doses of insulin to control their blood sugar. Smokers who have diabetes are more likely to have serious health problems, including:

- Heart and kidney disease

- Poor blood flow in the legs and feet that can lead to foot infections, ulcers, and possible amputation of toes or feet

- Retinopathy (an eye disease that can cause blindness)

- Peripheral neuropathy (damaged nerves to the arms and legs that cause numbness, pain, weakness, and poor coordination)

Even though no one exactly knows which smokers will develop type 2 diabetes, it is advisable that all diabetic smokers should quit smoking or using any type of tobacco product immediately. The health benefits of quitting begin right away. People with diabetes who quit have better control of their blood sugar. Studies have shown that insulin can start to become more effective at lowering blood sugar levels eight weeks after a smoker quits.

Section 20.3

Smoking and Digestive System

This section includes text excerpted from "Smoking and the Digestive System," National Institute of Diabetes and Digestive and Kidney Diseases (NIDDK), September 2013. Reviewed February 2019.

Smoking affects the entire body, increasing the risk of many life-threatening diseases—including lung cancer, emphysema, and heart disease. Smoking also contributes to many cancers and diseases of the digestive system. Estimates show that about one-fifth of all adults smoke, and each year at least 443,000 Americans die from diseases caused by cigarette smoking.

What Is the Digestive System?

The digestive system is made up of the gastrointestinal (GI) tract—also called the digestive tract—and the liver, pancreas, and gallbladder. The GI tract is a series of hollow organs joined in a long, twisting tube from the mouth to the anus. The hollow organs that make up the GI tract are the mouth, esophagus, stomach, small intestine, large intestine—which includes the colon and rectum—and anus. Food enters the mouth and passes to the anus through the hollow organs of the GI tract. The liver, pancreas, and gallbladder are the solid organs of the digestive system. The digestive system helps the body digest food, which includes breaking food down into nutrients the body needs. Nutrients are substances the body uses for energy, growth, and cell repair.

Does Smoking Increase the Risk of Cancers of the Digestive System?

Smoking has been found to increase the risk of cancers of the:

- Mouth
- Esophagus
- Stomach
- Pancreas

Research suggests that smoking may also increase the risk of cancers of the:

- Liver
- Colon
- Rectum

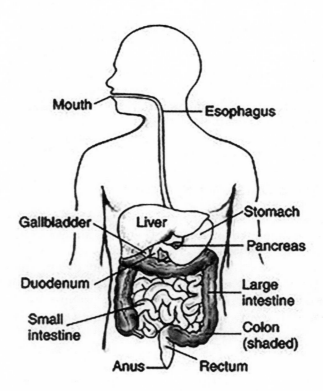

Figure 20.1. *How Smoking Affects the Digestive System*

How Does Smoking Affect Heartburn and Gastroesophageal Reflux Disease?

Smoking increases the risk of heartburn and gastroesophageal reflux disease (GERD). Heartburn is a painful, burning feeling in the chest caused by reflux, or stomach contents flowing back into the esophagus—the organ that connects the mouth to the stomach. Smoking weakens the lower esophageal sphincter, the muscle between the esophagus and stomach that keeps stomach contents from flowing back into the esophagus. The stomach is naturally protected from the acids it makes to help break down food. However, the esophagus is not protected from the acids. When the lower esophageal sphincter weakens, stomach contents may reflux into the esophagus, causing heartburn and possibly damaging the lining of the esophagus.

GERD is persistent reflux that occurs more than twice a week. Chronic, or long lasting, GERD can lead to serious health problems such as bleeding ulcers in the esophagus, narrowing of the esophagus that causes food to get stuck, and changes in esophageal cells that can lead to cancer.

How Does Smoking Affect Peptic Ulcers?

Smoking increases the risk of peptic ulcers. Peptic ulcers are sores on the inside lining of the stomach or duodenum, the first part of the small intestine. The two most common causes of peptic ulcers are infection with a bacterium called *Helicobacter pylori (H. pylori)* and long-term use of nonsteroidal anti-inflammatory drugs such as aspirin and ibuprofen.

Researchers are studying how smoking contributes to peptic ulcers. Studies suggest that smoking increases the risk of *H. pylori* infection, slows the healing of peptic ulcers, and increases the likelihood that peptic ulcers will recur. The stomach and duodenum contain acids, enzymes, and other substances that help digest food. However, these substances may also harm the lining of these organs. Smoking has not been shown to increase acid production. However, smoking does increase the production of other substances that may harm the lining, such as pepsin, an enzyme made in the stomach that breaks down proteins. Smoking also decreases factors that protect or heal the lining, including:

- Blood flow to the lining

- Secretion of mucus, a clear liquid that protects the lining from acid

263

- Production of sodium bicarbonate—a salt-like substance that neutralizes acid—by the pancreas

The increase in substances that may harm the lining and decrease in factors that protect or heal the lining may lead to peptic ulcers.

How Does Smoking Affect Liver Disease?

Smoking may worsen some liver diseases, including:

- Primary biliary cirrhosis, a chronic liver disease that slowly destroys the bile ducts in the liver

- Nonalcoholic fatty liver disease (NAFLD), a condition in which fat builds up in the liver

Researchers are still studying how smoking affects primary biliary cirrhosis, NAFLD, and other liver diseases.

Liver diseases may progress to cirrhosis, a condition in which the liver slowly deteriorates and malfunctions due to chronic injury. Scar tissue then replaces healthy liver tissue, partially blocking the flow of blood through the liver and impairing liver functions.

The liver is the largest organ in the digestive system. The liver carries out many functions, such as making important blood proteins and bile, changing food into energy, and filtering alcohol and poisons from the blood. Research has shown that smoking harms the liver's ability to process medications, alcohol, and other toxins and remove them from the body. In some cases, smoking may affect the dose of medication needed to treat an illness.

How Does Smoking Affect Crohn Disease?

Current and former smokers have a higher risk of developing Crohn disease than people who have never smoked.

Crohn disease is an inflammatory bowel disease that causes irritation in the GI tract. The disease, which typically causes pain and diarrhea, most often affects the lower part of the small intestine; however, it can occur anywhere in the GI tract. The severity of symptoms varies from person to person, and the symptoms come and go. Crohn disease may lead to complications such as blockages of the intestine and ulcers that tunnel through the affected area into surrounding tissues. Medications may control symptoms. However, many people with Crohn disease require surgery to remove the affected portion of the intestine.

Among people with Crohn disease, people who smoke are more likely to:

- Have more severe symptoms, more frequent symptoms, and more complications
- Need more medications to control their symptoms
- Require surgery
- Have symptoms recur after surgery

The effects of smoking are more pronounced in women with Crohn disease than in men with the disease.

Researchers are studying why smoking increases the risk of Crohn disease and makes the disease worse. Some researchers believe smoking might lower the intestines' defenses, decrease blood flow to the intestines, or cause immune system changes that result in inflammation. In people who inherit genes that make them susceptible to developing Crohn disease, smoking may affect how some of these genes work.

How Does Smoking Affect Colon Polyps?

People who smoke are more likely to develop colon polyps. Colon polyps are growths on the inside surface of the colon or rectum. Some polyps are benign, or noncancerous, while some are cancerous or may become cancerous.

Among people who develop colon polyps, those who smoke have polyps that are larger, more numerous, and more likely to recur.

How Does Smoking Affect Pancreatitis?

Smoking increases the risk of developing pancreatitis. Pancreatitis is inflammation of the pancreas, which is located behind the stomach and close to the duodenum. The pancreas secretes digestive enzymes that usually do not become active until they reach the small intestine. When the pancreas is inflamed, the digestive enzymes attack the tissues of the pancreas.

How Does Smoking Affect Gallstones?

Some studies have shown that smoking may increase the risk of developing gallstones. However, research results are not consistent and more study is needed.

Gallstones are small, hard particles that develop in the gallbladder, the organ that stores bile made by the liver. Gallstones can move into the ducts that carry digestive enzymes from the gallbladder, liver, and pancreas to the duodenum, causing inflammation, infection, and abdominal pain.

Can the Damage to the Digestive System from Smoking Be Reversed?

Quitting smoking can reverse some of the effects of smoking on the digestive system. For example, the balance between factors that harm and protect the stomach and duodenum lining returns to normal within a few hours of a person quitting smoking. The effects of smoking on how the liver handles medications also disappear when a person stops smoking. However, people who stop smoking continue to have a higher risk of some digestive diseases, such as colon polyps and pancreatitis, than people who have never smoked.

Quitting smoking can improve the symptoms of some digestive diseases or keep them from getting worse. For example, people with Crohn disease who quit smoking have less severe symptoms than smokers with the disease.

Eating, Diet, and Nutrition

Eating, diet, and nutrition can play a role in causing, preventing, and treating some of the diseases and disorders of the digestive system that are affected by smoking, including heartburn and GERD, liver diseases, Crohn disease, colon polyps, pancreatitis, and gallstones.

Section 20.4

Smoking and Bone Health

This section includes text excerpted from "Smoking and Bone Health," NIH Osteoporosis and Related Bone Diseases—National Resource Center (NIH ORBD—NRC), May 2016.

Many of the health problems caused by tobacco use are well known. Cigarette smoking causes heart disease, lung and esophageal cancer, and chronic lung disease. Additionally, several research studies have identified smoking as a risk factor for osteoporosis and bone fracture. According to the Centers for Disease Control and Prevention (CDC), more than 16 million Americans are living with a disease caused by smoking.

Facts about Osteoporosis

Osteoporosis is a condition in which bones weaken and are more likely to fracture. Fractures from osteoporosis can result in pain and disability. In the United States, more than 53 million people either already have osteoporosis or are at high risk due to low bone mass.

In addition to smoking, risk factors for developing osteoporosis include:

- Thinness or small frame

- Family history of the disease

- Being postmenopausal and particularly having had early menopause

- Abnormal absence of menstrual periods (amenorrhea)

- Prolonged use of certain medications, such as those used to treat lupus, asthma, thyroid deficiencies, and seizures

- Low calcium intake

- Lack of physical activity

- Excessive alcohol intake

Osteoporosis can often be prevented. It is known as a "silent" disease because, if undetected, bone loss can progress for many years without symptoms until a fracture occurs. It has been called a childhood disease with old age consequences because building healthy bones in youth helps prevent osteoporosis and fractures later in life. However, it is never too late to adopt new habits for healthy bones.

Smoking and Osteoporosis

Cigarette smoking was first identified as a risk factor for osteoporosis decades ago. Studies have shown a direct relationship between tobacco use and decreased bone density. Analyzing the impact of cigarette smoking on bone health is complicated. It is hard to determine whether a decrease in bone density is due to smoking itself or to other risk factors common among smokers. For example, in many cases smokers are thinner than nonsmokers, tend to drink more alcohol, may be less physically active, and have poor diets. Women who smoke also tend to have an earlier menopause than nonsmokers. These factors place many smokers at an increased risk for osteoporosis apart from their tobacco use.

In addition, studies on the effects of smoking suggest that smoking increases the risk of having a fracture. As well, smoking has been shown to have a negative impact on bone healing after fracture.

Osteoporosis Management Strategies

Start by quitting: The best thing smokers can do to protect their bones is to quit smoking. Smoking cessation, even later in life, may help limit smoking-related bone loss.

Eat a well-balanced diet rich in calcium and vitamin D: Good sources of calcium include low-fat dairy products; dark green, leafy vegetables; and calcium-fortified foods and beverages. Supplements can help ensure that you get adequate amounts of calcium each day, especially in people with a proven milk allergy. The Institute of Medicine recommends a daily calcium intake of 1,000 mg (milligrams) for men and women up to age 50. Women over age 50 and men over age 70 should increase their intake to 1,200 mg daily.

Vitamin D plays an important role in calcium absorption and bone health. Food sources of vitamin D include egg yolks, saltwater fish, and liver. Many people, especially those who are older, may need vitamin D supplements to achieve the recommended intake of 600 to 800 IU (International Units) each day.

Exercise for your bone health: Like muscle, bone is living tissue that responds to exercise by becoming stronger. Weight-bearing exercise that forces you to work against gravity is the best exercise for bone. Some examples include walking, climbing stairs, weight training, and dancing. Regular exercise, such as walking, may help prevent bone loss and will provide many other health benefits.

Avoid excessive use of alcohol: Chronic alcohol use has been linked to an increase in fractures of the hip, spine, and wrist. Drinking too much alcohol interferes with the balance of calcium in the body. It also affects the production of hormones, which have a protective effect on bone, and of vitamins, which we need to absorb calcium. Excessive alcohol consumption also can lead to more falls and related fractures.

Talk to your doctor about a bone density test: A bone mineral density (BMD) test measures bone density at various sites of the body. This safe and painless test can detect osteoporosis before a fracture occurs and can predict one's chances of fracturing in the future. If you are a current or former smoker, you may want to ask your healthcare provider whether you are a candidate for a BMD test, which can help determine whether medication should be considered.

See if medication is an option for you: There is no cure for osteoporosis. However, several medications are available to prevent and treat the disease in postmenopausal women and in men. Your doctor can help you decide whether medication might be right for you.

Section 20.5

Smoking and Reproduction

This section includes text excerpted from "Smoking and Reproduction," Centers for Disease Control and Prevention (CDC), October 15, 2014. Reviewed February 2019.

What You Should Know about Smoking and Reproduction

For many reasons, men and women who want to have children should not smoke. Studies have long shown that smoking and exposure to tobacco smoke are harmful to reproductive health. Some of the

Surgeon General's Report on smoking and health says that tobacco use during pregnancy remains a major preventable cause of disease and death of mother, fetus, and infant, and smoking before pregnancy can reduce fertility.

The Dangers of Smoking during Pregnancy

Maternal smoking and exposure to secondhand smoke endanger the health of the mother and the baby. Each year, about 400,000 infants born in the United States are exposed to the chemicals in cigarette smoke before birth because their mothers smoke. Since the first Surgeon General's Report on smoking and health was released in 1964, 100,000 babies have died from sudden infant death syndrome (SIDS), prematurity, low birth weight, or other complications caused by exposure to the dangerous chemicals in tobacco smoke. The following are ways smoking affects fertility and the health of a pregnancy:

- **Fertility:** Early Surgeon General's Reports linked smoking and reduced fertility in women. Studies suggest that smoking affects hormone production, which can make it more difficult for women smokers to become pregnant. Men who smoke are more likely to have damaged DNA in their sperm; this can also reduce fertility.

- **Pregnancy complications:** Smoking is known to cause ectopic pregnancy, a condition in which the fertilized egg fails to move to the uterus, or womb, and instead attaches to other organs outside the womb. Ectopic pregnancy almost always causes the fetus to die and is potentially fatal for the mother. Evidence also suggests that spontaneous abortion, or miscarriage, is caused by smoking during pregnancy.

- **Fetal growth:** Mothers who smoke during pregnancy are more likely to deliver babies with low birth weight, even if the babies are full term. Mothers who smoke during pregnancy are also more likely to deliver their babies early. Low birth weight and preterm delivery are leading causes of infant disability and death.

- **Fetal development:** Smoking during pregnancy can cause tissue damage in the fetus, especially in the lungs and brain. Carbon monoxide in tobacco smoke is a dangerous toxin that can harm the central nervous system and impair fetal growth. Damage from maternal smoking can last throughout childhood and into the teenage years.

- **Birth defects:** Each year, approximately 3 babies out of 100 in the United States are born with major birth defects. Women who smoke during pregnancy are more likely to deliver babies with cleft lip and/or cleft palate, where the lip or the roof of the mouth fails to form correctly. Babies born with cleft lip or cleft palate have problems with feeding, hearing, and speech development; dental problems, including missing teeth; and middle ear infections. Correction of cleft defects requires surgery.

- **Sudden infant death syndrome (SIDS):** Babies whose mothers smoked during pregnancy or who are exposed to secondhand smoke after birth are more likely to die of SIDS than are babies who are not exposed.

Male Sexual Function and Fertility

Smoking is also a cause of erectile dysfunction (ED), a condition that currently affects 18 million American men over age 20. ED is defined as the inability to maintain an erection that is adequate for satisfactory sexual performance. This can affect reproduction. Cigarette smoke alters blood flow needed for an erection, and smoking interferes with the healthy function of blood vessels in erectile tissue.

Quit Smoking—for Reproductive Health!

Although no one exactly knows which smokers might have reproduction complications because of smoking, it has been found that smoking can be a threat to the ability to become pregnant, to a safe and healthy pregnancy, and to healthy babies and mothers. Quitting smoking can improve your reproductive health. Women and men who want to have children can talk with their doctors about getting help to quit smoking before pregnancy.

Section 20.6

Smoking, Gum Disease, and Tooth Loss

This section includes text excerpted from
"Gum (Periodontal) Disease," Centers for Disease
Control and Prevention (CDC), March 22, 2018.

What Is Gum Disease?

Gum (periodontal) disease is an infection of the gums and can affect the bone structure that supports your teeth. In severe cases, it can make your teeth fall out. Smoking is an important cause of severe gum disease in the United States.

Gum disease starts with bacteria (germs) on your teeth that get under your gums. If the germs stay on your teeth for too long, layers of plaque (film) and tartar (hardened plaque) develop. This buildup leads to early gum disease, called gingivitis.

When gum disease gets worse, your gums can pull away from your teeth and form spaces that get infected. This is severe gum disease, also called periodontitis. The bone and tissue that hold your teeth in place can break down, and your teeth may loosen and need to be pulled out.

Warning Signs and Symptoms of Gum Disease

- Red or swollen gums
- Tender or bleeding gums
- Painful chewing
- Loose teeth
- Sensitive teeth
- Gums that have pulled away from your teeth

How Is Smoking Related to Gum Disease?

Smoking weakens your body's infection fighters (your immune system). This makes it harder to fight off a gum infection. Once you have gum damage, smoking also makes it harder for your gums to heal.

What Does This Mean for Me If I Am a Smoker?

- You have twice the risk for gum disease compared with a nonsmoker.

- The more cigarettes you smoke, the greater your risk for gum disease.

- The longer you smoke, the greater your risk for gum disease.

- Treatments for gum disease may not work as well for people who smoke.

Tobacco use in any form—cigarettes, pipes, and smokeless (spit) tobacco—raises your risk for gum disease.

How Can Gum Disease Be Prevented?

You can help avoid gum disease with good dental habits.

- Brush your teeth twice a day

- Floss often to remove plaque

- See a dentist regularly for checkups and professional cleanings

- Don't smoke. If you smoke, quit.

How Is Gum Disease Treated?

Regular cleanings at your dentist's office and daily brushing and flossing can help treat early gum disease (gingivitis).
More severe gum disease may require:

- Deep cleaning below the gum line

- Prescription mouth rinse or medicine

- Surgery to remove tartar deep under the gums

- Surgery to help heal bone or gums lost to periodontitis. Your dentist may use small bits of bone to fill places where bone has been lost. Or your dentist may move tissue from one place in your mouth to cover exposed tooth roots.

If you smoke or use spit tobacco, quitting will help your gums heal after treatment.

Section 20.7

Smoking and Human Immunodeficiency Virus

This section includes text excerpted from "Smoking and HIV," Centers for Disease Control and Prevention (CDC), March 22, 2018.

Smoking is a serious health threat for everyone, but it's especially dangerous for people living with human immunodeficiency virus (HIV). Smoking raises your risk for heart disease, cancer, serious lung diseases and infections such as pneumonia, and other illnesses. People with HIV are more likely to develop these harmful consequences of smoking than those without HIV.

Quitting smoking can help people with HIV have a better quality of life (QOL) and fewer HIV-related symptoms. When you quit, your risk goes down for many serious illnesses, including heart attacks and pneumonia.

What Is Human Immunodeficiency Virus?

HIV stands for human immunodeficiency virus. If HIV is not treated properly, it can lead to acquired immunodeficiency syndrome, or AIDS. In the United States, HIV is spread mainly by having unprotected sex or by sharing needles or other drug equipment (works) with someone who has HIV. Once you have HIV, you have it for life. Your body cannot get rid of HIV.

HIV harms your body's infection fighters—the cells of the immune system called CD4 cells, or T cells. Over time, HIV can destroy so many of these cells that your body can't fight off infections and disease. When this happens, HIV leads to AIDS.

How Is Smoking Related to Human Immunodeficiency Virus?

Smoking causes one in every five deaths in the United States each year. About one in five Americans smokes cigarettes.

If you smoke and have HIV, you're more likely to get HIV-related infections, including:

- Thrush (a mouth infection, also called oral candidiasis)

- Hairy leukoplakia (white mouth sores)

- Bacterial pneumonia

- Pneumocystis pneumonia, a dangerous lung infection

Smoking when you have HIV also makes you more likely to get other serious illnesses than nonsmokers with HIV. These illnesses can make you too sick to work (disabled) or even lead to an early death. They include:

- COPD (chronic obstructive pulmonary disease, a serious lung disease that causes severe breathing problems and includes emphysema and chronic bronchitis)

- Heart disease and stroke

- Lung cancer, head and neck cancer, cervical cancer, and anal cancer

How Can Human Immunodeficiency Virus Be Prevented?

Currently, there is no vaccine to prevent HIV infection and no cure for HIV infection. You can lower your risk of becoming infected with HIV or spreading the virus to other people with these steps:

- Get tested regularly for HIV

- Practice abstinence (not having sex)

- Remain faithful to your spouse or partner

- Consistently use male latex or female polyurethane condoms

- Do not share needles or other drug injection equipment (works)

- Limit the use of alcohol, and don't use drugs that affect your judgment

How Is Human Immunodeficiency Virus Treated?

HIV is treated with a mix of drugs called antiretroviral therapy (ART). You take these medicines daily to prevent the virus from multiplying and destroying your body's infection fighters—the CD4 cells. This, in turn, helps protect you from life-threatening illnesses, such as pneumonia and cancer.

ART can't cure HIV, but it helps people with HIV live longer, healthier lives.

It's very important to take your HIV medicines exactly as directed. If you don't, your CD4 count may go down and your viral load may go up. Not taking your medicines as directed can also make the HIV virus resistant, which means the medicines won't help you anymore. Smoking may limit how well HIV medicines work to fight the virus.

Tell your doctor if your medicines are making you sick. She or he may be able to help you deal with side effects and feel better. Don't just stop taking your medicines.

The Benefits of Quitting Smoking

If you smoke, quit. Quitting smoking has major and immediate health benefits for all tobacco users, especially those living with HIV. Quitting reduces your chances of developing disease, helps you feel better, and improves your quality of life.

Also, stay away from secondhand smoke, which is smoke in the air from other people smoking. Secondhand smoke has immediate harmful effects on your blood and blood vessels, which can raise your risk for a heart attack. People who already have heart disease are at especially high risk for a heart attack. Secondhand smoke can also cause a stroke or lead to lung cancer.

Talk with your healthcare provider about programs and products that can help you quit.

Section 20.8

Smoking and Pain

"The Effects of Cigarette Smoke on Pain,"
© 2017 Omnigraphics. Reviewed February 2019.

Many of smoking's harmful effects are well known, including its link to heart disease, respiratory disease, and cancer. Smoking is a factor in one among five deaths in the United States and is the greatest preventable cause of disease and death. However, what is less well known is its link to chronic pain. Many recent research studies have found that smoking makes pain worse and can interfere with effective pain management. Paradoxically, individuals with chronic pain tend to smoke more than the general public at large. While the overall smoking rate in the United States is 22 percent, more than half of patients who opt for pain management are found to be smokers.

What Is the Effect of Smoking on Pain?

Nicotine, the addictive ingredient in cigarettes, is a proven painkiller, but it is generally effective only in the short term. Despite this short-term effect, smokers have overall lower thresholds of pain compared to nonsmokers. The reasons are not fully understood, but scientists believe that contributing factors may include the effect of smoking on how the body regulates inflammation and perceives pain as well as the role smoking plays in the development of chronic, painful conditions. Smokers with chronic pain often fall into a cycle of smoking to cope with pain, which in turn interferes with pain management, leading the patient to smoke more when their pain threshold has invariably changed.

How Does Smoking Interfere with Pain Management?

Smoking affects the body and chronic pain in a number of different ways:

- **Aches and pains.** Smokers experience pain in the neck, arm, back, and legs and existing pain becomes worse with smoking.

- **Bones.** Smoking affects the musculoskeletal system, which includes the muscles, bones, tendons, and ligaments. Smoking acts on osteoblasts, the cells that develop into bone. It also interferes with the absorption of calcium from food and can lead to osteoporosis. Weakening of bones increases the incidence of fractures, which in turn produces acute and ongoing pain.

- **Decrease in oxygen supply to the body.** Smoking decreases the level of oxygen in the blood and also affects blood circulation. Healing of wounds takes longer in such conditions and can also have an effect on how the body regulates inflammation. Lack of enough oxygen in blood also affects the production of osteoblasts and increases healing time for fractures.

- **Chronic disease.** Multiple sclerosis, lupus, fibromyalgia, and arthritis are chronic and painful conditions that can be made worse by smoking.

- **Degenerative disc disease.** Smoking hardens arterial walls which impede blood flow in areas like bones and discs in the spinal cord that are fed by smaller vessels. The spine is starved of oxygen as a result so that after an injury or collapse, the spine becomes less capable of repairing itself and degenerates.

- **Increased perception of pain.** Research indicates that smoking affects brain circuits associated with pain. The nervous system becomes more sensitive to pain and perceives pain more acutely.

- **Effect on pain medication.** Smokers often require higher dosages of analgesics or narcotics for effective pain management according to research.

What Can Smokers Do?

Simply put, the most effective way for patients who smoke to overcome chronic pain is to quit. Smoking cessation has been repeatedly linked to the reduction of chronic pain and to an overall improvement in health. Smokers should be well prepared and armed with techniques to aid their decision to quit and succeed. Some strategies include:

- Designate a day for not smoking.

- Enlist the support of friends and relatives to help you quit.

- Find ways of becoming physically active.

- Stop stocking up on cigarettes and throw away ashtrays, matches, and lighters.

- Join a smoking cessation program.

- Change your daily routine and break associations linked to smoking.

- Use nicotine substitute products like inhalers, patches, and gum to reduce dependence.

- Take it slow by handling it one day at a time.

Patients can also consult a pain management specialist to determine how quitting smoking can work as part of an overall program to address their chronic pain.

References

1. Thompson, Dennis Jr. "Chronic Pain and Smoking," Everyday Health Media, LLC, 2017.

2. Ingraham, Paul. "Smoking and Chronic Pain," PainScience.com, November 18, 2016.

3. "Smokers Have More Aches and Pains," WebMD, LLC, January 8, 2003.

4. Whiteman, Honor. "Smoking Linked to Increased Risk of Chronic Back Pain," MedicalNewsToday, November 4, 2014.

5. "Smoking and Joint Pain—Why Smokers Have More Aches and Pains?" ConsumerHealthDigest, June 15, 2017.

6. "The Shocking Truth behind Smoking and Pain," Advanced Pain Management, January 17, 2017.

7. Macadaeg, Dr. "What You Don't Know about Smoking and Your Spine," Indiana Spine Group, n.d.

Part Three

Smoking Cessation

Chapter 21

Quitting Smoking

Chapter Contents

Section 21.1

Quitting Smoking: An Overview

This section contains text excerpted from the following sources: Text in this section begins with excerpts from "Quitting Smoking," Centers for Disease Control and Prevention (CDC), December 11, 2017; Text beginning with the heading "How to Quit" is excerpted from "Ways to Quit," Smokefree.gov, U.S. Department of Health and Human Services (HHS), September 6, 2018.

Tobacco use can lead to tobacco/nicotine dependence and serious health problems. Quitting smoking greatly reduces the risk of developing smoking-related diseases. Tobacco/nicotine dependence is a condition that often requires repeated treatments, but there are helpful treatments and resources for quitting. Smokers can and do quit smoking. In fact, today there are more former smokers than current smokers.

Nicotine Dependence

- Most smokers become addicted to nicotine, a drug that is found naturally in tobacco.

- More people in the United States are addicted to nicotine than to any other drug. Research suggests that nicotine may be as addictive as heroin, cocaine, or alcohol.

- Quitting smoking is hard and may require several attempts. People who stop smoking often start again because of withdrawal symptoms, stress, and weight gain.

- Nicotine withdrawal symptoms may include:
  - Feeling irritable, angry, or anxious
  - Having trouble thinking
  - Craving tobacco products
  - Feeling hungrier than usual

Health Benefits of Quitting

Tobacco smoke contains a deadly mix of more than 7,000 chemicals; hundreds are harmful, and about 70 can cause cancer. Smoking increases the risk for serious health problems, many diseases, and death.

People who stop smoking greatly reduce their risk for disease and early death. Although the health benefits are greater for people who stop at earlier ages, there are benefits at any age. You are never too old to quit.

Stopping smoking is associated with the following health benefits:

- Lowered risk for lung cancer and many other types of cancer

- Reduced risk for heart disease, stroke, and peripheral vascular disease (narrowing of the blood vessels outside your heart)

- Reduced heart disease risk within one to two years of quitting

- Reduced respiratory symptoms, such as coughing, wheezing, and shortness of breath. While these symptoms may not disappear, they do not continue to progress at the same rate among people who quit compared with those who continue to smoke.

- Reduced risk of developing some lung diseases (such as chronic obstructive pulmonary disease, also known as COPD, one of the leading causes of death in the United States).

- Reduced risk for infertility in women of childbearing age. Women who stop smoking during pregnancy also reduce their risk of having a low birth weight baby.

Smokers' Attempts to Quit

Among all current U.S. adult cigarette smokers, nearly 7 out of every 10 (68.0%) reported in 2015 that they wanted to quit completely.

- Since 2002, the number of former smokers has been greater than the number of current smokers.

Percentage of adult daily cigarette smokers who stopped smoking for more than 1 day in 2015 because they were trying to quit:

- More than 5 out of 10 (55.4%) of all adult smokers

- Nearly 7 out of 10 (66.7%) smokers aged 18 to 24 years

- Nearly 6 out of 10 (59.8%) smokers aged 25 to 44 years

- More than 4 out of 10 (49.6%) smokers aged 45 to 64 years

- About 4 out of 10 (47.2%) smokers aged 65 years or older

Percentage of high school cigarette smokers who tried to stop smoking in the past 12 months:

- More than 4 out of 10 (45.5%) of all high school students who smoke

Ways to Quit Smoking

Most former smokers quit without using one of the treatments that scientific research has shown can work. However, the following treatments are proven to be effective for smokers who want help to quit:

- Brief help by a doctor (such as when a doctor takes 10 minutes or less to give a patient advice and assistance about quitting)

- Individual, group, or telephone counseling

- Behavioral therapies (such as training in problem-solving)

- Treatments with more person-to-person contact and more intensity (such as more or longer counseling sessions)

- Programs to deliver treatments using mobile phones

Medications for quitting that have been found to be effective include the following:

- Nicotine replacement products

  - Over-the-counter (nicotine patch [which is also available by prescription], gum, lozenge)

  - Prescription (nicotine patch, inhaler, nasal spray)

- Prescription nonnicotine medications: bupropion SR (Zyban®), varenicline tartrate (Chantix®)

Counseling and medication are both effective for treating tobacco dependence, and using them together is more effective than using either one alone.

How to Quit

There are many medications, resources, and services available to help you quit smoking. There isn't one perfect way to quit, but using more than one quitting tool can increase your chances of quitting successfully.

Get Prepared

Boost your chances of becoming smoke-free for good with a plan. Here are a few things you can to do to prepare yourself.

- **Know why you're quitting:** Before you quit, it's helpful to know the reasons why you're doing it. Maybe it's because you want to look better, be healthier, save money, or have another good reason. When quitting gets hard, think about your reasons. They can inspire you to stop smoking for good.

- **Make a quit plan:** Creating your personalized quit plan can help you stay focused, confident, and motivated to quit.

- **Manage your quit day:** Having a plan to handle your quit day can give you the tools and confidence to take on your first smoke-free day—and help you stay quit for good.

Quit Smoking Medications

Nicotine replacement therapy (NRT) and other medications can double your chances of quitting for good. Some medications work better than others for some people. It can take a few tries to find the approach that is right for you. Ask your doctor' dentist' or pharmacist about what might work for you. Always use medications as prescribed or as on the packaging.

Section 21.2

Preparations for Smoking Cessation

This section includes text excerpted from "Take Steps to Quit," Centers for Disease Control and Prevention (CDC), March 21, 2018.

Prepare to Quit

Quitting is hard. But quitting can be a bit easier if you have a plan. When you think you're ready to quit, here are a few simple steps you can take to put your plan into action.

Commit to Your Quit

Set your quit date. The first step to becoming smoke-free is to choose when you want to quit. Here are some tips to help you pick a quit date:

- **Give yourself time to prepare.** Getting prepared can help you build the confidence and skills you will need to stay smoke-free.

- **Don't put it off for too long.** Picking a date too far away gives you time to change your mind or become less motivated. Choose a date that is no more than a week or two away.

- **Try quitting on a Monday.** Many people who quit on Mondays feel more confident about quitting—it can be the fresh start you need to begin your smoke-free journey. You can also use Mondays to recommit to your quit every week, set goals for the week, contact your support system, or reset your quit date if you slipped. If you're having a hard time setting a date, sign up for a Practice Quit (women.smokefree.gov/tools-tips/text-programs/practice-quit) text program to learn skills and become more comfortable with quitting for good. Practice quitting for one, three, or five days, and try the program as many times as you need.

- **Create your quit plan.** A personalized quit plan can help you stay focused, confident, and motivated to quit. Don't worry if you've already started your quit, because making a plan now can help you stay on track. Check in with your plan each Monday— it's a great way to recommit to your quit each week!

Know Why You're Quitting

Before you actually quit, it's important to know why you're doing it. Do you want to be healthier? Save money? Keep your family safe? If you're not sure, ask yourself these questions:

- What do I dislike about smoking?
- What do I miss out on when I smoke?
- How is smoking affecting my health?
- What will happen to me and my family if I keep smoking?
- How will my life get better when I quit?

Still not sure? Different people have different reasons for quitting smoking. Get ready to stop smoking by thinking about why you want to quit.

Learn How to Handle Your Triggers and Cravings

Triggers are specific persons, places, or activities that make you feel like smoking. Knowing your smoking triggers can help you learn to deal with them.

Cravings are short but intense urges to smoke. They usually only last a few minutes. Plan ahead and come up with a list of short activities you can do when you get a craving.

Find Ways to Handle Nicotine Withdrawal

During the first few weeks after you quit, you may feel uncomfortable and crave a cigarette. These unpleasant symptoms of quitting smoking are known as withdrawal. Withdrawal is common among smokers who quit, whether they are doing it cold turkey or with the help of medications, counseling, or other tools.

During withdrawal, your body is getting used to not having nicotine from cigarettes. For most people, the worst symptoms of withdrawal last a few days to a few weeks. During this time, you may:

- Feel a little depressed

- Be unable to sleep

- Become cranky, frustrated, or mad

- Feel anxious, nervous, or restless

- Have trouble thinking clearly

You may be tempted to smoke to relieve these feelings. Just remember that they are temporary, no matter how powerful they feel at the time.

One of the best ways to deal with nicotine withdrawal is to try nicotine replacement therapy (NRT). NRT can reduce withdrawal symptoms. And NRT can double your chances of quitting smoking for good. NRT comes in several different forms, including gum, patch, nasal spray, inhaler, and lozenge. Many are available without a prescription.

A lot of research has been done on NRT. It has been shown to be safe and effective for almost all smokers who want to quit. But teens, pregnant women, and people with severe medical conditions should talk to their doctor before using NRT.

If you plan to use NRT, remember to have it available on your quit day. Read the instructions on the NRT package and follow them carefully. NRT will give you the most benefit if you use it as recommended.

Tell Your Family and Friends You Plan to Quit

Quitting smoking is easier when the people in your life support you. Let them know you are planning to quit and explain how they can help. Here are a few tips:

- Tell your family and friends your reasons for quitting.

- Ask them to check in with you to see how things are going, especially on your quit date.

- Ask them to help you think of smoke-free activities you can do together (like going to the movies or a nice restaurant).

- Ask a friend or family member who smokes to quit with you, or at least not smoke around you.

- Ask your friends and family not to give you a cigarette—no matter what you say or do.

- Alert your friends and family that you may be in a bad mood while quitting. Ask them to be patient and help you through it.

Section 21.3

My Quit Plan

This section includes text excerpted from "Have You Built a Quit Plan?" Smokefree.gov, U.S. Department of Health and Human Services (HHS), July 30, 2013. Reviewed February 2019.

One of the keys to a successful quit is preparation. A great way to prepare to quit smoking is to create a quit plan. Quit plans:

- Combine quit smoking strategies to keep you focused, confident, and motivated to quit

- Help you identify challenges you will face as you quit and ways to overcome them

- Can improve your chances of quitting smoking for good

The following steps will help you to create your own customized quit plan. As you move through the steps, keep a record of your plan and have it readily available during your quit.

Pick a Quit Date

When it comes to choosing a quit date, sooner is better than later. Many smokers choose a date within two weeks to quit smoking. This will give you enough time to prepare. Really think about your quit date. Avoid choosing a day where you know you will be busy, stressed, or tempted to smoke (e.g., a night out with friends or days where you may smoke at work).

Next Step: Circle your quit day on your calendar. Write it out somewhere where you will see it every day. This will remind you of your decision to become smoke-free and give you time to prepare to quit.

Let Your Loved Ones Know You Are Quitting

Quitting smoking is easier with support from important people in your life. Let them know ahead of your quit date that you are planning to quit. Explain how they can help you quit. We all need different things, so be sure you let friends and family know exactly how they can help.

Next Step: Support is one of the keys to successfully quitting. However, it can be hard to ask for help, even from the people closest to you. Review tips on getting support to make sure you get the help you need.

Remove Reminders of Smoking

Getting rid of smoking reminders can keep you on track during your quit. Smoking reminders can include your cigarettes, matches, ashtrays, and lighters. It may also help to make things clean and fresh at work, in your car, and at home. Even the smell of cigarettes can cause a cigarette craving.

Next Step: Throw away all your cigarettes and matches. Give or throw away your lighters and ashtrays. Don't save one pack of cigarettes "just in case."

Identify Your Reasons to Quit Smoking

Everyone has their own reasons for quitting smoking. Maybe they want to be healthier, save some money, or keep their family safe. As you prepare to quit, think about your own reasons for quitting. Remind yourself of them every day. They can inspire you to stop smoking for good.

Next Step: Make a list of all the reasons you want to quit smoking. Keep it in a place where you can see it every day. Any time you feel the urge to smoke, review your list. It will keep you motivated to stay smoke-free.

Identify Your Smoking Triggers

When you smoke, it becomes tied to many parts of your life. Certain activities, feelings, and people are linked to your smoking. When you come across these things, they may "trigger" or turn on your urge to smoke. Try to anticipate these smoking triggers and develop ways to deal with them.

Next Step: Make a list of everything that makes you feel like smoking. Now, write down one way you can deal with or avoid each item on your list. Keep this list nearby during your quit.

Develop Coping Strategies

Nicotine is the chemical in cigarettes that makes you addicted to smoking. When you stop smoking, your body has to adjust to no longer having nicotine in its system. This is called withdrawal. Withdrawal can be unpleasant, but you can get through it. Developing strategies to cope with withdrawal ahead of your quit can help ensure you stay smoke-free for good!

Next Steps: Medications and behavior changes can help you manage the symptoms of withdrawal. Many quit-smoking medications are available over the counter. Make sure you have them on hand prior to your quit. While medications will help, they can't do all the work for you. Develop other quit smoking strategies to use with medications. Remember that withdrawal symptoms, including cravings, will fade with every day that you stay smoke-free.

Have Places You Can Turn to for Immediate Help

Quitting smoking is hardest during the first few weeks. You will deal with uncomfortable feelings, temptations to smoke, withdrawal symptoms, and cigarette cravings. Whether it is a quitline, support group, or good friend, make sure you have quit smoking support options available at all times.

Next Steps: Plan on using multiple quit smoking support options. Keep them handy in case you need them during your quit. Here a few options you may want to consider:

- **SmokefreeTXT:** A mobile text-messaging service designed for adults and young adults across the United States who are trying to quit smoking.

- **Quitlines:** If you want to talk to a quit-smoking counselor right away, call 800–QUIT–NOW (800–784–8669).

- **Quit Smoking Apps:** Mobile phone applications can help you prepare to quit, provide support, and track your progress.

- **Support Groups:** Visit your county or state government's website to see if they offer quit-smoking programs in your area.

- **Friends and Family:** Getting support from the important people in your life can make a big difference during your quit.

- **Medications:** If you are using a quit-smoking medication, such as the patch, gum, or lozenges, make sure you have them on hand.

Set up Rewards for Quit Milestones

Quitting smoking happens one minute, one hour, one day at a time. Reward yourself throughout your quit. Celebrate individual milestones, including being 24 hours smoke-free, one week smoke-free, and one month smoke-free. Quitting smoking is hard, be proud of your accomplishments.

Next Steps: You should be proud every time you hit a quit-smoking milestone. Treat yourself with a nice dinner, day at the movies, or any other smoke-free activity. Plan out your milestones ahead of time and set up a smoke-free reward for each one.

Section 21.4

Managing Your Quit Day

This section contains text excerpted from the following sources:
Text under the heading "Steps to Manage Quit Day" is excerpted
from "Manage Your Quit Day," Centers for Disease Control and
Prevention (CDC), March 21, 2018; Text under the heading "Within
20 Minutes of Quitting" is excerpted from "Within 20 Minutes of
Quitting," Centers for Disease Control and Prevention (CDC), July
15, 2015. Reviewed February 2019.

Steps to Manage Quit Day

You've decided to quit smoking. Congratulations! Your first day
without cigarettes can be difficult. Here are five steps you can take to
handle quit day and be confident about being able to stay quit.

1. **Make a quit plan**

   Having a plan can make your quit day easier. A quit plan gives
   you ways to stay focused, confident, and motivated to quit.
   You can build your own quit plan or find a quit program that
   works for you. If you don't know what quit method might be
   right for you, you can explore different quit methods. No single
   approach to quitting works for everyone. Be honest about your
   needs. If using nicotine replacement therapy is part of your
   plan, be sure to start using it first thing in the morning.

2. **Stay busy**

   Keeping busy is a great way to stay smoke-free on your quit
   day. Being busy will help you keep your mind off smoking and
   distract you from cravings. Think about trying some of these
   activities:

   • Exercise

   • Get out of the house for a walk

   • Chew gum or hard candy

   • Keep your hands busy with a pen or toothpick, or play a
     game in the QuitGuide app

   • Drink lots of water

   • Relax with deep breathing

- Go to a movie

- Spend time with nonsmoking friends and family

- Go to dinner at your favorite smoke-free restaurant

3. **Avoid smoking triggers**

Triggers are the people, places, things, and situations that set off your urge to smoke. On your quit day, try to avoid all your triggers. Here are some tips to help you outsmart some common smoking triggers:

- Throw away your cigarettes, lighters, and ashtrays if you haven't already

- Avoid caffeine, which can make you feel jittery. Try drinking water instead

- Spend time with nonsmokers

- Go to places where smoking isn't allowed

- Get plenty of rest and eat healthy. Being tired can trigger you to smoke

- Change your routine to avoid the things you might associate with smoking

4. **Stay positive**

Quitting smoking is difficult. It happens one minute . . . one hour . . . one day at a time. Try not to think of quitting as forever. Pay attention to today and the time will add up. It helps to stay positive. Your quit day might not be perfect, but all that matters is that you don't smoke—not even one puff. Reward yourself for being smoke-free for 24 hours. You deserve it. And if you're not feeling ready to quit today, set a quit date that makes sense for you. It's okay if you need a few more days to prepare to quit smoking.

5. **Ask for help**

You don't need to rely on willpower alone to be smoke-free. Tell your family and friends when your quit day is. Ask them for support on quit day and in the first few days and weeks after. They can help you get through the rough spots. Let them know exactly how they can support you. Don't assume they'll know.

Within 20 Minutes of Quitting

Within 20 minutes after you smoke that last cigarette, your body begins a series of changes that continue for years.

- **20 minutes after quitting**

Your heart rate drops.

- **12 hours after quitting**

Carbon monoxide level in your blood drops to normal.

- **2 weeks to 3 months after quitting**

Your heart attack risk begins to drop.

Your lung function begins to improve.

- **1 to 9 months after quitting**

Your coughing and shortness of breath decrease.

- **1 year after quitting**

Your added risk of coronary heart disease is half that of a smoker's.

- **5 years after quitting**

Your stroke risk is reduced to that of a nonsmoker's 5 to 15 years after quitting.

- **10 years after quitting**

Your lung cancer death rate is about half that of a smoker's.

Your risk of cancers of the mouth, throat, esophagus, bladder, kidney, and pancreas decreases.

- **15 years after quitting**

Your risk of coronary heart disease is back to that of a nonsmoker.

Section 21.5

Questions and Answers about Smoking Cessation

This section contains text excerpted from "FAQs about Quitting," Smokefree.gov, U.S. Department of Health and Human Services (HHS), February 3, 2017.

Lots of people want to quit smoking, but they have questions about it. Here are some answers to common questions.

1. "I Don't Think I Can Live without Cigarettes. What Can I Do?"

It's common for people to think they can't function without smoking. And it's also normal to be worried about or afraid of quitting. Many people view smoking as a central part of their lives—so quitting can seem overwhelming. Instead of trying to stop all at once, you might try to cut back on smoking over time as you get closer to your quit date.

Make a list of the reasons why you want to quit. Keep it handy for motivation. Do you want to be healthier? Save money? Keep your family safe? Focus on the ways that your life will be better when you are a nonsmoker. Pick a quit day in the future, and work toward it. Get tips to manage your quit day.

2. "I Have a Lot of Stress in My Life Right Now. Shouldn't I Wait to Quit When I'm Not as Stressed?"

There is no drug in cigarettes that magically gets rid of stress. Stress is something that you will have to deal with for the rest of your life. And in the long run, quitting smoking can reduce stress.

Talk to your healthcare provider about ways to manage stress. You can also try exercise or a meditation class to learn how to deal with stress. Learn ways to cope with your emotions without smoking.

3. "Well, My Doctor Didn't Say Anything about My Smoking, So . . . It Can't Be That Bad, Right?"

Unfortunately, many physicians don't address tobacco use for a variety of reasons. But don't take that as their approval that you should continue to smoke. The scientific evidence is very clear: smoking is the leading cause of disease and death.

If your healthcare provider doesn't bring up your smoking during your next visit, ask how quitting can improve your health and ask what they can do to help you quit.

4. "The Last Time I Quit, My Depression Got Worse. I'm Just Starting to Feel Good Now. Will I Backslide?"

Depression is one symptom of nicotine withdrawal in the early weeks and months after quitting. But here's the good news: you can quit without depression coming back. And after you quit, you're likely to see improvements in your mood and quality of life (QOL).

If you have a history of depression, it's important that you work with your doctor. This way, your medication levels can be monitored. Talk about your quit plan with your healthcare provider before you get started.

5. "Will the Stress of Quitting Increase My Chances of Drinking Again?"

People with substance-use problems who quit smoking are more likely to stay sober than those who keeping smoking.

Lots of veterans have successfully quit-smoking and drinking. Want to talk to someone about it? You can join a quit-smoking group at your local VA, or find supportive friends or family that have quit. And you can call 855-784-8838 to make a plan with a quit-smoking counselor.

6. "How Can I Quit When Everyone Else in My Group Home Smokes?"

Many people who live with other smokers quit every day. You can create a plan so that you will be able to be comfortable in your home. Here are some suggestions for dealing with this situation:

- Have a meeting with your housemates to tell them you're quitting.

- Talk with them about where they will and won't smoke.

- Ask them not to leave cigarettes or dirty ashtrays where you can find them. Explain that these might trigger you to smoke.

- Talk with your housemates—they may be interested in quitting, too.

Chapter 22

Products to Quit Smoking

Quitting smoking can be hard, but it is possible. In fact, every time you put out a cigarette is a new chance to try quitting again, according to the U.S. Food and Drug Administration's (FDA) newest tobacco education campaign, "Every Try Counts."

If you want to quit—almost 70 percent of adult smokers say they do—you may want to use a "smoking cessation" product proven to help. Data has shown that using FDA-approved cessation medicine can double your chances of quitting successfully.

Some products contain nicotine as an active ingredient and others do not. These products include over-the-counter (OTC) options like skin patches, lozenges, and gum, as well as prescription medicines.

Smoking cessation products are intended to help you quit smoking. They are regulated through the FDA's Center for Drug Evaluation and Research (CDER), which ensures that the products are safe and effective and that their benefits outweigh any known associated risks.

What to Know about Smoking Cessation Products

Understanding how smoking cessation products work—and what side effects they may cause—can help you determine which product may be best for you.

This chapter includes text excerpted from "Want to Quit Smoking? FDA-Approved Products Can Help," U.S. Food and Drug Administration (FDA), December 12, 2017.

If you're considering one of these products, reading labels and talking to your pharmacist and other healthcare providers are good first steps to take.

And remember to weigh each product's benefits and risks, among other considerations.

Nicotine Replacement Therapy

Nicotine is the substance primarily responsible for causing addiction to tobacco products. Tobacco users who are addicted to nicotine are used to having nicotine in their bodies. As you try to quit smoking, you may have symptoms of nicotine withdrawal. When you quit, this withdrawal may cause symptoms like cravings, or urges, to smoke; depression; trouble sleeping; irritability; anxiety; and increased appetite. Nicotine withdrawal can discourage some smokers from continuing with a quit attempt. But the FDA has approved several smoking-cessation products designed to help users gradually withdraw from smoking (that is, "wean" themselves from smoking) by using specific amounts of nicotine that decrease over time. This type of product is called a "nicotine replacement therapy," or NRT. It supplies nicotine in controlled amounts while sparing you from other chemicals found in tobacco products.

NRTs are available over the counter and by prescription. You should generally use them only for a short time to help you manage nicotine cravings and withdrawal. However, the FDA recognizes that some people may need to use these products longer to stay smoke-free. Talk to your healthcare provider to determine the best course of treatment for you.

Over-the-counter NRTs are approved for sale to people age 18 and older. They are available under various brand names and sometimes as generic products. They include:

- **Skin patches** (also called "transdermal nicotine patches"). These patches are placed on the skin, similar to how you would apply an adhesive bandage.

- **Chewing gum** (also called "nicotine gum"). This gum must be chewed according to the labeled instructions to be effective.

- **Lozenges** (also called "nicotine lozenges"). You use these products by dissolving them in your mouth.

For OTC products, it's important to follow the instructions on the Drug Facts Label (DFL) and to read the enclosed User's Guide for complete directions and other important information. Ask your health-care provider if you have questions.

Prescription nicotine replacement therapy is available only under the brand name Nicotrol, and is available both as a nasal spray and an oral inhaler. The products are FDA-approved only for use by adults.

If you are under age 18 and want to quit smoking, talk to a health-care professional about whether you should use nicotine replacement therapies.

Important Advice for People Considering Nicotine Replacement Therapy

Women who are pregnant or breastfeeding should talk to their healthcare providers and use nicotine replacement products only if the healthcare providers approve.

Also talk to your healthcare provider before using these products if you have:

- Diabetes, heart disease, asthma, or stomach ulcers
- Had a recent heart attack
- High blood pressure that is not controlled with medicine
- A history of irregular heartbeat
- Been prescribed medication to help you quit smoking

If you take prescription medication for depression or asthma, tell your healthcare provider if you are quitting smoking because she or he may need to change your prescription dose.

Stop using a nicotine replacement product and call your healthcare professional if you have any of the following symptoms:

- Nausea
- Dizziness
- Weakness
- Vomiting
- Fast or irregular heartbeat
- Mouth problems with the lozenge or gum
- Redness or swelling of the skin around the patch that does not go away

Prescription Cessation Medicines without Nicotine

The FDA has approved two smoking-cessation products that do not contain nicotine. They are Chantix (varenicline tartrate) and Zyban® (bupropion hydrochloride). Both are available in tablet form and by prescription only.

Chantix acts at sites in the brain affected by nicotine by reducing the rewarding effects of nicotine. The precise way that Zyban® helps with smoking cessation is unknown.

As with other prescription products, the FDA has evaluated these medicines and found that the benefits outweigh the risks. For users taking these products, risks include changes in behavior, depressed mood, hostility, aggression, and suicidal thoughts or actions.

The most common side effects of Chantix include nausea; constipation; gas; vomiting; and trouble sleeping or vivid, unusual, or strange dreams. Chantix also may change how you react to alcohol, so talk to your healthcare provider about your drinking habits (if you drink alcohol) and whether these habits need to change. Chantix is not recommended for people under the age of 18.

The most commonly observed side effects consistently associated with the use of Zyban are dry mouth and insomnia.

Because Zyban® contains the same active ingredient as the antidepressant Wellbutrin (bupropion), the FDA encourages people who use Zyban®—and those who are considering it—to talk to their healthcare providers about the risks of treatment with antidepressant medicines. Zyban® has not been studied in children under the age of 18 and is not approved for use in children and teenagers.

Note: If your healthcare provider prescribes Chantix or Zyban®, please read the product's patient medication guide in its entirety. These guides offer important information on side effects, risks, warnings, product ingredients, and what you should talk about with your healthcare provider before taking the products.

Finally, if you ever have any side effects related to any smoking cessation products, or have any other problems related to your treatment, the FDA would like to hear from you. Please consider making a voluntary and confidential report to the FDA's MedWatch program.

Chapter 23

Smokeless Tobacco: Quitting

Quitting smokeless tobacco is hard. But it helps if you have a plan. Here are some steps you can take to quit.

Pick a Quit Date

Every day is a good day to quit. Try to pick a date within the next two weeks so you can prepare. A time with low stress works well for many people.

Understand Withdrawal

After you quit, you may feel uncomfortable and crave a chew or dip. This is withdrawal. Your body is getting used to not having nicotine. For most people, the worst symptoms of withdrawal last a few days to a few weeks. During this time, you may:

- Feel a little depressed
- Be unable to sleep
- Become cranky, frustrated, or mad
- Feel anxious, nervous, or restless
- Have trouble thinking clearly

This chapter includes text excerpted from "How to Quit Smokeless Tobacco," Smokefree.gov, U.S. Department of Health and Human Services (HHS), January 14, 2017.

You may be tempted to chew or dip when these feelings hit. They are temporary, no matter how powerful they feel at the time.

Learn How to Handle Triggers and Cravings

Triggers are specific people, places, or activities that make you want to chew or dip. Knowing your triggers can help you learn to deal with them. Different people have different triggers. Some common ones are:

- Waking up during the night or having nightmares
- After meals or during breaks
- Driving
- Stress or pain
- Feeling anxious, angry, impatient, or bored
- Seeing someone else smoke, chew, or dip
- Drinking coffee or alcohol
- After sex
- Before bedtime
- Watching TV or a live game
- Playing a sport

Triggers can cause cravings. Cravings are short but intense urges to chew or dip. Most cravings last for 15 to 20 minutes.

Use Nicotine Replacement Therapy

One way to deal with nicotine withdrawal is to try nicotine replacement therapy (NRT). A lot of research has been done on NRT. It has been shown to be safe and effective for almost all tobacco users who want to quit.

NRT can reduce withdrawal symptoms. And NRT can double your chances of quitting for good. Explore the different types of NRT that can help you quit smokeless tobacco. NRT is available from your Veterans Affairs (VA) provider at a low cost to you or at your neighborhood drugstore without a prescription.

NRT comes in several different forms, including patch, gum, and lozenge. The patch is a long-acting form of NRT that releases a small, steady amount of nicotine through the skin. This small amount of nicotine helps satisfy your craving for nicotine. But you may still have cravings while on the patch. Doctors recommend using the patch along with a short-acting form of NRT, such as the gum or the lozenge. This will help you fight these strong cravings. That's because using the patch with either the gum or lozenge will work better than using any of these on their own. If you have a severe medical condition or are pregnant, talk to your doctor about using NRT.

If you plan to use NRT, have it available on your quit day. Read the instructions on the NRT package and follow them carefully. NRT will give you the most benefit if you use it as recommended.

Get Help

It is hard to quit chewing or dipping on your own. Quitting "cold turkey" is not your only choice. There are lots of resources to help you quit.

Check out the VA programs that can help you quit:

- **Tobacco Quitline:** Call 855-QUIT-VET (855-784-8838) to talk with a trained counselor who can help you come up with a quit plan that works for your life. She or he can also give you support to stay smoke-free.

- **SmokefreeVET:** This text messaging program sends you support, advice, and encouragement while you are quitting. Sign up for the program in English by texting the word VET to 47848 or go to smokefree.gov/VET. For Spanish, text VETesp to 47848 or go to smokefree.gov/VETespanol.

There are also medications that can help you quit. Medications can double your chances of quitting for good. Using medication doesn't mean you're not strong enough to quit on your own. Using medication can help you keep committed to quitting for yourself and others.

Talk to your VA healthcare provider about medications to help you quit.

Celebrate Successes

An hour, a day, or any time without chewing or dipping is something to celebrate. Get together with the people who've supported your quit journey. Take time to feel proud of yourself. Check out tobacco-free ways to celebrate your successes.

Chapter 24

Quitting as a Woman

Chapter Contents

Section 24.1

How Quitting Can Be Different for Women

This section includes text excerpted from "How Quitting Can Be
Different for Women," Smokefree Women, U.S. Department of
Health and Human Services (HHS), September 6, 2018.

Women can face different challenges when quitting. Learning what
makes quitting unique for some women can set you up for smoke-free
success.

Everyone's path to becoming smoke-free is different. But, many
women have similar challenges when quitting. Knowing what to expect
can help you prepare to quit.

Triggers

Negative emotions and reminders of smoking are powerful triggers for
many smokers. Some women are very sensitive to these types of triggers.
This means that feeling stressed or sad can make it harder to quit and
stay quit. Seeing or smelling a cigarette can also make you want to smoke.

- Learn how to handle the hard times without using cigarettes—
  this can help you avoid slips and stay smoke-free.

- Create an environment that will set you up for success by
  surrounding yourself with supportive people who want to help
  you become smoke-free.

Worries about Weight Gain

If you're worried about gaining weight after quitting, you're not
alone. Many women who smoke have this concern. Think about all the
things quitting will do for you—such as improving how you look and
feel. Becoming smoke-free can give you the confidence to reach other
health goals. Don't let worries about weight gain stop you from trying
to quit. Make becoming smoke-free your priority.

Relationships

Relationships are important in reaching smoke-free success—especially for women. Loved ones and partners can help you face challenges
and celebrate milestones. But, being around people who aren't supportive of your quit can lead to slips or relapses. Planning ahead will help
you manage situations with the people in your life while you try to quit.

- Identify people who will support your decision to quit and ask them for help.

- Take some space (temporarily) from people who don't support your quit.

- Buddy up with someone you are close with who smokes, and quit together. You'll have a built-in support system—and someone who understands the challenges of becoming smoke-free.

Being a Smoker

Many women worry that their life will change for the worse when they quit smoking. Some women who have quit return to smoking because they miss being a smoker. This might be because smoking, and being a smoker, became an important part of your life—especially if you've been smoking for a long time. Thinking about why you smoke and what good things happen when you quit can help you mentally prepare to become smoke-free.

Nicotine

Nicotine is the main addictive substance in cigarettes and other forms of tobacco. It affects many parts of your body, and up to 90 percent of people who smoke regularly are addicted to nicotine. Nicotine increases levels of dopamine—a substance that makes the body feel good. But, when you stop smoking' your body has to get used to not having nicotine. That's called withdrawal.

More research is needed, but women seem to have a slight advantage when it comes to nicotine addiction. When you smoke, dopamine is released, and because dopamine makes you feel good, you start to think of smoking as a reward. This seems to affect men who smoke more than women. It might be the reason women are more likely to smoke when dealing with stress or managing their moods, and men are more likely to smoke because of nicotine-related triggers (such as craving and withdrawal). Learning ways to cope with stress and bad moods can help you avoid a slip or relapse after you quit.

Quitting for Two

Many women successfully quit smoking when they're pregnant or trying to have a baby. Some women go back to smoking before or after their baby is born. The stresses of parenting and dealing with the changes in your life, relationships, and body can make it hard to

309

quit before, during, or after pregnancy. Quitting smoking—at any time—gives you and your child important health benefits. Preparing to quit while pregnant and staying smoke-free after your baby is born will help you and your baby be happier and healthier.

Quit Attempts

The unique challenges women face may contribute to why they typically try to quit more times than men, before they quit for good. It might take you a few tries to become smoke-free—and, that's okay. Each time you try, you learn what does and does not work for you, plus you'll be living healthier and one step closer to a life without cigarettes.

Section 24.2

Quitting and Women's Health

This section includes text excerpted from "How Quitting Helps Women's Health," Smokefree Woman, U.S. Department of Health and Human Services (HHS), September 6, 2018.

Physical Benefits of Quitting

Within 20 minutes of quitting, nicotine starts to leave your body and it begins to heal. You'll get some of these benefits right away. Quitting also improves your health in the years ahead and greatly reduces your risk of smoking-related illness.

It's never too late to quit—no matter your age or how long you've been smoking. But the earlier you quit, the better.

Healthier Blood, Heart, and Lungs

- You'll breathe easier. Within two weeks of quitting, you might notice it's easier to walk up the stairs because you're less short of breath.

- Your "smoker's cough" starts to go away. You might cough more than usual when you first quit. But, this is a sign that the cilia

310

in your lungs are growing—they're one of the first things in your body to heal after you quit.

- Quitting can prevent permanent damage to your lungs. Scarring of the lungs is not reversible. But, if you do have lung disease (such as chronic obstructive pulmonary disease (COPD)), quitting can prevent symptoms from getting worse.

- The oxygen in your blood rises to a normal level. This will make it easier for your heart to pump blood to important parts of your body.

- Your heart rate and blood pressure lower. This puts less stress on your heart and lowers your risk of heart disease, including heart attacks.

Stronger Body

- Quitting can keep your bones strong and healthy. Quitting can reduce the risk of fractures now and later in life. This is important because both women and smokers are more likely than men and nonsmokers to get osteoporosis (a disease when your bones become weak and more likely to break).

- Your immune system will become stronger. When you have a strong immune system, you'll be less likely to get sick.

- Your muscles will become stronger and healthier. Quitting smoking will help increase the availability of oxygen in your blood, and your muscles will become stronger and healthier.

Changes You Can See

- Your skin will look healthier. Quitting can help clear up blemishes and protect your skin from premature aging and wrinkling. It will also help wounds heal better.

- You'll have a cleaner mouth. Quitting will make your teeth brighter and gums healthier. Your breath will also smell better.

- Quitting can reduce belly fat and lower your risk of diabetes. If you already have diabetes, quitting can help you manage your blood sugar levels.

- Your sense of taste and smell will improve. When you quit you'll taste foods better and you'll smell things such as foods, flowers, and other scents better.

311

Fewer Fertility Problems and Pregnancy Risks

- Your estrogen levels will gradually return to normal. Low estrogen can cause many problems, such as dry skin, thinning hair, and mood swings.

- Quitting smoking will increase your chances of having a healthy pregnancy. Women of a childbearing age who quit smoking are less likely to have problems becoming pregnant. You'll also lower your risk of certain pregnancy complications that can harm you and your baby, such as preeclampsia and placenta previa.

- Quitting now will increase your chances of having a healthy baby. Quitting at any time during your pregnancy lowers your risk of having a miscarriage. Quitting smoking also gives your baby benefits, such as lowering their risk of being born too early, having certain birth defects, or dying from sudden infant death syndrome (SIDS).

In the Long Run, You Can Lower Your Chance of . . .

- **Heart disease.** Heart disease can lead to heart attacks, chest pain, and stroke. Within 10 to 15 years of quitting smoking, your risk of heart disease may be the same as nonsmokers.

- **Stroke.** The longer you are smoke-free the more your risk of stroke goes down. Within 5 to 15 years of quitting, your risk of stroke may be the same as nonsmokers.

- **Lung cancer.** The longer you are smoke-free the more your risk of lung cancer goes down. Within 10 years of quitting, your risk of dying from lung cancer will have decreased by half.

- **Dying early.** People who smoke die about 10 years earlier than people who don't. Quitting can lower your risk of dying from smoking-related illness and diseases.

If You Have Been Diagnosed with Cancer, It Is Not Too Late to Benefit from Quitting

- Patients with some cancers increase their chances for survival if they quit when they are diagnosed with cancer.

- For those having surgery, chemotherapy, or other treatments, quitting smoking helps improve the body's ability to heal and respond to therapy. You will be less likely to develop pneumonia or respiratory failure.

- Quitting smoking may also reduce the chances that the cancer will recur (come back), that another cancer will develop, or that you will die from the cancer.

Emotional Benefits of Quitting

Quitting can improve your quality of life (QOL) and boost your mood. After you become smoke-free you will experience many positive changes. Below are some things you may experience after quitting.

- You'll feel more in control of your life. Being smoke-free means that you won't have to plan your life around smoking, worry about finding places to smoke, or worry about bothering others.

- Your hair, clothes, home, car, and breath won't smell like smoke anymore.

- You'll have more money.

- You'll have more energy to walk, play with your kids, or be active.

- Your loved ones will be proud of you.

- You'll feel empowered and proud of your success.

Section 24.3

Quitting While Pregnant

This section contains text excerpted from the following sources:
Text in this section begins with excerpts from "Smoking and Your
Baby," Smokefree Women, U.S. Department of Health and Human
Services (HHS), September 6, 2018; Text under the heading "Myths:
Smoking and Pregnancy" is excerpted from "Myths: Smoking and
Pregnancy," Smokefree Women, U.S. Department of Health and
Human Services (HHS), September 6, 2018.

Smoking before, during, and after pregnancy is harmful to both you
and your baby. It's never too late to gift yourself and your new baby a
healthier future and quit.

Smoking can make it harder for you to get pregnant. And, if you
get pregnant, continuing to smoke while you're pregnant can cause
serious harm to your baby's health. Even if you smoked during a past
pregnancy and had a healthy baby, every pregnancy is different. Quitting smoking at any time during pregnancy—especially early on—can
give your baby a healthier start in life.

After your baby is born, exposing your baby to secondhand smoke
also causes harm to your child's health and development. Understand the risks of continuing to smoke before and after your baby is
born. Quitting will give you and your baby a brighter and healthier
future.

Smoking While Pregnant

Smoking can cause complications during your pregnancy, such as:

- **Miscarriage:** An unexpected loss of the baby.

- **Ectopic pregnancy:** When the fertilized egg implants outside
  the uterus (where the baby usually develops). The result is
  usually a miscarriage.

- **Placental abruption:** When the placenta (which provides
  oxygen and nutrients to the baby) separates too early from the
  uterus. This can cause bleeding in the mother. It can also lower
  or stop the baby's supply of oxygen and food.

- **Placenta previa:** When the placenta covers the cervix (the
  lower end and opening of the uterus). This can cause severe
  bleeding during pregnancy and delivery.

- **Preeclampsia:** When the mother experiences high blood pressure and swollen feet, legs, and hands. This can have severe complications for both the mother and baby, such as placental abruption and having a premature baby.

- **Preterm birth:** When the baby is born too early. Babies who survive early delivery can have problems breathing, digestion issues, and bleeding in their brains. As the child gets older, she or he may experience developmental delays (when a baby or child is behind in language, thinking, or movement skills) and not do as well in school.

Have a Healthy Baby

Quitting before you give birth will give your baby these benefits:

- Less risk of being born too early

- Less risk of being born with birth defects, such as cleft lip or cleft palate

- Higher chance of having a healthy birth weight (more than 5.5 pounds) and growing on track

- More likely to come home from the hospital with you. Babies who are too small or who need care for health problems may need to stay in the hospital until they are healthy enough to go home

- Increased chance lungs will develop well

- More likely to have normal brain development before birth and through early childhood

- Less likely to die from sudden infant death syndrome (SIDS)

Staying Smokefree after Baby

Once you bring your baby home, it's just as important to stay smoke-free and protect your baby from cigarettes or secondhand smoke. Babies who are exposed to secondhand smoke after birth are more likely to have certain health problems or die from SIDS.

Protect your baby from secondhand smoke by avoiding smoking around your baby, spending time in places that do not allow smoking, and making your house and car smoke-free. This is important because

315

babies and young children living in smoke-free environments typically have:

- Fewer coughs and chest colds
- Lower chances of getting bronchitis or pneumonia
- Fewer ear infections
- Less frequent and less severe asthma, if your baby has asthma
- Fewer missed school days because of asthma attacks and breathing illnesses
- Less of a chance of becoming a smoker as an adult
- Lower chances of dying from SIDS

Breastfeeding and Smoking

Don't stop breastfeeding if you smoke. Breastfeeding is good for your baby, so it's better to do it than not, even if you're still smoking. Here's why breastfeeding is good for both you and your baby:

- Gives your baby nutrients and antibodies that help her or him grow and develop
- Protects your baby from illnesses, such as ear infections and diarrhea
- Lowers your baby's risk of diabetes and SIDS
- Lowers your risk for diabetes, breast cancer, ovarian cancer, and depression
- Can help you form a special bond with your baby
- Can help you lose weight during the first year after your baby is born

Smoking while breastfeeding does expose the baby to nicotine and other harmful substances through breastmilk. If it's not the right time for you to quit, you can help your baby by following a plan to reduce your baby's exposure to cigarette smoke.

- Smoke right after breastfeeding and avoid smoking right before. Your body will have more time to clear the nicotine from breast milk.
- Make your house and car smoke-free to keep your baby away from secondhand smoke.

- Try cutting down on the number of cigarettes you smoke.

- Don't smoke while you're feeding your baby. She or he will inhale your smoke or could get burned by the cigarette.

- Sign up for the SmokefreeMOM text program for help cutting back on cigarettes and eventually becoming smoke-free.

Myths: Smoking and Pregnancy

Smoking anytime during your pregnancy is dangerous. Quitting is the best thing you can do for you and your baby. A list of some common myths about quitting and pregnancy follows.

Quitting smoking—at any point during your pregnancy—is one of the best things you can do for your baby.

Myth 1: I'm Pregnant and Have Been Smoking, So There Is No Point in Stopping Now

Fact: Quitting smoking at any stage of your pregnancy has health benefits for you and your baby. Even after just one day of not smoking, your baby will get more oxygen. This will help your baby's lungs develop well. Quitting now also lowers your chances of having a baby with low birth weight.

Myth 2: Quitting Smoking Will Be Too Stressful on My Baby

Fact: Quitting smoking doesn't put extra stress on your baby. It's one of the best things that you can do for your health and your baby's health during pregnancy—and after the baby is born. By quitting smoking now, you will be protecting your infant from the dangers of secondhand smoke and reducing the risk of sudden infant death syndrome.

Myth 3: Smoking Fewer Cigarettes or Switching to E-Cigarettes during Pregnancy Is Okay

Fact: There is no safe amount of smoking. Every puff of a cigarette releases harmful chemicals that will reach your baby and affect your health too. E-cigarettes are also not harmless. Although there is still much to learn about e-cigarettes, pregnant women should not use them. The nicotine in e-cigarettes is harmful for developing babies.

317

Myth 4: Smoking Relaxes Me, and Being Relaxed Is Better for Me and My Baby

Fact: Smoking may make you feel calmer, but it hurts your body more than it helps. The relaxed feeling is only temporary and whatever is causing your stress will likely return. Smoking speeds up your heart rate and increases your blood pressure. It also increases the carbon monoxide in your bloodstream, which means your baby gets less oxygen.

Myth 5: There Is Nothing Wrong with Having a Small Baby

Fact: Smoking during pregnancy increases the chances of having a low-birth-weight (LBW) baby. Babies with low birth weight are more likely to have serious health problems than normal-weight babies. These problems can affect your baby's health now, throughout their childhood, and into adulthood.

Myth 6: The Only Way to Quit Smoking Is Cold Turkey

Fact: Pregnant women have other ways to quit smoking besides "cold turkey," which is quitting without any preparation or counseling. Smokefree women offers many resources that can help you quit. Try signing up for a text message program, such as SmokefreeMom. While some smokers try medications to help them quit, the risks of using them while pregnant are not fully known. If you're having trouble quitting after trying different methods, be open with your doctor about your challenges and ask if medication may be right for you and your baby.

Myth 7: I Smoked during My Last Pregnancy and Had a Healthy Baby, so This Next Baby Will Be Healthy, Too

Fact: Every time you smoke during pregnancy, you put your baby's health at risk. If you smoked and had a healthy pregnancy in the past that does not mean your next one will be healthy, too.

Myth 8: I Smoke, So I Should Not Breastfeed My Baby

Fact: According to the American Academy of Pediatrics (AAP), mothers who smoke are encouraged to quit smoking, but can breastfeed their baby if they continue to smoke. Breast milk is good for your baby. It provides your baby with what she or he needs for healthy growth and development. Not smoking while you are breastfeeding your baby, waiting to smoke until after you breastfeed, and making your car and home smoke-free are important ways to protect your baby from the effects of nicotine and secondhand smoke.

Chapter 25

Medications for Smoking Cessation

Medications can make it easier to quit smoking by reducing cravings and withdrawal symptoms. Learn about the different types of medications and see which ones might work best for you.

Nicotine Replacement Therapy

Nicotine replacement therapy (NRT) is the most used quit smoking medication and comes in many forms, like gum and patches. It works by giving you a small, controlled amount of nicotine, which is the main addictive substance in cigarettes and other tobacco products. NRT doesn't have any of the other dangerous chemicals found in cigarettes. This small amount of nicotine helps satisfy cravings for nicotine and reduces your urge to smoke. Some smokers have mild to moderate side effects. However, research shows that NRT is safe and works. NRT is not recommended for women who are pregnant or trying to become pregnant.

Most types of NRT are available over-the-counter (OTC), which means you can buy them without a prescription. Before you start, follow the directions on the package so you use it properly.

This chapter includes text excerpted from "Using Medications Can Help You Quit," Smokefree Women, U.S. Department of Health and Human Services (HHS), September 6, 2018.

Nonnicotine Prescription Medications

There are two common prescription medications that help smokers quit: Bupropion SR and Varenicline. You'll need a prescription from your doctor to buy these medications. Many insurance plans cover quit-smoking medications. Check with your insurance plan to see if you are covered.

Bupropion SR

Bupropion SR (often referred to as Wellbutrin), is a medicine without nicotine. It may help with withdrawal symptoms and the urge to smoke. This medicine may not be right for pregnant women, people who have seizures, people who have eating disorders, or heavy alcohol users.

Varenicline

Varenicline (often referred to as Chantix) does not have nicotine. This drug may help you quit by improving withdrawal symptoms and making nicotine from cigarettes less effective if you start smoking again. This medicine may not be right for people with kidney problems and women who are pregnant, planning to become pregnant, or are breastfeeding.

Ask your doctor, dentist, or pharmacist if these medicines are right for you. Always use them as prescribed or as listed on the packaging.

Chapter 26

Using Bupropion (Zyban®)

How Bupropion Will Help You

Bupropion (Zyban®) is used to help people stop smoking by reducing the urge to smoke and decreasing withdrawal symptoms. Also, many patients report that cigarettes do not taste as good after starting this medication.

Instructions for Using Bupropion

1. Follow the directions on your prescription label carefully, and ask your primary-care provider (PCP) or pharmacist to explain any part you do not understand. Take bupropion exactly as directed.

2. Bupropion comes as a tablet to take by mouth. It is usually taken one to two times a day and may be taken with or without food. Do not crush, chew, or divide bupropion tablets. Take it at the same time each day.

3. Bupropion for smoking cessation is usually started 7 to 14 days before your smoking Quit Date (the date you plan to stop using tobacco).

This chapter includes text excerpted from "Instructions for Using Bupropion (Zyban)," Mental Illness Research, Education and Clinical Centers (MIRECC), U.S. Department of Veterans Affairs (VA), July 2013. Reviewed February 2019.

4. If you take bupropion two times per day, take the missed dose as soon as you remember it and take the second dose for that day at least eight hours later. Do not take an extra tablet to make up for the dose you forgot. However, if it is almost time for the next dose, skip the one you forgot and continue your regular dosing schedule.

5. Do not take a larger dose, or take it more often or for a longer period of time than your PCP tells you.

6. Keep this medication in a tightly closed container, and out of reach of children. Store it at room temperature and away from excess heat and moisture (not in the bathroom). Throw away any medication that is outdated or no longer needed. Talk to your pharmacist about the proper disposal of your medication.

Precautions

There is a chance of having a seizure (convulsion or fit) with bupropion if it is taken in large doses. Take bupropion precisely as instructed to minimize the risk of seizures.

Some people have had changes in behavior, hostility or anger, agitation, depression, suicidal thoughts or actions while taking bupropion to help them quit smoking. These symptoms can develop during treatment with bupropion or after stopping treatment with bupropion. If you, your family member, or your caregiver notices any of these symptoms, call your PCP right away. Tell your PCP if you have or have ever had depression, suicidal thoughts or actions, or other mental-health problems.

Tell your PCP if you have or have ever had seizures, brain tumors, head injuries, or an eating disorder or are taking antipsychotics, tranquilizers, or other antidepressants.

Before taking bupropion, tell your PCP and pharmacist if you are allergic to bupropion or any other drugs. Tell your PCP and pharmacist what other prescription and nonprescription drugs you are taking, or have taken in the last two weeks, especially anticoagulants [warfarin (Coumadin)]; antihistamines; antipsychotics; carbamazepine (Tegretol); cimetidine (Tagamet); diet pills; insulin; levodopa (Sinemet, Larodopa); lithium (Eskalith, Lithobid); monoamine oxidase inhibitors (MAO) inhibitors [phenelzine (Nardil), tranylcypromine (Parnate)]; medication for high blood pressure, seizures, asthma, colds, or allergies; methylphenidate (Ritalin); oral antidiabetic medications; other antidepressants; ritonavir (Norvir); sedatives; sleeping pills;

theophylline (Theobid, TheoDur, others), thyroid medications; tranquilizers; and vitamins.

You should not take Wellbutrin or Wellbutrin SR and Zyban® together.

Also tell your PCP if you drink large amounts of alcohol or are addicted to cocaine, opiates, or other drugs so that you can properly taper and end their use. If you stop using these things all of a sudden it can increase your chance of having a seizure. Let your PCP know if you have or have ever had glaucoma, an enlarged prostate, difficulty urinating, an overactive thyroid gland, anorexia or bulimia, or liver, or kidney or heart disease.

Tell your PCP if you are pregnant, plan to become pregnant, or are breastfeeding. If you become pregnant while taking bupropion, call your PCP immediately. If you are having surgery, including dental surgery, tell the doctor or dentist that you are taking bupropion.

Side effects from bupropion are common and may include:

- Dry mouth
- Insomnia (trouble sleeping)
- Weakness or tiredness
- Excitement or anxiety (nervousness)
- Nightmares
- Change in appetite or weight

Tell your PCP if any of these symptoms are severe or do not go away:

- Frequent urination
- Difficulty urinating
- Constipation
- Blurred vision
- Change in sex drive or ability
- Excessive sweating

Chapter 27

Genetics and Quitting

A study shows that genetic information can help predict whether medications will be likely to help a person quit smoking. The study brings healthcare providers a step closer to providing individualized treatment plans to help smokers kick the habit.

Tobacco use is the most preventable cause of disease, disability, and death in the United States. Smoking and exposure to secondhand smoke results in more than 440,000 preventable deaths each year—about one in five U.S. deaths overall. Another 8.6 million people live with a serious illness caused by smoking. Yet over 46 million U.S. adults continue to smoke cigarettes.

Many smokers would like to quit, but find that it can be a difficult and frustrating process. Previous studies have found that certain genetic variations influence the chances of successfully quitting. In a new study, a team of researchers led by Dr. Li-Shiun Chen of Washington University tested whether genetics can be used to personalize therapeutic strategies. The work was supported by several National Institutes of Health (NIH) components, including the National Institute on Drug Abuse (NIDA), National Human Genome Research Institute (NHGRI) and National Cancer Institute (NCI). The study appeared online on May 30, 2012, in the *American Journal of Psychiatry*.

This chapter includes text excerpted from "Genetics May Guide Ways to Quit Smoking," National Institutes of Health (NIH), June 25, 2012. Reviewed February 2019.

Genetics May Guide Ways to Quit Smoking

The scientists focused on specific variations in a cluster of nicotinic receptor genes, CHRNA5-CHRNA3-CHRNB4. Prior studies had shown that this cluster contributes to nicotine dependence and a pattern of heavy smoking. Using data on over 5,000 smokers from a previous study supported by NIH's National Heart Lung and Blood Institute (NHLBI), the researchers found that people carrying the high-risk form of the gene cluster reported a two-year delay in the median quit age compared to those with the low-risk version.

The researchers next carried out a clinical trial. They randomly divided more than 1,000 smokers into six groups. One received only a series of brief counseling sessions. The others received the sessions plus one of five treatments: nicotine patch, nicotine lozenge, the medication bupropion (Zyban®), nicotine patch and nicotine lozenge, or bupropion and nicotine lozenge.

Eight weeks after quitting, nearly half the participants had abstained from smoking. When given only counseling, those with the high-risk gene cluster were more likely to fail in their quit attempts than those with the low-risk cluster. However, medications increased the likelihood of success in the high-risk groups. Those with the highest risk had a three-fold higher success rate with active treatment than with counseling only. In the low-risk group, medication added no significant benefits over counseling only. The impact of genetic variation on success was similar when compared across the other treatment options.

"We found that the effects of smoking cessation medications depend on a person's genes," Chen says. "If smokers have the risk genes, they don't quit easily on their own and will benefit greatly from the medications. If smokers don't have the risk genes, they are likely to quit successfully without the help of medications such as nicotine replacement or bupropion."

"This study builds on our knowledge of genetic vulnerability to nicotine dependence, and will help us tailor smoking cessation strategies accordingly," comments NIDA Director Dr. Nora D. Volkow.

Chapter 28

Overcoming Challenges When Quitting

Chapter Contents

Section 28.1

Know Your Triggers

This section includes text excerpted from "Know Your Triggers,"
Smokefree Women, U.S. Department of Health and Human
Services (HHS), September 6, 2018.

A trigger is something that makes you want to smoke. It can be a stressful situation, sipping coffee, going to a party, or smelling cigarette smoke. Different people have different triggers, and there are some triggers that have a bigger effect on women than they do on men. Knowing your triggers can help you control cravings and make it less likely that you will smoke.

Emotional Triggers

Feeling intense emotions, either good or bad, can make you want a smoke. These emotional triggers are common among all smokers. Women, more than men, are more likely to want to smoke when they experience negative emotions, such as sadness, anger, frustration, and stress. This means that bad moods and stress can make it harder for women to stay smoke-free. Smoking may help you feel better, but it will only be temporary. It won't make whatever is bothering you go away for good. There are ways you can manage your stress and mood without cigarettes. Learning healthy ways to handle these emotions is important for preventing slips and relapses.

Environmental Triggers

When you stop smoking, you also have to get used to dealing with the things around you that remind you of smoking. Environmental triggers can be related to activities you do every day or in your social life. These reminders can be a trigger for any smoker, but they may affect women more than men.

Pattern Triggers

A pattern trigger is an activity where you typically include smoking. Some examples of these include:

- Drinking alcohol or coffee
- Watching television

- Driving

- Finishing a meal

- Relaxing after having sex

- One way to beat pattern triggers is to break the connection to smoking by replacing it with another activity. Try changing your routine or finding a nonsmoking replacement, such as chewing gum or eating sugar-free candy.

Social Triggers

Social triggers are events or activities where other people are smoking. Here are some examples:

- Going to a bar or concert

- Going to a party or other social event

- Seeing someone else smoke

- Being with friends who smoke

- Celebrating a big event

You might be worried that you won't be able to stay smoke-free with all the things that remind you of smoking. But there are small steps you can take that can help a lot. Start by telling your friends and family about your decision to quit. Ask them for their support and to not smoke around you. When you first quit, it is best to avoid places where people smoke. But over time, it will get easier to be around smokers.

Withdrawal Triggers

If you've been a long-time smoker, your body is used to getting nicotine from cigarettes. When you quit, withdrawal symptoms will make you want to keep smoking. Some examples of withdrawal triggers include:

- Craving the taste of a cigarette

- Smelling cigarette smoke

- Handling cigarettes, lighters, and matches

- Needing to do something with your hands or mouth

- Feeling restless or having other withdrawal symptoms

There are ways to deal with withdrawal triggers. Distract yourself by finding something to take your mind off the craving. Try nicotine replacement medication.

Section 28.2

Fight Cravings

This section includes text excerpted from "Fight Cravings," Smokefree Women, U.S. Department of Health and Human Services (HHS), September 6, 2018.

It's no secret that quitting is hard. You won't be able to avoid all of your triggers. And learning how to deal with triggers takes practice. So, when a craving is triggered, it's important to have a plan to beat that urge to smoke.

Triggers lead to cravings to smoke. Cravings are uncomfortable, and the urge to smoke can be strong. The good news is that cravings are temporary, and there are things you can do to make yourself feel better during this time. Make a list of things to do when a craving hits. Here are some ideas:

- **Keep your mouth and hands busy.** Chew gum. Suck on a straw. Squeeze a handball. Do beading or needlework.

- **Exercise.** Go for a walk. Ride a bike. Go swimming. Exercising can distract you from smoking.

- **Change your routine.** For example, try drinking your coffee at a different time, or brushing your teeth right after you eat a meal.

- **Use nicotine replacement therapy.** These medications can double your chances of quitting for good. Experiment with combinations of medications to see what works best for you.

- **Tell others that you are quitting.** Let the people in your life know you are quitting and ask them to avoid smoking around you. This will allow them to support you and might prevent them from offering you a cigarette.

- **Prepare to handle places where people smoke.** In the weeks or months after you quit, you might need to stay away from places where people smoke. The smell of cigarettes and other reminders of smoking can be a trigger, which might lead to a slip.

- **Take deep breaths.** Your breaths can become small and shallow when you're stressed. This makes the physical symptoms of stress worse. Take a few deep breaths to help your body relax.

- **Let your feelings out.** Talk, laugh, cry, or write to express your negative feelings. You don't have to deal with your stress and problems alone. It is okay to reach out to your family, friends, counselor, or loved ones for support.

Section 28.3

Dealing with Withdrawal

This section includes text excerpted from "Dealing with Withdrawal," Smokefree Women, U.S. Department of Health and Human Services (HHS), September 6, 2018.

When you stop smoking, your body needs to get used to not having nicotine. This can be uncomfortable, but there are ways you can manage nicotine withdrawal.

Understanding Withdrawal

Withdrawal is the collection of uncomfortable symptoms you have when you stop getting nicotine. Nicotine is the main addictive drug in cigarettes and other tobacco products. For most smokers, the worst withdrawal symptoms last from a few days up to two weeks. Every smoker's withdrawal experience is different. Take this short quiz to better understand your withdrawal symptoms.

Common withdrawal symptoms include:

- Having intense cravings for cigarettes
- Feeling sad, irritable, or restless
- Having trouble sleeping
- Having trouble thinking clearly and concentrating
- Feeling hungrier or gaining weight

Withdrawal and Emotional Triggers

Negative feelings and mood changes can happen during withdrawal—and they can also be powerful smoking triggers for women. Learning how to manage emotional symptoms of withdrawal is very important for women to stay smoke-free. Try the tips below to help you get through it.

- **Build healthy coping skills.** Learn how to handle your emotional triggers without using cigarettes.

- **Consider using medication.** If you are not pregnant, or trying to become pregnant, try nicotine replacement therapy to help manage these symptoms and withdrawal-related cravings. You can also talk to your doctor about prescription medications that can help you quit.

- **Explore other ways to beat withdrawal.** The more tools you have at your disposal, the more successful you will be at dealing with triggers, cravings, and withdrawal.

Get Support

Having symptoms of withdrawal is a major challenge smokers face while quitting. Many people will start smoking again to feel better. There are ways to get support and avoid a slip or relapse.

- **Lean on your loved ones.** Let your friends and family know that quitting smoking may affect your mood and patience. Ask them to be understanding and supportive of the challenging time you're going through.

- **Try SmokefreeTXT.** This mobile text-messaging service offers 24/7 encouragement, advice, and tips to help smokers quit smoking and stay quit. Sign up online or text START to 47848.

- **Connect with a trained quit-smoking counselor.** Call 800-QUIT-NOW (800-784-8669) or chat online at LiveHelp. Both services are available Monday through Friday, 8:00 a.m. to 11:00 p.m. EST. Also available in Spanish.

Withdrawal and Depression

It's normal to go through negative mood changes after you quit smoking. This can be part of withdrawal. Some people feel sadder. You might also be irritable or restless. But if you're feeling very down for more than two weeks, it could be depression.

If you think you have depression, try not to smoke. Talk to your doctor about your options for treating depression while you stay smokefree.

Find Help 24/7

If you need help now, call a 24-hour crisis center at 800-273-TALK (800-273-8255) or 800-SUICIDE (800-784-2433) for free, private help, or dial 911.

Sometimes people who are feeling depressed think about hurting themselves or dying. If you or someone you know is having these feelings, get help now.

The Substance Abuse and Mental Health Services Administration (SAMHSA)—a part of the U.S. Department of Health and Human Services (HHS)—runs both crisis centers.

Section 28.4

Weight Gain and Quitting

This section includes text excerpted from "Weight Gain and Quitting," Smokefree Women, U.S. Department of Health and Human Services (HHS), September 6, 2018.

Fear of gaining weight keeps some people from trying to quit smoking. When you make quitting your priority, it's the best step you can take for a healthier you.

Smoking and Weight

It's true that some people gain weight after quitting. On average, people who gain weight after quitting gain about 5 to 10 pounds. Here's why:

- Smoking lowers your appetite. Smoking cigarettes will make you feel less hungry. When you quit smoking, you might feel hungrier and eat more.

- Eating can be a substitute for smoking. Smoking gave you something to do with your hands, and you're used to putting a cigarette in your mouth. For a lot of people, food replaces cigarettes, so it's more likely you'll gain weight after quitting.

- Eating may be a new way for you to deal with negative emotions. Maybe smoking was your go-to when you feel stressed. When you quit, you may eat to feel better or deal with stress. This can cause weight gain.

If you are worried about gaining weight after you quit, try to remember:

- Quitting smoking is the best thing you can do for your health right now.

- You can do things to help prevent weight gain while you become smokefree. Take small steps, such as healthy snacking, watching portion sizes, or finding ten minutes a few times a day to move more.

- If you do gain weight after quitting, it doesn't have to be forever. There are healthy ways you can lose the weight.

Set Realistic Goals

When you try to lose weight, think of steps that are manageable, positive, under your control, and specific. Every step toward your goal is a step to a healthier you. If some of these steps aren't as small as you thought or if your current plan isn't working, change it. It may take a few tries to find what works best with quitting smoking, and making other healthy lifestyle choices is no different.

Healthy Weight Loss and Body Image

Becoming and staying smoke-free will give you more energy and confidence, which will help you lose any weight you might have gained after quitting. There is no magic number for healthy weight. Everyone is different. Trying to lose weight with extreme diets or over-the-top workouts doesn't work long term. The key is slow and steady weight loss. Find what works for you and recognize your successes. Celebrate milestones like losing five percent of your body weight, fitting into a pair of jeans that used to be too tight, or improving your jogging pace.

If weight gain is a big concern, think about why. Many women struggle with poor body image and unrealistic expectations for how they should look. Our bodies are capable of amazing things and the better we treat them, the more they can do. Try to be easier on yourself and find things to love about your body.

Chapter 29

Lifestyle Modifications and Smoking Cessation

Chapter Contents

Section 29.1

Nutrition and Appetite While Quitting

This section includes text excerpted from "Nutrition and Appetite While Quitting," Smokefree.gov, U.S. Department of Health and Human Services (HHS), October 13, 2016.

It's normal for your appetite to increase and your sense of taste to change after quitting smoking. Avoid overeating, weight gain, and poor nutrition by staying in control of your food choices.

Increased appetite is a common withdrawal symptom after quitting. It tends to last longer than other symptoms. When you quit smoking, your senses of taste and smell improve and return to normal. Food is more enjoyable. This may also increase your appetite.

After quitting, your food likes and dislikes might change. It is common for people to say that before quitting they didn't have a sweet tooth but now eat sweet foods. Studies show that people want more sweet and fatty foods after quitting. Sweet and fatty foods usually are high in calories.

Snack Smart

When you quit smoking, you may miss the feeling of having something to do with your mouth and hands. Eating or snacking is like the action of smoking. This need to have something in your mouth goes away over time. Try to keep your hands and mouth busy with a toothpick or straw. Or you can chew on healthy foods such as carrots and celery, or even sugar-free mints.

Eat Mindfully

Research shows that people tend to use food in the same ways they used cigarettes. They use food to deal with stress or boredom, reward themselves, pass time, or help them be social. It is important to know why you eat even when you are not hungry. You can eat mindfully by eating slower. Eating slower is healthier too. Take time to enjoy your food. You are more likely to notice when you are full. Try turning off the television and shutting off your phone while you eat. Going gadget-free helps you pay attention to the details of your food. You might notice flavors and textures you otherwise would have missed.

Control Weight

Changes in your diet or eating more food might cause you to gain weight. It is possible to reduce the chance you will gain weight after you quit smoking. If you have a plan to deal with appetite and nutrition changes, you can manage your weight. Exercise can be an important part of that plan to help you reduce cravings, reduce your appetite, and handle stress.

Section 29.2

Dealing with Weight Gain

This section includes text excerpted from "Dealing with Weight Gain," Smokefree.gov, U.S. Department of Health and Human Services (HHS), May 16, 2016.

Many people delay quitting smoking because they are worried about weight gain. While many smokers gain some weight after they quit, it is better for your health if you quit as soon as possible.

Once you quit, you can begin to build healthy habits for exercise, nutrition, and—if necessary—weight loss.

How to Control Weight Gain after Quitting

Control your appetite. Smoking cigarettes makes you feel less hungry. When you quit smoking, you might feel hungrier and eat more than you used to eat. Portion control can help you prevent overeating. And make sure you stay hydrated. It will keep you from eating when you are actually thirsty.

Start exercising regularly. Smoking cigarettes increases your metabolism, so you burn more calories. When you stop smoking, you may burn fewer calories, which can lead to weight gain. Regular exercise can help you continue to burn those calories. Just 10 minutes of exercise a day can make a difference.

Learn how to eat mindfully. Eating can become a substitute for smoking. It doesn't have to. Learn how to eat mindfully. Mindful eating means preparing, choosing, and eating with awareness. Here are some tips to help you eat healthier and enjoy your food more:

- Avoid distractions. Turn off the phone, television, and computer.

- Eat at your dinner table so you can focus on eating.

- Think carefully about your portion sizes. Consider starting with a smaller portion.

- Avoid serving food family style. If you still feel hungry, you will have to get up from the table for seconds.

- Do a belly check. Are you really hungry or eating for some other reason?

Learn to cope with emotions without cigarettes. When you stop smoking, you may want to turn to eating when you feel bad or life is stressful. But you can learn other ways to cope without smoking.

You can take charge of your weight even while quitting smoking. Sign up for HealthyYouTXT, a text messaging service that provides 24/7 encouragement, advice, and tips to help you achieve your health goals.

Section 29.3

Fit for Life

This section includes text excerpted from "Be Fit for Life," Smokefree.gov, U.S. Department of Health and Human Services (HHS), January 10, 2017.

Including exercise in your quit-smoking plan can help you get through withdrawal and cravings.

Making physical activity part of your routine gives other important benefits too, such as:

- Reduced risk of heart disease, diabetes, and cancer

- More positive mood

- Increased energy

- Better sleep

- Improved stress management

How much physical activity do you need to get the most benefits? That may depend on what you want to achieve. To reach and stay at a healthy level of fitness, experts recommend.

Get Your Heart Pumping

Cardiovascular exercise is an important part of any exercise routine. Aim for at least 30 minutes of moderate-intensity physical activity five or more days a week. Or at least 20 minutes of vigorous-intensity physical activity three or more days a week. Don't have 30 minutes a day? Studies show that exercising for 10 minutes three times a day gives the same benefits as 30 minutes of nonstop exercise. You don't need fancy equipment for cardio exercises. Plenty of cardio exercises rely only on your body, such as jumping jacks. Brisk walking, running, dancing, jumping rope, swimming, and cycling are other cardio exercises you might try—indoors or outdoors, alone or with friends.

Grow Stronger

Improve your muscle tone with strength training. Aim for 8 to 10 resistance exercises on two or more days a week. Resistance exercises use an opposing force to increase muscle size and strength. Dumbbells, or barbells, are commonly used for resistance exercises. You also can use your own body weight, bricks, bottles of water, or any other object that causes your muscles to contract. Include a set of exercises for each major muscle group: arms, legs, back, and chest. Each exercise set should include 8 to 12 repetitions. That means lifting and lowering the weights 8 to 12 times slowly. As you get stronger, you might want to work up to two or three sets of exercises for each major muscle group.

Stretch and Connect with Your Body

Stretching can improve your flexibility and balance. It also can help you perform better and reduce your risk of injury during other activities. When done safely and properly, stretching can help relax tense muscles and create a sense of well-being. You can stretch anytime, anywhere. Aim for 20 to 30 minutes of stretching activities two or three days a week. Stretch each of the major muscle groups.

Take a Stand for Health

Some studies show that sitting for a long time is bad for your health, even if you exercise regularly. Sitting too much has been linked to heart attacks, heart disease, and death from cancer, although experts aren't sure why. Here are a few simple ways you can reduce your sitting time and maintain your health:

- Break up sitting times. Try to get up and move around about every 30 minutes or so.

- Stand when you can. If you work in an office, try a standing desk. More workplaces are making standing desks available to employees.

- Give yourself reminders to sit less. At home, get out of your chair briefly at every television commercial.

- At work, use a smaller coffee cup or glass so you need to make more trips for refills. Or schedule several walking or standing meetings a week.

Section 29.4

Physical Activity and Cessation

This section includes text excerpted from "Be SMART about
Physical Activity," Smokefree.gov, U.S. Department of
Health and Human Services (HHS), April 12, 2017.

Everyone can gain health benefits from starting or increasing physical activity. Setting SMART goals can help you get the most out of moving your body.

Regular physical activity is good for your health. Being active reduces the risk of heart disease, stroke, high blood pressure, some cancers, and type two diabetes. It also helps control weight and may help ease stress. Starting an exercise routine or taking your current workout to the next level is a great way to gain these health benefits. Set SMART goals to stay motivated and on track. SMART goals also can help you make weight loss and healthier eating part of an overall healthy lifestyle.

Specific

"I'll move more" or "I'll watch less TV" are too general to be helpful goals. Be specific about what activities you'll do and for how long. Your goals should be clear and easy to understand—for example, "I will walk five mornings a week for 30 minutes each day." Have a back-up goal for times when your schedule, weather, or other issues prevent you from staying with your plan: "If I can't walk in the morning, I'll walk after dinner." Also, be specific about how you'll reward yourself when you reach your goals, such as "I'll buy myself new sneakers if I stay with my plan for a month."

Measurable

Including a number in your goal helps you track your progress. For example, you might start with brisk walking for 10 minutes and over time build up to 30 minutes. Keep a record of your progress in an exercise journal. You might try wearing a fitness tracker or getting a health and fitness app for your smartphone to see how your workouts are going. This can be important for staying motivated if you set some of your goals for months ahead.

Attainable

Some people set their physical activity goals too high. Goals that are too hard could make you give up. Try creating a series of small challenges for yourself, especially if you haven't been active in a while. Make your goals realistic and try not to change too much at once. For example, you might set a small goal, such as "Call a friend this week to arrange a hike together," to reach a broader goal, "Go on a hike this month." On the other hand, goals that are too easy may not keep you interested or give the results you're working toward. Finding a healthy balance can help you create exercise goals that you stick with and feel good about when you reach them.

Relevant

Setting goals should be enjoyable. Your goals should mean something to you and where you are in your life. Avoid setting goals that someone else wants you to do. Physical activity should be something you like doing or it will not be something you keep up. Try including different activities. If one kind of activity becomes boring, switch to another type.

Time-Oriented

Pick a time frame for completing small goals to help you reach a larger goal. For example, "Work in my garden for an hour every Saturday for the next two months." Setting a start date and deadline for small goals may help you hold yourself accountable. Plus, it easily lets you see if you've achieved your goal. Once you complete a goal, you might want to do it again, making it a healthy habit.

Chapter 30

Health Benefits of Quitting

The health benefits of quitting smoking can help most of the major parts of your body: from your brain to your deoxyribonucleic acid (DNA).

Brain
Broken Addiction Cycle

Quitting smoking can re-wire your brain and help break the cycle of addiction. The large number of nicotine receptors in your brain will return to normal levels after about a month of being quit.

Head and Face
Sharp Hearing

Quitting smoking will keep your hearing sharp. Remember, even mild hearing loss can cause problems.

Better Vision

Stopping smoking will improve your night vision and help preserve your overall vision by stopping the damage that smoking does to your eyes.

This chapter includes text excerpted from "Benefits of Quitting," Smokefree. gov, U.S. Department of Health and Human Services (HHS), May 31, 2018.

Clean Mouth

Nobody likes a dirty mouth. After a few days without cigarettes, your smile will be brighter. Not smoking now will keep your mouth healthy for years to come.

Clear Skin

Quitting smoking is better than antiaging lotion. Quitting can help clear up blemishes and protect your skin from premature aging and wrinkling.

Heart
Decreased Heart Risks

Smoking is the leading cause of heart attacks and heart disease. But many of these heart risks can be reversed simply by quitting smoking. Quitting can lower your blood pressure and heart rate almost immediately. Your risk of a heart attack declines within 24 hours.

Thin Blood

Another effect of quitting smoking is that your blood will become thinner and less likely to form dangerous blood clots. Your heart will also have less work to do, because it will be able to move the blood around your body more easily.

Lower Cholesterol

Quitting smoking will not get rid of the fatty deposits that are already there, but it will lower the levels of cholesterol and fats circulating in your blood, which will help to slow the buildup of new fatty deposits in your arteries.

Lungs
Stop Lung Damage

Scarring of the lungs is not reversible. This is why it is important to quit smoking before you do permanent damage to your lungs. Within two weeks of quitting, you might notice it's easier to walk up the stairs because you may be less short of breath. Don't wait until later; quit today!

Prevent Emphysema

There is no cure for emphysema. But quitting when you are young, before you have done years of damage to the delicate air sacs in your lungs, will help protect you from developing emphysema later.

Return of Cilia

Cilia start to regrow and regain normal function very quickly after you quit smoking. They are one of the first things in your body to heal. People sometimes notice that they cough more than usual when they first quit smoking. This is a sign that the cilia are coming back to life. But you're more likely to fight off colds and infections when you're cilia are working properly.

Deoxyribonucleic Acid
Lower Cancer Risk

Quitting smoking will prevent new DNA damage from happening and can even help repair the damage that has already been done. Quitting smoking immediately is the best way to lower your risk of getting cancer.

Stomach and Hormones
Smaller Belly

Quitting smoking will reduce your belly fat and lower your risk of diabetes. If you already have diabetes, quitting can help you keep your blood sugar levels in check.

Normal Estrogen Levels

If you're a woman, your estrogen levels will gradually return to normal after you quit smoking. And if you hope to have children someday, quitting smoking right now will increase your chances of a healthy pregnancy in the future.

Erectile Dysfunction
Sexual Healing

If you quit smoking now, you can lower your chances of erectile dysfunction (ED) and improve your chances of having a healthy sexual life.

Blood and the Immune System
Normal White Blood Cell Count

When you quit smoking, your body will begin to heal from the injuries that smoking caused. Eventually, your white blood cell counts will return to normal and will no longer be on the defensive.

Proper Healing

Quitting smoking will improve blood flow to wounds, allowing important nutrients, minerals, and oxygen to reach the wound and help it heal properly.

Stronger Immune System

When you quit smoking, your immune system is no longer exposed to tar and nicotine. It will become stronger, and you will be less likely to get sick.

Muscles and Bones
Strong Muscles

Quitting smoking will help increase the availability of oxygen in your blood, and your muscles will become stronger and healthier.

Stronger Bones

Quitting smoking can reduce your risk of fractures, both now and later in life. Keep your bones strong and healthy by quitting now.

Chapter 31

Quitting Smoking and Mood Changes

Chapter Contents

Section 31.1

Anxiety and Smoking

This section includes text excerpted from "Anxiety and Smoking," Smokefree.gov, U.S. Department of Health and Human Services (HHS), June 7, 2017.

It is common for people to think that smoking is a way to calm your nerves and deal with feelings of anxiety. But the truth is, nicotine can cause anxiety symptoms or make them worse.

Nicotine and mood are connected. Researchers know that nicotine in cigarettes affects your brain, including your mood.

Anxiety is feeling frightened, nervous, or panicky. Most people feel anxiety from time to time in difficult situations, but feel better when the situation ends. Anxiety can be a problem if it continues. You might feel sad or depressed and have trouble sleeping or concentrating. Your heart might race or you could feel faint or have stomach problems.

Some regular smokers believe smoking eases anxiety and they report this is a reason they continue to smoke. However, that's because smoking relieves their nicotine withdrawal symptoms. This relief is only temporary. Unless they deal with what's bothering them, anxiety is likely to return and the cycle will continue.

There are many smoke-free ways to handle stress and anxiety. Explore these ways to find one or more that works for you.

Section 31.2

Stress and Smoking

This section contains text excerpted from the following
sources: Text in this section begins with excerpts from "Stress and
Smoking," Smokefree.gov, U.S. Department of Health and Human
Services (HHS), July 21, 2016; Text under the heading "Coping with
Stress without Smoking" is excerpted from "Coping with Stress
without Smoking," Smokefree.gov, U.S. Department of
Health and Human Services (HHS), May 31, 2018.

Some people smoke when they feel stressed. They use smoking as a
way to cope. There are many problems with using cigarettes as a way
to cope with stress or other unpleasant feelings.

- Smoking isn't a long-term stress reliever. In the time it takes
  to smoke a cigarette, you could do something else that's more
  effective—like take a short walk or try a relaxation exercise.

- Smoking doesn't solve the problem that's giving you stress. Your
  stress will return.

- Nicotine addiction causes stress. Cravings for nicotine feel
  stressful because your body begins to go through withdrawal.

Some smokers find it hard to give up cigarettes as a way to cope
with stress. It's important to find healthy ways to handle stress and
take care of yourself without smoking. There are many other ways to
cope with stress that don't involve smoking.

Coping with Stress without Smoking

Stress is a normal part of life—in moderation it can help you reach
your goals, but too much stress creates more problems. Managing
stress is a key part of quitting smoking.

You may have learned to deal with stress by smoking. But there
are ways to handle stress without smoking. Here are a few ideas you
might find helpful. Some of these tips may take practice, but others
you can do right away. Try one or more to learn what works for you.

Relax

Our bodies respond to stress by releasing hormones that increase
your heart rate and raise your blood pressure. Practicing relaxation

techniques, like the ones below, may improve your health and help you handle your stress in positive ways.

Breathe

Take a few slow, deep breaths—in through your nose, out through your mouth. You will feel your body start to relax.

Locate Your Stress

Take a minute to figure out how stress affects your body. Where do you feel tension in your body? Finding ways to reduce that tension will also help your mental stress. A warm bath, a massage, or stretching can help you release built-up tension.

Visualize

Think of a place where you feel safe, comfortable, and relaxed. Picture it as clearly as you can, including imagining what you would feel, hear, and maybe even smell if you were in that relaxing place. Let yourself enjoy being there for a few minutes.

Exercise

Being active sends out natural chemicals that help your mood and reduce your stress. Sometimes a short walk is all it takes to relieve stress. And walking is free!

Talk

You don't have to deal with stress alone. Share your feelings with friends, family, and other important people in your life who are able to support you in staying smoke-free.

Focus

Life can sometimes be overwhelming. Try not to get caught up in worrying about what's next. Instead, try to focus on what is happening now, not what you might have to deal with in the future.

Care

Make an extra effort to take care of yourself. This includes basic things such as eating a balanced diet, drinking lots of water, and getting enough sleep.

Do Good

Doing something nice for others can make your day a little better too. Being caring toward others helps you reduce your own stress.

Decaffeinate

Caffeine can help you stay awake, but it also can make you feel tense, jittery, and stressed. Cutting back or even doing away with caffeine can help reduce your feelings of stress. Switching to herbal tea or even hot water with lemon gives you a chance to enjoy a hot beverage but without the caffeine.

Accept

Life is full of twists and turns. You'll always have some stress in your life. It helps to understand that there will be good days and bad days.

Section 31.3

Depression and Smoking

This section includes text excerpted from "Smoking and Depression," Smokefree.gov, U.S. Department of Health and Human Services (HHS), June 5, 2016.

Smokers are more likely to have depression than nonsmokers. Nobody knows for sure why this is. People who have depression might smoke to feel better. Or smokers might get depression more easily because they smoke. No matter what the cause, there are treatments that work for both depression and smoking.

Mood changes are common after quitting smoking. Some people feel increased sadness. You might be irritable, restless, or feel down or blue. Changes in mood from quitting smoking may be part of withdrawal. Withdrawal is your body getting used to not having nicotine. Mood changes from nicotine withdrawal usually get better in a week or two. If mood changes do not get better in a couple of weeks, you should talk to your doctor. Something else, like depression, could be the reason.

Coping with Depression

Smoking may seem to help you with depression. You might feel better in the moment. But there are many problems with using cigarettes to cope with depression. There are other things you can try to lift your mood:

- **Exercise.** Being physically active can help. Start small and build up over time. This can be hard to do when you're depressed. But your efforts will pay off.

- **Structure your day.** Make a plan to stay busy. Get out of the house if you can.

- **Be with other people.** Many people who are depressed are cut off from other people. Being in touch or talking with others every day can help your mood.

- **Reward yourself.** Do things you enjoy. Even small things add up and help you feel better.

- **Get support.** If you are feeling down after quitting smoking, it may help to talk about this with friends and family. Your doctor also can help.

Section 31.4

Postpartum Depression and Smoking

This section includes text excerpted from "Postpartum Depression and Smoking," Smokefree.gov, U.S. Department of Health and Human Services (HHS), September 6, 2018.

Some women experience depression when they are pregnant or after they give birth—this is called "postpartum depression." About 15 percent of pregnant women and new moms experience it. Any woman can develop postpartum depression, but women who smoke are at a higher risk than women who don't smoke. Knowing the signs and how to get support will help protect you and your baby, and help you stay smoke-free.

Know What to Expect

If you have postpartum depression, you may experience symptoms of depression, like feeling sad, empty, or worthless. You may also:

- Have trouble sleeping when your baby sleeps
- Feel numb or disconnected from your baby
- Have scary or bad thoughts about your baby
- Have thoughts about hurting yourself or your baby
- Feel guilty about not being a good mom or embarrassed that you can't care for your baby

It is common to feel sad or down during your pregnancy or after you give birth. But postpartum depression is more than the "baby blues." People can experience symptoms daily, or most days—which can last at least two weeks, usually longer. It interferes with your everyday life and can stop you from doing things that are important to you.

If you're feeling very down after childbirth, talk to your doctor or other health professional to find out if you might have postpartum depression. If it's not treated, postpartum depression can:

- Affect your ability to care for yourself and your baby
- Disturb your baby's sleeping, eating, and behavior
- Impact your bond with your baby
- Make it harder for you to stay smoke free, especially if you've used smoking to deal with sadness or anxiety in the past

There are treatments for both depression and smoking. And, even though dealing with depression can be hard, women with postpartum depression can still successfully quit smoking.

Section 31.5

Boost Your Mood

This section includes text excerpted from "Boost Your Mood," Smokefree.gov, U.S. Department of Health and Human Services (HHS), July 8, 2014. Reviewed February 2019.

Mood changes are common after quitting smoking. You might be irritable, restless, or feel down or blue. If you have these feelings after quitting smoking, there are things you can do to help lift your mood.

- **Stay active:** Any kind of exercise can help—taking a walk, going to the gym, or joining a team sport are a few you might try. If you need to, start small and build up over time. This can be hard to do when you are feeling down, but making the effort can pay off.

- **Structure your day:** Create a plan to stay busy. Try to get out of the house whenever you can.

- **Do things with other people:** Some people who are feeling down are cut off from others. Having regular contact with other people can help your mood. Try to connect with people regularly, whether it's in person, over the phone, or via text message.

- **Build rewards into your life:** For many people who are sad, rewards and fun activities are missing from their lives. Finding ways to reward yourself can help lift your mood. Even small things, such as reading a magazine or listening to music, can add up and help you feel better.

- **Do what used to be fun:** One of the common signs of depression is not wanting to do activities that used to be fun. It may be hard, but you might try doing fun activities again to help improve your mood. Try making a list of activities or events that you enjoy and plan to do one a day.

- **Get support:** You don't have to deal with negative moods alone. Your friends, family, and others who are important to you can support you. Let them know what they can do to help.

Chapter 32

Relationships and Quitting Smoking

If you've been smoking for a while, people may know you as a smoker. Deciding to quit doesn't mean you won't be able to spend time with your friends and family—even if they still smoke.

Relationships are common reasons why people start or continue to smoke. Maybe you grew up around smokers and it seemed natural for you to smoke, too. Maybe your partner smokes and it's a way that you spend time together. The important people in your life can also be a key to your success in quitting. You might even be surprised by how much help they can be.

Tell Them You're Quitting

Quitting is hard. It's okay to ask for help. If you let people close to you know you're trying to quit, you can tell them what will help you be successful. Be specific—say exactly what you need from them. Here are some suggestions about what to ask of your friends, family, and co-workers:

- **Be supportive:** Some people may be able to quit on their own, but many need help and support. Ask friends, family, and loved ones to be there for you during this time.

This chapter includes text excerpted from "Relationships and Quitting," Smokefree Women, U.S. Department of Health and Human Services (HHS), September 6, 2018.

- **Understand your change in mood:** Mood changes are common after people quit smoking. Remind the people around you that this is temporary. You can say, "The longer I go without smoking, the sooner I'll be back to my old self."

- **Offer distractions:** Most people have cravings when they quit smoking. Your friends and family can distract you until the craving passes. Ask them to help you come up with a list of things to do.

- **Celebrate with you:** Ask friends and family to help you celebrate your smoke-free successes—such as throwing away your cigarettes, setting a quit date, and reaching milestones such as being smoke-free for three days. Even if you slip or relapse, you can still celebrate all the cigarettes you didn't smoke!

- **Quit with you:** Does someone close to you smoke? Quitting can be easier with support, so ask them to quit with you. If they're not ready, ask them not to smoke around you or let you bum a cigarette.

The Power of a Partner

Romantic partners can have a big effect on whether people become smoke-free. People who get support from their partners have an easier time quitting. But some partners may not be supportive, or may try to control the smoke-free process. This can lower a person's confidence and make it harder to quit. Every couple is different, but relationships tend to follow four patterns when it comes to how partners relate to each other about smoking:

- **Involved:** Both partners see smoking as a problem because of health, financial, or other reasons, but also know that trying to quit can be hard. Regardless of who is smoking in the relationship, they try to be supportive and understanding.

- **Accommodating:** Both partners see smoking as okay and make sure there are chances to smoke during their everyday activities. This is true even if only one partner smokes.

- **Disengaged:** Both partners see smoking as a choice each partner makes. Even if they both smoke, they often smoke when they aren't with each other.

- **Conflictual:** One partner criticizes their partner's smoking habits or their decision to continue or quit smoking. This can cause tension, arguments, and hurt feelings. The stress on the relationship might drive someone to slip up and smoke or even go back to smoking regularly. This may especially be true if smoke breaks have been used in the past to find peace. If this sounds like your relationship, ask for help from people who can be more supportive.

Knowing your partner's beliefs about smoking can help you figure out what support you need in order to become smoke-free. It can also help you prepare for how your partner might react to your quit attempt. If you think your partner might not understand your quit attempt or won't want you to quit smoking, there are other ways to find the support you need.

Chapter 33

Relapse: What If You Have a Cigarette?

Slips and Relapses

Many people try quitting smoking often before they finally succeed. Slips are a common part of quitting.

A slip is one or two cigarettes after you quit. For most people, even "just one puff" counts. If you have a slip ("I'll have just one"), it could be harder for you to stay smoke-free. But a slip is different from a relapse. A relapse means going back to smoking regularly.

You can learn from slips. Remind yourself that you've had a temporary setback. You have not failed and you're not back to square one. A slip doesn't make you a smoker again. It also isn't an excuse to relapse and go back to smoking regularly.

Prepare yourself to recover from slips and avoid a relapse. Think about how you will avoid or handle your triggers and deal with cravings.

This chapter contains text excerpted from the following sources: Text under the heading "Slips and Relapses" is excerpted from "Slips and Relapses," Smoke-free.gov, U.S. Department of Health and Human Services (HHS), June 6, 2016; Text under the heading "Tips for Slips" is excerpted from "Tips for Slips," Smokefree.gov, U.S. Department of Health and Human Services (HHS), May 31, 2018.

Tips for Slips

Many smokers slip and smoke one or two cigarettes while they're quitting smoking. You're not alone. Don't use a slip as an excuse to start smoking again.

If you slip, you might try these ways to get back on track:

- Slips are common, so don't be too hard on yourself. A slip doesn't make you a failure or mean your relapsing. It doesn't mean you can't quit for good.

- Feel proud of the time you went without smoking cigarettes. Think about ways you avoid your triggers and beat cravings. Try to use those ways to cope again.

- It's important to restart quitting right away—today or tomorrow at the latest. Don't give up on your goal of no cigarettes at all.

- If quitting forever seems too hard right now, try a text message program to help you prepare to quit in the future. These programs help you build skills for dealing with cravings, triggers, and stressful situations. You can try a Practice Quit (www.smokefree.gov/tools-tips/text-programs/practice-quitting/practice-quit) for a few days or do a week of Daily Challenges (www.smokefree.gov/tools-tips/text-programs/practice-quitting/daily-challenge) without quitting.

- Use nicotine replacement therapy (NRT). You don't need to stop using NRT after you slip and smoke one or two cigarettes. Using NRT increases your chances of staying smoke-free for good.

- Get support. If you slip, talk to family or friends. Ask them for help to stay smoke-free. You don't have to do it alone.

Think about what you learned when you were not smoking. What helped you to stay smoke-free and what caused you to have a slip? What can you do differently now to help yourself be smoke-free again?

Part Four

Tobacco-Related Research

Chapter 34

Link between Maternal Smoking and Child Behavior

Many studies have established that a pregnant woman's smoking raises her child's risk of disruptive behavior disorders and of delinquency in the teen and young adult years, but its behavioral effects in early life have been difficult to trace. Now, however, National Institute on Drug Abuse (NIDA)-funded researchers have revealed associations between a child's in utero exposure to smoking and specific patterns of aberrant behavior as a toddler, at school age, and as a teen. The researchers propose that these patterns form a continuum, united by an underlying theme of disrupted social information processing.

An Early Start to Disruptive Behavior

In an initial study, Dr. Lauren Wakschlag of the Institute for Juvenile Research at the University of Illinois at Chicago (UIC) and her colleagues, Dr. Rolf Loeber of the University of Pittsburgh and Dr. Kate Pickett of The University of York in England, analyzed disruptive

This chapter contains text excerpted from the following sources: Text in this chapter begins with excerpts from "Behavioral Problems Related to Maternal Smoking During Pregnancy Manifest Early in Childhood," National Institute on Drug Abuse (NIDA), June 1, 2008. Reviewed February 2019; Text beginning with the heading "Smoking in Pregnancy" is excerpted from "Smoking in Pregnancy: A Possible Risk for ADHD," Centers for Disease Control and Prevention (CDC), December 5, 2017.

behavior patterns in first graders and subsequent problems that have been associated with later delinquency. Data were derived from the first-grade cohort of the Pittsburgh Youth Study (PYS), a community sample of boys at risk for delinquency who were followed over several decades under the direction of Dr. Loeber.

The researchers concentrated on 448 boys, who were roughly age seven when the PYS study began. One hundred and sixty-six boys in this group had mothers who smoked during pregnancy. These boys developed the antisocial behavior pattern known as oppositional defiant disorder (ODD) at more than double the rate of the rest. Children with ODD demonstrate defiant, disobedient, and hostile behavior toward authority figures that persists for at least six months, and they are touchy, easily angered, and resentful. ODD is often considered a developmental precursor of conduct disorder (CD), a condition in older children and adolescents characterized by persistent antisocial behaviors such as lying, truancy, vandalism, and aggression.

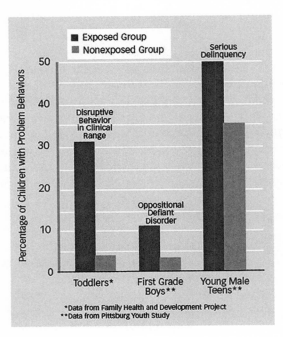

Figure 34.1. *Pathway to Trouble*

Children of mothers who smoked during pregnancy had higher rates of disruptive behavior throughout development. Toddlers were evaluated for disruptive behavior; first-grade boys for oppositional defiant disorder; and young teenage boys for serious delinquency.

Boys whose mothers smoked while pregnant did not have a higher incidence of attention deficit hyperactivity disorder (ADHD) without ODD than the nonexposed boys. However, the incidence of co-occurring ODD and ADHD—a combination that often results in chronic disruptive behavior problems—was nearly twice as high in the exposed group as in the nonexposed group. As the boys entered and traversed their teens, delinquent behavior began earlier and was more severe in the exposed group.

"All the children with ODD in the PYS study were diagnosed in first grade, meaning the disorder developed in the first five or six years of life. This provides evidence of a coherent developmental pathway from prenatal exposure to cigarettes to a subsequent sequence of conduct problems," Dr. Wakschlag says. "While previous research established a link between prenatal exposure to cigarettes and CD in older children, this study is the first to establish connections to ODD and to do so as early as first grade."

Toddlers with Troubles

To look for exposure-related behavioral abnormalities at even younger ages, Dr. Wakschlag's team conducted the Family Health and Development Project (FHDP), in collaboration with colleagues from the University of Illinois, The University of York, the National Institute of Mental Health (NIMH), and the University of Massachusetts—Boston. The researchers recruited 96 expectant mothers, age 18 and older, at several clinics. The women were predominantly white and working class. Along with the women's self-reports, the researchers collected biological data, such as measurements of the nicotine metabolite cotinine in urine samples, to assess fetal exposure to maternal smoking. These measurements, taken three times during pregnancy, indicated that 47 percent of the women smoked throughout their pregnancies. Ninety-three infants and their mothers completed the study's developmental phase, which lasted until the babies were 24 months old.

The babies were evaluated every 6 months. At the 12-, 18-, and 24-month evaluations, each mother filled out the Infant–Toddler Social Emotional Assessment (ITSEA). During 20-minute laboratory observations of the toddlers and their mothers interacting at 24 months, the researchers rated specific components of the toddlers' behavior using codes from the Disruptive Behavior Diagnostic Observation Schedule.

The results indicated that toddlers whose mothers had smoked during pregnancy demonstrated a high and escalating pattern of disruptive behavior from 12 to 24 months, whereas nonexposed toddlers exhibited a relatively stable pattern. A mother's smoking during pregnancy increased the likelihood of the observed atypical trajectory of behavior independent of several associated risk factors, including parental antisocial behavior, quality of parenting, and postnatal exposure to tobacco smoke. At 24 months, toddlers whose mothers had smoked while pregnant were more than 11 times as likely as nonexposed peers to exhibit clinically significant patterns of disruptive behavior, shown on the ITSEA.

To more precisely determine the nature of the boys' behavior problems, the researchers examined four components of disruptive behavior, each of which is considered a precursor to disruptive behavior patterns seen at later ages:

- Aggressive/destructive behavior, including threatening, hitting, and throwing or smashing toys

- Dysregulated negative effect, characterized by persistent, uncontrolled outbursts of anger with loud yelling, intense crying, and temper tantrums

- Stubborn defiance, marked by obstructive behavior that persists after the mother has increased expressions of support for her child and has tried several strategies to change her child's behavior and

- Low social competence, where the child misses social cues and exhibits low social interest or concern

These four behaviors, while viewed as normal in toddlers, are considered precursors to clinical problems if they are severe or pervasive.

The children whose mothers had smoked during pregnancy displayed lower social competence than other children and significantly higher levels of aggressive/destructive behavior and stubborn defiance. They were not more likely to exhibit dysregulated negative effect.

"Dr. Wakschlag has teased out some components of disruptive behavior problems when they first emerge between 18 and 24 months of age," says Dr. Nicolette Borek of NIDA's Division of Clinical Neuroscience and Behavioral Research (DCNBR). "This gives us a way to identify at-risk children early and raises interesting questions about the role of brain development in later-stage behavioral issues."

On to Adolescence

Dr. Wakschlag and colleagues have hypothesized that the resistant, hostile, and unresponsive patterns of behavior demonstrated in FHDP, PYS, and similar studies may reflect disruptions in social information processing that resulted from prenatal exposure to cigarette smoke. To test this hypothesis, the team is conducting the NIDA-funded East Boston Family Study (EBFS), which includes 272 adolescents and is a followup to the Maternal–Infant Smoking Study of East Boston (MISSEB). Dr. Wakschlag and her colleagues are also examining the influence of genetic makeup on exposure-related disruptive behavior among these young people. The researchers are using maternal exposure data originally collected by MISSEB but applying more sophisticated methods to measure prenatal exposure to cigarette smoke. These new techniques, which combine maternal self-report and biological data, were developed from FHDP-derived data by Dr. Vanja Dukic at the University of Chicago in collaboration with Dr. Neal Benowitz of the University of California, San Francisco, and Dr. Wakschlag.

"Maternal self-reports are affected by memory lapses and social pressure not to smoke, and biological methods can be inaccurate because the smoke-derived chemicals have a short half-life and rates of metabolism differ among individuals," says Dr. Wakschlag. "In addition, we know that smoking levels fluctuate throughout a pregnancy. The new technique incorporates the unique information from both of these methods to provide a more precise estimate of prenatal exposure to cigarettes."

Smoking in Pregnancy

Attention deficit hyperactivity disorder (ADHD) is one of the most common neurobehavioral disorders of childhood. About 1 in 10 children 4 to 17 years of age in the United States have been diagnosed with ADHD, based on parent reports. The Centers for Disease Control and Prevention (CDC) is working to understand the risk factors for ADHD, including smoking during pregnancy, so that more can be done to prevent the disorder.

Attention Deficit Hyperactivity Disorder and Smoking in Pregnancy

The year 2014 marked the 50th anniversary of the first Surgeon General's report on the risks associated with smoking. For the 2014 report, researchers at the CDC reviewed the evidence for any associations between prenatal smoking and a set of childhood mental disorders, including ADHD, oppositional defiant disorder (ODD), conduct disorder (CD), anxiety disorders, depression, Tourette syndrome (TS), schizophrenia, and intellectual disability (ID).

The evidence reviewed suggests that:

- Prenatal exposure to smoking is associated with disruptive behavioral disorders in children, including ADHD, ODD, and CD.

- The ways that prenatal exposure to smoking is linked to these behavioral outcomes is not yet understood.

- There is limited data and the results were mixed for the relationship between prenatal exposure to smoking and other neurobehavioral disorders (anxiety disorders, depression, TS, schizophrenia, and ID).

We know that there is a link, but we do not yet know how smoking during pregnancy is related to ADHD in childhood. Many of the studies relied on retrospective reports of smoking and many studies had small sample sizes, especially for less common conditions such as Tuberous sclerosis (TS) and schizophrenia. There are also many other risks that are often present along with maternal smoking that might explain the link with neurobehavioral disorders. Thus, more research studies are necessary to determine how prenatal exposure to smoking is related to developmental outcomes in children.

This evidence adds to the concerns about the negative effects of smoking that were already included in the previous Surgeon General's reports. In addition to the well-known risks to the health of the person who is smoking, there was consistent evidence that the toxins in tobacco from maternal smoking have negative effects on reproductive and developmental outcomes—for example, premature birth which can lead to death, disability, and disease among newborns.

What Can Healthcare Providers Do?

There are many risks from smoking before and during pregnancy, so it is especially important that women do not smoke during their reproductive years. The CDC has gathered many resources that healthcare providers can use to help women quit smoking before or during pregnancy and uses a public-health approach to eliminate tobacco use and exposure during the reproductive years.

What Can You Do If You Are Concerned about Smoking and Pregnancy?

Quitting smoking can be hard, but it is one of the best ways a woman can protect herself and her baby's health. For support with quitting, including free quit coaching, a free quit plan, free educational materials, and referrals to local resources, please call 800-QUIT-NOW (800-784-8669); Toll-Free TTY 800-332-8615.

What Is Attention Deficit Hyperactivity Disorder?

Many children have trouble focusing and behaving at one time or another. These symptoms may reach the level of a disorder if they continue over time and cause difficulty at school, at home, or with friends. ADHD may put a child at risk for other concerns and conditions. The impact of ADHD can continue into adulthood.

Children with ADHD:

- Often have other behavioral disorders
- May have problems making or keeping friends
- May show risky behavior that can lead to injury
- May have problems succeeding in school

What Is the Centers for Disease Control and Prevention Doing about Attention Deficit Hyperactivity Disorder?

The CDC is gathering information about ADHD and how it impacts children and families, so that researchers can do more about prevention and treatment. Currently, research studies are underway to:

- Understand the number of children with ADHD and how they are being treated
- Follow children with symptoms of ADHD over time, to learn about changes in the symptoms and their effects and changes in treatment; gather information about children's mental, emotional, and behavioral disorders, including ADHD, from multiple sites around the United States, and
- Gather evidence on factors that increase the likelihood that a child will develop ADHD

Chapter 35

Are Health Risks Different for Flavored Tobacco Products?

Vanilla, cherry, chocolate, grape, apple, cotton candy, banana split, bubble gum. These could be flavors from your local ice-cream shop. They (and many others) are also flavors used in many e-cigarette fluids (often called "e-liquid" or "e-juice") as well as popular tobacco products such as hookahs.

Something that comes in a variety of fun-sounding flavors may seem like it should be okay to consume. How dangerous could they really be?

This chapter contains text excerpted from the following sources: Text in this chapter begins with excerpts from "How Does Addiction Taste? Flavored Tobacco Products and E-Cigs Are No Safer Than Unflavored," National Institute on Drug Abuse (NIDA) for Teens, October 23, 2015. Reviewed February 2019; Text under the heading "The U.S. Food and Drug Administration Unveils New Steps to Protect Youth by Preventing Access to Flavored Tobacco" is excerpted from "FDA Unveils New Steps to Protect Youth by Preventing Access to Flavored Tobacco Products, Announces Plans to Ban Menthol in Cigarettes and Cigars," U.S. Food and Drug Administration (FDA), November 15, 2018; Text under the heading "Econometric Research on Regulating Menthol Cigarettes and Smoking Cessation" is excerpted from "Econometric Research on Regulating Menthol Cigarettes and Smoking Cessation," U.S. Food and Drug Administration (FDA), January 10, 2018.

That depends. Many of the e-cigarette companies are adding these flavors to liquid nicotine to make them appealing to young people. Will you fall for that? Liquid nicotine with flavors is just as unhealthy as its plain cousin. But if a flavored product says it contains no nicotine, is it safe? We simply don't know. It's still made with chemicals, and why risk inhaling them into your lungs?

Known—and Unknown—Risks

How many teens are trying the flavored nicotine e-cigs? According to a new study from the Centers for Disease Control and Prevention (CDC) and the U.S. Food and Drug Administration (FDA), approximately 70 percent of middle-school and high-school students who have used a tobacco product or an e-cigarette in the past 30 days have used at least one flavored product during that time. E-cigarettes were the most commonly used, followed by hookahs and cigars.

Even though the jury is still out on how harmful e-cigarettes are, they have definite health consequences, whatever the taste. For starters, research has found they may be as addictive as regular cigarettes. Another study found that teens who use e-cigarettes are more likely than others to smoke regular cigarettes and use other tobacco products when they're older.

Also, the aerosol (vapor) created by e-cigarettes contains chemicals at concentrations (amounts) that are toxic. We do know that the liquid nicotine in e-cigarettes also can cause nicotine poisoning if a person accidentally drinks, sniffs, or touches it.

Not So Tasty

E-cigarette manufacturers don't usually state the levels of specific chemicals they include in the liquid nicotine. This means that the flavors in e-cigarettes are like a smokeable version of "mystery meat."

What about flavored hookah products? With or without added flavors, hookahs are nothing more than inhaled tobacco smoke—same as cigarettes. Again, companies might use the flavoring to entice you, but it's up to you whether you take the bait.

Several makers of e-cigarettes are branding their flavors so that one brand has a different name for, say, its cherry-flavored product than a different brand does. But whatever the name, the risks are the same. Flavors and catchy brand names are just a gimmick to sell you the same old addictive product.

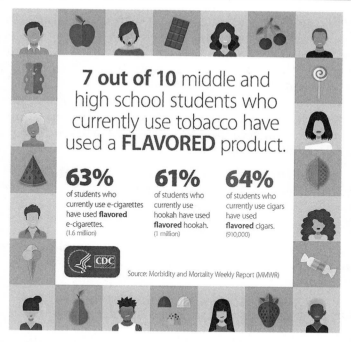

Figure 35.1. *High-School Students Use of Tobacco* (Source: Morbidity and Mortality Weekly Report (MMWR).)

The U.S. Food and Drug Administration Unveils New Steps to Protect Youth by Preventing Access to Flavored Tobacco

When Commissioner Gottlieb announced the FDA's Comprehensive Plan for Tobacco and Nicotine Regulation in 2017, the deadlines for certain deemed products were extended, in part, to allow for the potential benefits of innovative new products to be studied more thoroughly. This deadline was extended under the condition that the youth e-cigarette use numbers would not change for the worse. At the time, the FDA was seeing change in the right direction as high school current e-cigarette use had decreased from its peak of 16.0 percent in 2015 to 11.3 percent in 2016 and held steady in 2017.

Unfortunately, the situation changed drastically in 2018. According to new findings from the 2018 National Youth Tobacco Survey (NYTS), there has been a dramatic increase in youth use of e-cigarettes and other electronic nicotine delivery systems (ENDS): From 2017 to

2018, there was a 78 percent increase in current e-cigarette use among high-school students and a 48 percent increase among middle-school students. This startling surge in youth e-cigarette use is beyond troubling and the FDA will not sit idly by while facing the threat of a whole generation becoming addicted to these products.

Therefore, the FDA is outlining a new policy framework to address what appear to be the central problems—youth appeal and youth access to flavored tobacco products. The FDA will be taking steps on the following product categories:

- Flavored ENDS products (other than tobacco, mint, and menthol flavors or nonflavored products) that are not sold in an age-restricted, in-person location

- Flavored ENDS products (other than tobacco, mint, and menthol flavors or nonflavored products) that are sold online without heightened age-verification processes

- Flavored cigars

- ENDS products that are marketed to kids, and

- Menthol in combustible tobacco products, including cigarettes and cigars

The FDA intends to provide additional details soon, including what the agency might consider an "age-restricted" location, what it might consider "heightened" age-verification online, and timelines for when the FDA intends to implement these policies. Commissioner Gottlieb has stated that he hopes that in the next 90 days, manufacturers will choose to remove flavored ENDS products from stores where kids can access them and from online sites that do not have sufficient age-verification procedures.

The FDA also intends to move forward with proposed rulemaking to ban menthol in cigarettes and all flavors, including menthol, in cigars, on an expedited timeline.

The framework reflects the FDA's aim of providing the right balance between closing the on-ramp for kids to become addicted to nicotine while maintaining access to potentially less harmful forms of nicotine delivery for adult smokers seeking to transition away from combustible tobacco products.

This policy framework is an important step toward reversing the epidemic that is underway. The FDA will continue to base its actions on the best available science and will continue to take aggressive actions to protect our youth through a wide range of prevention and enforcement actions.

Econometric Research on Regulating Menthol Cigarettes and Smoking Cessation

An estimated 18.1 million United States adults, or about 30 percent of adult smokers, smoke menthol cigarettes. Some evidence suggests that the availability of menthol cigarettes increases the number of smokers, partly by reducing cessation. The goal of this project is to use econometric methods to study relationships between menthol cigarette use and smoking cessation. Specific aims are to:

1. Study the relationship between menthol cigarette use and smoking cessation

2. Investigate the economics of consumer choices about menthol cigarettes and smoking cessation

3. Investigate the economics of consumer choices about menthol cigarettes over the life course

To accomplish Aim 1, researchers will analyze data from the 2003, 2006–2007, and 2010–2011 cycles of the Tobacco Use Supplements to the Current Population Survey (TUS-CPS) and data from the Simmons National Consumer Survey (NCS) from 1995 onward, which together provide data on nearly 80,000 current and past-year smokers. To accomplish Aim 2, researchers will use geocode information to merge the individual-level data from the TUS-CPS and NCS with market-level demand influences (e.g., price and advertising), and will use information on respondents' magazine-reading habits to create individual-level measures of potential exposure to cigarette advertisements. This will enable researchers to investigate the extent to which smokers perceive menthol and nonmenthol cigarettes to be close substitutes and evaluate the role of advertising in consumer choices. To accomplish Aim 3, researchers will use TUS-CPS data (e.g., ever use of menthol cigarettes; whether smokers used menthol cigarettes all or nearly all, most, half, or less than half the years they smoked) to build approximate lifetime histories of menthol use. By evaluating the impact of menthol use on smoking cessation, this research may inform regulatory actions related to menthol.

Chapter 36

Combination of Genes and Prenatal Exposure to Smoking Increases Disruptive Behavior Risk

A study funded by the National Institute on Drug Abuse (NIDA), a component of the National Institutes of Health (NIH), shows that prenatal exposure to smoking combined with a specific genetic variant places children at greatest risk for behavioral problems. Many studies have established that there is an increased risk of aggressive behavior in children exposed to cigarette smoke before birth, a significant problem given that many women still smoke during pregnancies. According to the National Survey on Drug Use and Health (NSDUH), in 2006–2007 slightly more than 16 percent of pregnant women aged 15 to 44 (426,000) were current cigarette smokers.

Study Shows Different Gene Variants Influence Risk for Girls and Boys

A team of researchers led by the Institute for Juvenile Research, University of Illinois at Chicago (UIC), identified a long-lasting

---

This chapter includes text excerpted from "Combination of Genes and Prenatal Exposure to Smoking Increases Teens' Risk of Disruptive Behavior," National Institutes of Health (NIH), March 4, 2009. Reviewed February 2019.

influence on behavior of the monoamine oxidase A (MAOA) gene variant following tobacco exposure before birth. MAOA is an enzyme which regulates key neurotransmitters, or chemical messengers in the brain. Strikingly, the genetic variant that confers this increased risk differs between boys and girls.

"These findings illuminate how the interaction between genes and the environment can mold behavioral patterns very early in development," said NIDA Director Dr. Nora Volkow. "This research provides a foundation for studies of the impact of these interactions on brain development during pregnancy."

The researchers studied 176 non-Hispanic white youth, whose average age was 15, and their biologic mothers. In contrast with previous studies of prenatal smoking that have measured exposure based on the mother's recollection of past smoking, this study obtained accurate measurements of smoking behaviors during the actual pregnancy.

In boys, with the low-activity MAOA (MAOA-L) gene variant, exposure to prenatal smoking was associated with increased disruptive social interactions, including aggressive behaviors and serious rule-violating. "Low activity" means that the gene produces less of its product, the enzyme monoamine oxidase A. In contrast, exposure to prenatal smoking was associated with increased disruptive behavior in girls who had the high-activity MAOA (MAOA-H) variant. For both boys and girls, the more their mothers had smoked during pregnancy, the higher the risk of disruptive behavior.

Additionally, on computerized tasks, girls with both the MAOA-H variant and prenatal exposure to smoking had a greater tendency to perceive anger in a range of facial expressions, a tendency that researchers term "hostile attribution bias." This effect was not seen among boys.

"The tendency to overperceive anger suggests the possibility that the combination of prenatal tobacco exposure and the MAOA risk variant affects the brain's processing of emotional cues," said the study's principal investigator, Dr. Lauren S. Wakschlag, associate professor of psychiatry at the Institute for Juvenile Research, University of Illinois at Chicago. "Individuals with a greater tendency to perceive hostility in others are more likely to respond aggressively. These findings provide us with clues to the possible mechanism by which prenatal exposure may exert its effects on brain and behavior. Clearly, close attention to sex differences in these patterns will be critical for future studies."

Dr. Wakschlag led the study in collaboration with colleagues from the Institute for Juvenile Research as well as researchers from the

National Institute of Mental Health (NIMH) (which is also a component of the National Institutes of Health); the University of Chicago; Harvard University Medical School (HMS); and the University of York, England.

Chapter 37

Research on Tobacco Use

Chapter Contents

Section 37.1

Global Tobacco Research

This chapter includes text excerpted from "Global Tobacco Research,"
Research Portfolio Online Reporting Tools (RePORT), National
Institutes of Health (NIH), June 30, 2018.

Yesterday

- By the 1960s, evidence of the adverse health effects of cigarette smoking was undeniable. By 1980, evidence of the burden of tobacco-related illnesses in low- and middle-income countries (LMICs) began to emerge.

- Since 1980, LMICs have faced a rapidly escalating epidemic of tobacco use due to many factors, including globalization, urbanization, and demographic shifts in populations.

- Historically, research and prevention efforts were primarily focused on developed countries, where the tobacco epidemic was well established. There were few coordinated global efforts to address tobacco-control research and practice.

Today

- Today, science-based research confirms that all tobacco products, including cigars, smokeless tobacco, bidis (a thin, hand-rolled unfiltered cigarette made with sun-dried tobacco), and waterpipes are hazardous. Additionally, the addictive nature of tobacco and the harmful effects of exposure to secondhand smoke on nonsmokers are clearly established.

- Tobacco use is one of the gravest public-health challenges the world has ever confronted. More than five million people die of tobacco-caused illness each year—an average of one person every six seconds—and tobacco use accounts for 10 percent of adult mortality globally. Because tobacco-related disease strikes people in the prime of their working lives, it also negatively impacts economic development.

- Tobacco use is increasing in LMICs, which will bear the brunt of the tobacco epidemic in the 21st century. Without a significant shift in worldwide prevalence patterns, smoking is projected to cause roughly eight million deaths annually by 2030; notably, more than 80 percent of these deaths will occur in LMICs.

- Women now comprise 20 percent of the world's more than one billion smokers. In addition to tobacco-related cancer, heart disease, and respiratory disease that affect both genders, tobacco also causes additional female-specific cancers and negatively affects pregnancy and reproductive health. Understanding and controlling the tobacco epidemic among women is a critical part of the tobacco-control research agenda.

- The vast majority of smokers begin using tobacco products well before 18 years of age. Adolescent smoking is a major concern in both the developed world and LMICs. Therefore, the development of effective youth interventions is essential in controlling the tobacco epidemic.

- Successful implementation of tobacco-control strategies is informed by scientific evidence. However, a major obstacle to tobacco control in LMICs is the lack of capacity for local surveillance, research, and evaluation of interventions. In addition, globalization of the tobacco epidemic requires coordinated global-health efforts and collaboration between researchers from high-income countries and their counterparts in LMICs.

- The International Tobacco and Health Research and Capacity Building Program (www.fic.nih.gov/programs/research_grants/tobacco) addresses the critical role of research and local research capacity in reducing the burden of tobacco consumption in LMICs and the need to generate a solid evidence base that can inform effective local tobacco control strategies and policies. The program supports epidemiological and behavioral research, as well as prevention, treatment, communications, health services, and policy research.

Tomorrow

By training the next generation of scientists, forging research partnerships, and creating a global network of scientists, the National Institutes of Health (NIH) will continue to make an important contribution to global tobacco-control efforts. NIH will support research that addresses key priority areas in tobacco control and build research capacity in LMICs. For example:

- Researchers at the University of Alabama at Birmingham are collaborating with Brazilian scientists to develop a Network for Tobacco Control among Women in order to promote gender-relevant tobacco-control efforts among Brazilian women. The "Network" is intended to reduce tobacco use and exposure to environmental tobacco smoke among Brazilian women, and to develop a cadre of well-trained researchers in tobacco control.

- In Delhi, India, researchers are testing the efficacy and cost-effectiveness of a comprehensive, community-based behavioral intervention for tobacco cessation among disadvantaged youth living in low-income communities. Studies such as this can also inform efforts to curb adolescent smoking in the United States.

- Researchers in California are working with Turkish scientists to design and evaluate a text-messaging-based smoking intervention that harnesses mobile technologies adopted by adult smokers in Turkey. Given the widespread use of mobile phones, evidence from studies that utilize such technologies to promote behavior change can be utilized in the design of interventions in the United States and other countries around the world.

Section 37.2

Setting the Stage for the Next Decade of Tobacco-Control Research

This section includes text excerpted from "Setting the Stage for the Next Decade of Tobacco Control Research," National Cancer Institute (NCI), July 5, 2016.

One of the biggest dangers we face in public health is prematurely declaring victory over a major health threat. Nowhere is that more true than in the case of tobacco use.

Smoking rates have dropped precipitously over the past several decades—a monumental achievement that resulted from implementing evidence-based policies and programs, such as increasing tobacco taxes, implementing comprehensive smoke-free laws, and efforts to help smokers quit.

But, with 40 million Americans still smoking, there is still a lot to be done. The data speak for themselves: tobacco use causes at least 13 different types of cancer and results in one-third of cancer deaths. It also causes heart disease, stroke, and serious lung disease. Indeed, smoking harms nearly every organ in the body.

In addition to the cost in lives and suffering, tobacco use also exacts a staggering financial toll: $170 billion in direct medical costs and $156 billion in lost productivity in the United States annually. Furthermore, proliferation and use of emerging tobacco products, such as e-cigarettes and hookahs, are growing. In 2014, e-cigarettes were the most commonly used tobacco product among middle- (4%) and high- (13%) school students.

In 2015, in recognition of tobacco's role as the leading cause of preventable cancer deaths, National Cancer Institute (NCI) Acting Director Doug Lowy, M.D., convened a working group of prominent tobacco-control experts to formulate research priorities to help to shape NCI's future tobacco-control research portfolio.

Dr. Lowy specifically charged the working group with identifying and highlighting research priorities that NCI's Tobacco-Control Research Branch (TCRB) is particularly well-positioned to address over the next decade, with an emphasis on areas with the greatest potential to help end the tobacco epidemic and the diseases it causes.

To ensure that it would have a comprehensive view of the current tobacco control landscape, the working group heard from leaders from other federal agencies and nonprofit groups working in tobacco control to better understand those groups' priority research areas and efforts.

In spring 2016, the working group completed its report and presented it to NCI's Board of Scientific Advisors. As the report notes, "the elimination of tobacco use and its unequaled harms is now within our grasp," but "achieving this goal will require significant scientific progress, including advances led by the NCI."

Further, the report says that the research community must consider key factors that help maintain continued tobacco use, including a changing population of tobacco users, evolving patterns of tobacco use, new and different tobacco products, and a complex and rapidly evolving policy environment.

The working group's report provides a comprehensive guide that helps direct NCI research efforts moving forward. For example, it recommends a focus on improving the efficacy of new and existing smoking prevention and cessation interventions, including making them more accessible, whether that's via an app for a smartphone or in nontraditional care settings such as emergency departments and mental-health and substance-abuse treatment centers.

Another priority area identified by the working group is reaching out to vulnerable populations. The working group specifically identified adolescents and young adults as an important target population. And recent studies show why. In 2016, for example, an NCI-supported study showed that high-school juniors and seniors who had never smoked conventional cigarettes, but had begun using e-cigarettes, were much more likely to begin smoking conventional cigarettes during the following 16 months than students who had not used e-cigarettes.

The working group also identified research to better address racial and financial disparities as a top priority, because the evidence clearly demonstrates that tobacco's harms are not equal across all populations. For instance, people who only have a high-school education or less smoke at more than twice the rate of those with a college degree or more. Smoking rates are also substantially higher among people whose income is below the poverty level, in certain racial/ethnic groups, and with mental-health or substance-abuse conditions.

Existing NCI-supported research is well positioned to help jump-start NCI's response to these recommendations. These research efforts include the State and Community Tobacco Control (SCTC) research initiative, the Adolescent Brain Cognitive Development Study (in collaboration with many federal partners), and a new funding opportunity aimed at promoting smoking cessation among low-income populations using scalable interventions.

In the mid-1960s, approximately 50 percent of all men and 33 percent of women in the United States smoked. In 2016 those figures are approximately 17 percent and 14 percent. That's a tremendous achievement by any measure. But we can and must do more.

NCI is committed to working with many partners to help end the tobacco-use epidemic. These recommendations—and the ability of NCI to follow through on them—is an important step in that direction.

Section 37.3

Adolescents Who Wouldn't Have Smoked May Be Drawn to E-Cigarettes

This section includes text excerpted from "Adolescents Who Wouldn't Have Smoked May Be Drawn to E-Cigarettes," National Cancer Institute (NCI), August 11, 2016.

Some adolescents who otherwise would never have smoked are using e-cigarettes, according to a study published July 11, 2016, in the journal *Pediatrics*. The findings suggest that adolescents are not just using e-cigarettes as a substitute for conventional cigarettes but that e-cigarettes are attracting new users to tobacco products.

E-cigarettes are electronic devices that create an aerosol by heating a liquid solution that often contains nicotine and flavorings, as well as other chemicals. They allow users to simulate smoking conventional cigarettes by inhaling the aerosol, which mimics combustible cigarette smoking. The U.S. Food and Drug Administration (FDA) finalized a rule extending its regulation of tobacco products to include e-cigarettes. The rule went into effect in August 2016.

The new findings come from the Children's Health Study (CHS), which includes periodic surveys that assess smoking behavior among high-school juniors and seniors in southern California.

Although use of conventional cigarettes in this population has declined steadily over the past two decades, the study showed an increase in the percentage of students who reported using cigarettes and/or e-cigarettes in 2014 (the most recent year data were collected) compared with the percentage of students using cigarettes in 2004. This increase in overall use of tobacco products threatens to reverse the downward trend of smoking among adolescents in southern California, the researchers suggested.

Two Decades of Declining Tobacco Use

Researchers from the University of Southern California (USC) Keck School of Medicine have been conducting the Children's Health Study, a longitudinal prospective cohort study of the effects of air pollution on schoolchildren in southern California, since 1992. As part of the study, they have surveyed eleventh- and twelfth-grade students at high schools in southern California about their use of cigarettes at different time periods, starting in 1994 and most recently in 2014. A total of 5,490 students have reported their history of cigarette use in individually administered questionnaires.

The study has documented a two-decades-long decline in cigarette smoking among adolescents in the area. In 1994, 19.1 percent of twelfth-grade students reported current cigarette use. By 2004, the prevalence of current cigarette use among twelfth graders had dropped to 9 percent. Data collected in 2014 show that the drop in current cigarette use has continued, with only 7.8 percent of twelfth-grade students reporting current cigarette smoking.

E-Cigarettes Reverse the Trend

However, the researchers found that use of cigarettes and e-cigarettes combined has increased among high-school juniors and seniors since the 2007 introduction of e-cigarettes in the United States. In 2014, 13.7 percent of twelfth-grade students in the study reported current use of cigarettes and/or e-cigarettes, a prevalence only slightly less than the current use of cigarettes that was reported in 2001 (14.7%) and much higher than current cigarette smoking reported in 2004 (9%). This increased prevalence suggests that e-cigarettes are being used by adolescents who likely would not have initiated smoking conventional cigarettes, they explained.

"The combined e-cigarette and cigarette use in 2014 far exceeded what we would have expected if teens were simply substituting cigarettes with e-cigarettes. The data suggest that at least some of the teens who are [using e-cigarettes] would not have smoked cigarettes," lead author Jessica L. Barrington-Trimis, Ph.D., of the University of Southern California, said in a news release.

Known and Unknown Harms

Although e-cigarettes do not carry the same dangers as conventional tobacco products, e-cigarette aerosol contains a number of harmful and potentially harmful chemicals as well as possible dangers that are not yet understood, the authors noted.

Nicotine is often included in the liquid used in e-cigarettes. The risk of nicotine addiction associated with e-cigarette use makes it likely that e-cigarettes could serve as a gateway device to conventional tobacco smoking, the authors contend, and some research already supports this hypothesis.

"The use of e-cigarettes by youth is a serious concern," explained Elizabeth Ginexi, Ph.D., of National Cancer Institute's (NCI) Tobacco

Control Research Branch (TCRB), who was not involved in the study. "Research has revealed that nicotine exposure during adolescence may have long-term effects on the developing brain, including putting the adolescent at increased risk for addiction later in life."

For adolescents who have never used other tobacco products, nicotine experimentation with e-cigarettes could lead to nicotine addiction, Dr. Barrington-Trimis said.

And other studies have suggested that e-cigarette use may normalize behaviors associated with cigarette smoking, such as seeing clouds of smoke in public or gathering in a common place to "vape," which may make smoking initiation more likely.

Potential Solutions

In an accompanying editorial, Jonathan P. Winickoff, M.D., M.P.H., of the Massachusetts General Hospital Tobacco Research and Treatment Center, and Sarah E. Winickoff, a Brookline, MA, public-school student, proposed a number of strategies to address e-cigarette use by high-school students. They suggest implementing school-based initiatives that emphasize the potential harms of e-cigarettes, including addiction, inhalation of toxins, and increased likelihood of smoking.

They also recommended that health professionals screen their adolescent patients for e-cigarette use and educate them about the effects of nicotine on the developing brain. And they called for policy makers to extend current prohibitions on traditional tobacco sales to adolescents and young adults to the purchase of e-cigarettes, as well as prohibiting use of e-cigarettes anywhere cigarette smoking is currently prohibited.

"Moving beyond the documentation of e-cigarette use toward vigorously pursuing possible solutions (to prevent such use) could help many adolescents escape addiction to nicotine," these experts wrote.

Implications for Public Health

A major strength of this study is its use of more than 20 years of data collected from five cohorts in the same southern California communities, Dr. Ginexi said. The cohorts were drawn from entire classrooms of students, giving the researchers large samples of ethnically diverse students who were representative of youth in the region during this time period.

The authors noted that more research is needed to determine whether current users of e-cigarettes in the study population who have not used traditional cigarettes continue to use e-cigarettes exclusively, go on to smoke combustible tobacco products, become users of both, or stop using tobacco products altogether.

"Although the adverse health effects of e-cigarettes may be less than those of cigarettes, the long-term consequences of e-cigarette use are not known because these products have been on the market for less than a decade," they wrote.

Section 37.4

Stronger Nicotine Dependency Associated with Higher Risk of Lung Cancer

This section includes text excerpted from "Study Finds Stronger Nicotine Dependency Associated with Higher Risk of Lung Cancer," National Cancer Institute (NCI), June 19, 2014. Reviewed February 2019.

People who are highly addicted to nicotine—those who smoke their first cigarette within five minutes after awakening—are at higher risk of developing lung cancer than those who wait for an hour or more to smoke. Researchers at National Cancer Institute (NCI) found this simple measure of nicotine dependency improved lung-cancer risk prediction beyond standard smoking measures, such as cigarettes per day, age, gender, history of chronic obstructive pulmonary disease and other lung cancer risk factors. The results of this study were published online in the *Journal of the National Cancer Institute*, June 19, 2014.

Neil E. Caporaso, M.D., senior investigator, and Fangyi Gu, M.Med., Sc.D., postdoctoral fellow, Genetic Epidemiology Branch, Division of Cancer Epidemiology and Genetics (DCEG), NCI, and their colleagues, analyzed data from about 1,800 lung cancer patients and 1,400 people without cancer from five cities in the Lombardy region in Italy. The researchers observed that people who had smoked their first cigarette within five minutes of waking up had more than three times the risk of lung cancer compared with those who waited longer than an hour before smoking their first cigarette of the day, after taking into account other smoking characteristics, age, and gender. The increased risks were observed for each category of cigarette per day, as well as each category of smoking history. Assessing time to a first cigarette may help clinicians quickly assess lung-cancer risk. The findings underscore the need for even light smokers to quit, because even light smokers who are or who become dependent smokers, can be at substantial risk for developing lung cancer. The researchers plan to investigate the effect of nicotine addiction on risk for other health outcomes in future studies.

Section 37.5

What Research Is Being Done on Tobacco Use?

This section contains text excerpted from the following sources: Text in this section begins with excerpts from "Tobacco, Nicotine, and E-Cigarettes," National Institute on Drug Abuse (NIDA), January 2018; Text under the heading "Research Priorities" is excerpted from "Research Priorities," U.S. Food and Drug Administration (FDA), October 28, 2018.

New scientific developments can improve our understanding of nicotine addiction and spur the development of better prevention and treatment strategies.

Genetics and Epigenetics

An estimated 50 to 75 percent of the risk for nicotine addiction is attributable to genetic factors. A cluster of genes (CHRNA5-CHRNA3-CHRNB4) on chromosome 15 that encode the $\alpha5$, $\alpha3$, and $\beta4$ protein subunits that make up the brain receptor for nicotine are particularly implicated in nicotine dependence and smoking among people of European descent. Variation in the *CHRNA5* gene influences the effectiveness of combination nicotine replacement therapy (NRT), but not varenicline. Other research has identified genes that influence nicotine metabolism, and therefore, the number of cigarettes smoked, responsiveness to medication, and chances of successfully quitting.

For example, the therapeutic response to varenicline is associated with variants for the *CHRNB2*, *CHRNA5*, and *CHRNA4* genes, while bupropion-related cessation is linked with variation in genes that affect nicotine metabolism.

Smoking can also lead to persistent changes in gene expression (epigenetic changes), which may contribute to associated medical consequences over the long term, even following cessation. Epigenetic changes may serve as a potential biomarker for prenatal tobacco smoke exposure. Researchers found tobacco-specific changes at 26 sites on the epigenome, and this pattern predicted prenatal exposure with 81 percent accuracy. A large-scale meta-analysis of data on epigenetic changes associated with prenatal exposure to cigarette smoke also identified many epigenetic changes that persisted into later childhood. More research is needed to understand the long-term health impacts of these changes.

Neuroimaging

Cutting-edge neuroimaging technologies have identified brain changes associated with nicotine dependence and smoking. Using functional magnetic resonance imaging (fMRI), scientists can visualize smokers' brains as they respond to cigarette-associated cues that can trigger craving and relapse. Such research may lead to a biomarker for relapse risk and for monitoring treatment progress, as well as point to regions of the brain involved in the development of nicotine addiction.

A neuroimaging technology called "default-mode" or "resting-state fMRI" (rs-fMRI) reveals intrinsic brain activity when people are alert but not performing a particular task. Using this technique, researchers are examining the neurobiological profile associated with withdrawal and how nicotine impacts cognition. Comparisons between smokers and nonsmokers suggest that chronic nicotine may weaken connectivity within brain circuits involved in planning, paying attention, and behavioral control—possibly contributing to difficulty with quitting. fMRI studies also reveal the impact of smoking cessation medications on the brain—particularly how they modulate the activity of different brain regions to alleviate withdrawal symptoms and reduce smoking. A review of these studies suggested that NRT enhances cognition during withdrawal by modulating activity in default-network regions, but may not affect neural circuits associated with nicotine addiction.

Some imaging techniques allow researchers to visualize neurotransmitters and their receptors, further informing our understanding of nicotine addiction and its treatment. Using these techniques, researchers have established that smoking increases the number of brain receptors for nicotine. Individuals who show greater receptor upregulation are less likely to stop smoking. Combining neuroimaging and genetics may yield particularly useful information for improving and tailoring treatment. For example, nonsmoking adolescents with a particular variant in the CHRNA5-CHRNA3-CHRNB4 gene cluster (which is associated with nicotine dependence and smoking) showed reduced brain activity in response to reward in the striatum as well as the orbitofrontal and anterior cingulate cortex. This finding suggests that genetics can influence how the brain processes rewards which may influence vulnerability to nicotine dependence. Neuroimaging genetics also shows that other genes, including ones that influence dopamine neurotransmission, influence reward sensitivity and risk for addiction to nicotine.

Research Priorities

The Family Smoking Prevention and Tobacco Control Act (TCA) gives the U.S. Food and Drug Administration (FDA) broad authority to regulate tobacco-product manufacturing, distribution, and marketing. Although a vast and sound science base exists with regard to numerous areas of the TCA, new research will provide scientific evidence in several areas. Consistent with the Center's mission, Center for Tobacco Products (CTP) encourages research that addresses the following scientific domains:

Toxicity

Understanding how tobacco products and changes to tobacco product characteristics affect their potential to cause morbidity and mortality, including animal and cell-culture models, as well as novel alternative toxicology approaches that test the toxicity of tobacco smoke, aerosols, or specific constituents in tobacco. Priorities include:

- Toxicological assays (*in vivo* and *in vitro*) to compare toxicity across different types of tobacco products within the same class, including electronic nicotine delivery systems (ENDS), cigars, waterpipes, and smokeless tobacco

- How product-design characteristics (and changes in those characteristics) impact constituent exposure and toxicity from tobacco products

- Biomarkers to assess exposure, as well as biomarkers to assess harm or toxicity of noncigarette tobacco products, including ENDS

Addiction

Understanding the effect of tobacco-product characteristics on addiction and abuse liability. Priorities include:

- Impact of changes in tobacco-product characteristics (such as flavors or product design) on dependence

- Differences in dependence and tobacco-use patterns with use of low-nicotine-content cigarettes in context with other tobacco products

- The amounts of nicotine delivered to ENDS users during experimentation, regular ENDS use, dual use of ENDS and cigarettes, and cigarette-smoking quit attempts;

- Correlation of ENDS use behaviors with pharmacokinetic and pharmacodynamic effects of nicotine and other HPHCs delivered by ENDS

Health Effects

Understanding the short- and long-term health effects of tobacco products. Highest priority areas include cardiovascular or respiratory health effects, including inflammation. Other health effects including cancer, oral health, or reproductive health may be included within projects, but should not be the primary focus of the TCORS. Priorities include:

- Impact of changes in tobacco-product characteristics (such as flavors or product design) on human health

- Biomarkers to assess short- and long-term effects of noncigarette tobacco products

- Clinical evaluations to distinguish changes in cell function/ physiology specific to tobacco exposure (e.g., ENDS aerosol exposure) known to indicate longer-term disease development and progression

Behavior

Understanding the knowledge, attitudes, and behaviors related to tobacco-product use and changes in tobacco-product characteristics. Priorities include:

- Changes in tobacco-product characteristics (such as flavors, product design, or packaging) impact on tobacco-use behaviors, including experimentation, initiation, dual/poly use, transition to nonflavored products, and cessation

- Innovative methods and measures to assess tobacco use behaviors

- Measures, methods, or study designs to assess the likely impact of novel and/or potential modified-risk tobacco products on tobacco behavior, including perceptions, susceptibility, experimentation, adoption, switching, and use (including dual use)

- Measures (e.g., attitudes, perceptions, intentions) to best predict future behaviors of noncigarette tobacco-product use, including current and established users of cigars, waterpipe, and ENDS

Communications

Understanding how to effectively communicate to the public and vulnerable populations regarding nicotine and the health effects of tobacco products, including media campaigns and digital media. Priorities include:

- Messages to effectively communicate about nicotine and the harms of noncigarette tobacco-product use

- Methods and messages for communicating complex scientific concepts to the general public, including risk and harms of tobacco use, taking into account unintended consequences

- Effectiveness of text and graphic warnings for tobacco products other than cigarettes

Marketing Influences

Understanding why people become susceptible to using tobacco products (both classes of products and products within classes) and to transitions between experimentation and initiation to regular use and dual use. Topics may include tobacco industry marketing such as advertising, point-of-sale, digital media, and promotions. Priorities include:

- Methods, measures, and study designs to best assess the impact of tobacco-product advertising and promotion restrictions on users and nonusers of tobacco, including marketing of novel and/ or potential modified risk tobacco products

- Impact of potential marketing restrictions on youth experimentation, initiation, use, and cessation

Impact Analysis

Understanding the impact of potential FDA regulatory actions. Priorities include:

- Evaluation of policies at the state and community level that fall within FDA CTP regulatory authorities

- Methods and measures (e.g., behavioral economics, population modeling) to estimate the range of potential impacts on behavior and health of potential FDA regulatory actions such as products standards addressing toxicity, appeal, and addiction

Chapter 38

ASSIST: Americans Stop Smoking Intervention Study

Americans Stop Smoking Intervention Study (ASSIST) was a demonstration project for tobacco-use prevention and control conducted by the National Cancer Institute (NCI), the American Cancer Society (ACS), and 17 state health departments. The goal of ASSIST was to change the social, cultural, economic, and environmental factors that promote tobacco use by using policy, mass media, and program-services interventions. The four policy strategies were as follows:

- Raising excise taxes to increase the price of tobacco products
- Eliminating exposure to environmental tobacco smoke
- Limiting tobacco advertising and promotion
- Reducing minors' access to tobacco products

Shaping the Future of Tobacco Prevention and Control

The strategies for ASSIST were developed and implemented by state and local tobacco-control coalitions using population-based

This chapter includes text excerpted from "The American Stop Smoking Intervention Study for Cancer Prevention (ASSIST)," Division of Cancer Control and Population Sciences (DCCPS), National Cancer Institute (NCI), May 15, 2005. Reviewed February 2019.

research, public-health practices, policy development, and media advocacy. The concepts of building on a strong evidence base; designing interventions with broad population impacts; changing social norms in pursuit of greater justice; developing strong partnerships based on common goals and mutual respect; maintaining a determination not to be swayed or pushed off target by one's adversaries; and ensuring a serious commitment to evaluation, self-reflection, and midcourse correction were crucial components of ASSIST.

Evaluating ASSIST: A Blueprint for Understanding State-level Tobacco Control (NCI Tobacco Control Monograph 17), and the preceding one in this series, Monograph 16, *ASSIST: Shaping the Future of Tobacco Prevention and Control*, are designed as companion documents.

Monograph 16 provides in-depth descriptions of intervention processes, examples of materials and best practices, and resource lists and guidance for activities such as media-advocacy campaigns. Numerous case studies are presented, not in the form of formal social research, but as stories and vignettes from state and local public-health staff and volunteers who describe their efforts, the barriers they encountered, the lessons they learned, and insights they gained. These case studies show ASSIST as it was experienced by the many committed and diverse people responsible for its success.

Below are the major topics addressed in Monograph 16:

- The historical context and conceptual framework of ASSIST

- The national partners and state agencies and their respective roles, and communication linkages among all the structural units that promoted collaborative decision-making and were essential for the program to function

- National, state, and local capacity-building by mobilizing communities, establishing coalitions, promoting participatory planning, and providing training and technical assistance

- Descriptions of strategies and intervention methods, insights, and lessons learned for the three ASSIST intervention channels—policy development, mass media and media advocacy, and program services

- The tobacco industry challenge to ASSIST and the ASSIST response

- Strategic planning for a national tobacco-use prevention and control program

402

- The processes and challenges in maintaining capacity built by the ASSIST demonstration project, disseminating best practices, and building a comprehensive national tobacco-use prevention and control program

- Contributions of ASSIST to tobacco-use prevention and control and to other behavioral health programs

The insights and lessons learned from ASSIST have advanced our understanding of how research studies can be successfully translated and disseminated as demonstration projects, while illustrating how sustained funding builds effective tobacco-use prevention and control programs. The ASSIST legacy endures in the infrastructure that continues to support tobacco-use prevention and control interventions. As the first major public-health intervention grounded in ecological theory, ASSIST remains an exemplar for modern systems-level public-health programs.

Chapter 39

Low-Income, Rural Kids at Higher Risk for Second- or Third-Hand Smoke Exposure

Infants and toddlers in low-income, rural areas may be at higher risk for second- and third-hand smoke than previously reported, according to a study supported by the National Institutes of Health (NIH). Approximately 15 percent of children in the study tested positive for cotinine, a byproduct formed when the body breaks down nicotine, at levels comparable to those of adult smokers. About 63 percent of children in the study had detectable levels of cotinine, suggesting widespread exposure to smoke. The study appears in *Nicotine and Tobacco Research*.

"Few studies have explored the risks of very young children, especially infants, for second- or third-hand exposure to smoking," said James A. Griffin, Ph.D., deputy chief of the Child Development and Behavior Branch (CDBB) at NIH's *Eunice Kennedy Shriver* National Institute of Child Health and Human Development (NICHD), which funded the research. "The current study suggests that moving frequently, having more adults in the home and spending less time in center-based daycare facilities may increase a child's exposure to smoke or smoke residue."

This chapter includes text excerpted from "Low-Income, Rural Kids at Higher Risk for Second- or Third-Hand Smoke Exposure," National Institutes of Health (NIH), December 6, 2018.

The researchers analyzed data from the Family Life Project, a long-term study of rural poverty in North Carolina and Pennsylvania. For the study, saliva samples of 1,218 children were tested for cotinine. The samples were collected from children at age 6 months, 15 months, 2 years and 4 years. The presence of cotinine indicates that the child was exposed to second- or third-hand smoke. Second-hand smoke comes from a lit tobacco product, an electronic-smoking device or the smoker. Third-hand smoke is an invisible residue from smoke that settles onto floors, furniture, and clothing.

The researchers classified the children into three groups based on their cotinine levels. Fifteen percent of the children were in the high-exposure group, with cotinine levels comparable to active adult smokers (12 ng/mL or higher), 48 percent were in the moderate-exposure group (0.46 to 12 ng/mL) and 37 percent were in the low-exposure group (less than or equal to 0.46ng/mL). These values are higher than those seen in data previously reported in the National Health and Nutrition Examination Survey (NHANES), which found that only one-third to one-half of children's blood samples had detectable cotinine.

"We found that infants had higher cotinine levels compared to toddlers," said Lisa M. Gatzke-Kopp, Ph.D., a professor at Pennsylvania State University and the lead author of the study. "Because infants often put objects into their mouths and crawl on floors, they may be more likely to ingest smoke residue or get it on their skin, compared to older children." The study team evaluated independent factors that may influence a child's probability of being in one of the three exposure groups. They found that lower income, less education, frequent residential moves, and fluctuations in the number of adults within the home were associated with high smoke exposure, whereas time spent at a center-based daycare was associated with lower smoke exposure.

Factors influencing cotinine levels included the following:

- When a caregiver had at least a high-school degree, a child was 85 percent less likely to be in the high-exposure group, compared to the other two groups.

- Each residential move increased a child's odds of being in the high-exposure group, compared to the low-exposure group, by 43 percent.

- Each adult moving into or out of the home increased this risk by 11 percent.

- A child who spent time in a center-based daycare was 81 percent less likely to be in the high-exposure group, compared to the low-exposure group.

"Our results, if supported by future studies, can help educate parents and caregivers, as well as improve prevention programs that seek to reduce children's smoke exposure," said Clancy Blair, Ph.D., M.P.H., a professor at New York University's (NYU) Steinhardt School of Culture, Education and Human Development and the senior author of the study. "For instance, nonsmoking families may not be aware that nicotine can be present in their child's environment if their home was previously occupied by a smoker or if smoking is permitted at the workplace."

Funding for the current analysis was provided by NICHD, the National Institute on Drug Abuse (NIDA), and the Environmental influences on Child Health Outcomes (ECHO) program, all part of NIH.

Chapter 40

Population Assessment of Tobacco and Health

The Population Assessment of Tobacco and Health (PATH) Study is a uniquely large, long-term study of tobacco use and health in the United States. A collaboration between the U.S. Food and Drug Administration (FDA) Center for Tobacco Products and the National Institutes of Health (NIH) National Institute on Drug Abuse (NIDA), the study was launched in 2011, started the first wave of data collection in 2013, and is currently in its fourth wave.

About the Population Assessment of Tobacco and Health Study

By following study participants over time, the PATH Study helps scientists learn how and why people start using tobacco, quit using it, and start using it again after they've quit, as well as how different tobacco products affect health (such as cardiovascular and respiratory health) over time. Findings from the study may also inform the FDA's actions related to tobacco products, thereby helping to achieve the goals of the Family Smoking Prevention and Tobacco Control Act.

This chapter includes text excerpted from "FDA and NIH Study: Population Assessment of Tobacco and Health," U.S. Food and Drug Administration (FDA), December 14, 2018.

The PATH Study is conducted via a contract awarded to Westat and involves researchers from:

- Center for Tobacco Products, FDA
- National Institute on Drug Abuse, NIH
- Centers for Disease Control and Prevention
- Roswell Park Cancer Institute
- Dartmouth College
- Truth Initiative (formerly Legacy)
- The Medical University of South Carolina
- The University of California, San Diego
- The University of Waterloo
- The University of Minnesota

Research Goals for the Population Assessment of Tobacco and Health Study

By monitoring and assessing behaviors, attitudes, biomarkers, and health outcomes associated with tobacco use in the United States, the PATH Study helps enhance the evidence base available to inform the FDA's regulatory activities related to tobacco. Specifically, the study aims to:

- Examine what makes people susceptible to using a tobacco product
- Evaluate initiation and use patterns, including the:
  - Use of newer products, such as e-cigarettes or ENDS (electronic nicotine delivery systems)
  - Use of multiple products
  - Switching from one product to another
- Study patterns of tobacco product use, cessation, and relapse
- Track potential behavioral and health impacts, including biomarkers of exposure and harm
- Assess differences in tobacco-related attitudes, behaviors, and health conditions among racial/ethnic, gender, and age subgroups

Findings from the Population Assessment of Tobacco and Health Study

About 46,000 people aged 12 years and older, including tobacco users and nonusers, are included in the first wave of the PATH Study.

Initial data on adult and youth tobacco use, published January 2017 in the *New England Journal of Medicine (NEJM)*, showed that more than 25 percent of American adults were current users of tobacco in 2013–2014 and roughly 9 percent of youth reported using tobacco in the past 30 days. Multiple product use was common among tobacco users, accounting for roughly 40 percent of adult and youth tobacco users, with cigarettes and e-cigarettes being the most common combination among both age groups.

Among tobacco users who reported using more than one product:

- 23 percent of adults and 15 percent of youth used cigarettes and e-cigarettes

- 6 percent of adults and 4 percent of youth used cigarettes and hookah

- 2 percent of adults and 5 percent of youth used e-cigarettes and hookah

- 5 percent of adults and 10 percent of youth used cigarettes and cigarillos

The study reports prevalence for more product combinations used by study participants.

Data Access and Availability

Data and documentation (questionnaires, codebooks) related to the PATH Study are available on the National Addiction & HIV Data Archive Program (NAHDAP), including:

- Public-use files (PUFs) from Wave 1 (September 2013–December 2014), Wave 2 (October 2014–2015), and Wave 3 (October 2015–2016) of data collection

- Restricted-use files (RUFs) disclaimer icon from Wave 1 (September 2013–December 2014), Wave 2 (October 2014–October 2015), and Wave 3 (October 2015–October 2016). Qualified researchers are encouraged to apply for access through the NAHDAP.

- Biomarker Restricted-use files (BRUFs) disclaimer icon from Wave 1 (September 2013–December 2014)

- Biospecimen Access Program (BAP) disclaimer icon, which provides the research community with access to urine, serum, and plasma collected from adult PATH Study participants during Wave 1 (September 2013–December 2014)

Researchers interested in the PATH Study are encouraged to create an account to join the PATH Study Data User Forum. The forum enables researchers using PATH Study data to submit and answer questions. Announcements, data releases and updates, new publications, upcoming events, and other information for PATH Study data users are also posted to the forum.

Part Five

Tobacco Control and Use Prevention

Chapter 41

Preventing Tobacco Use

Chapter Contents

415

Section 41.1

How Can We Prevent Tobacco Use?

This section contains text excerpted from the following sources:
Text in this section begins with excerpts from "Tobacco, Nicotine, and
E-Cigarettes," National Institute on Drug Abuse (NIDA), January
2018; Text under the heading "Working Together, We Can End
the Tobacco Epidemic" is excerpted from "A Report of the Surgeon
General: Preventing Tobacco Use among Youth and Young Adults,"
Centers for Disease Control and Prevention (CDC),
March 8, 2012. Reviewed February 2019.

The medical consequences of tobacco use—including secondhand
exposure—make tobacco control and smoking prevention crucial parts
of any public-health strategy. Since the first Surgeon General's Report
on Smoking and Health in 1964, states and communities have made
efforts to reduce initiation of smoking, decrease exposure to smoke, and
increase cessation. Researchers estimate that these tobacco-control
efforts are associated with averting an estimated 8 million premature
deaths and extending the average life expectancy of men by 2.3 years
and of women by 1.6 years. But there is a long way yet to go: roughly
5.6 million adolescents under age 18 are expected to die prematurely
as a result of an illness related to smoking.

Prevention can take the form of policy-level measures, such as
increased taxation of tobacco products; stricter laws (and enforcement of laws) regulating who can purchase tobacco products; how and
where they can be purchased; where and when they can be used (i.e.,
smoke-free policies in restaurants, bars, and other public places); and
restrictions on advertising and mandatory health warnings on packages. Over 100 studies have shown that higher taxes on cigarettes, for
example, produce significant reductions in smoking, especially among
youth and lower-income individuals. Smoke-free workplace laws and
restrictions on advertising have also shown benefits.

Prevention can also take place at the school or community level.
Merely educating potential smokers about the health risks has not
proven effective. Successful evidence-based interventions aim to
reduce or delay initiation of smoking, alcohol use, and illicit drug
use, and otherwise improve outcomes for children and teens by
reducing or mitigating modifiable risk factors and bolstering protective factors. Risk factors for smoking include having family members or peers who smoke, being in a lower socioeconomic status,
living in a neighborhood with high density of tobacco outlets, not

participating in team sports, being exposed to smoking in movies, and being sensation-seeking. Although older teens are more likely to smoke than younger teens, the earlier a person starts smoking or using any addictive substance, the more likely they are to develop an addiction. Males are also more likely to take up smoking in adolescence than females.

Some evidence-based interventions show lasting effects on reducing smoking initiation. For instance, communities utilizing the intervention-delivery system, Communities That Care (CTC) for students aged 10 to 14 show sustained reduction in male cigarette initiation up to 9 years after the end of the intervention.

Working Together, We Can End the Tobacco Epidemic

If we choose to, we can end the tobacco epidemic in this country. But it's going to take all of us—parents, teachers, healthcare providers, communities, states, schools, and policymakers—supporting policies, programs, and media campaigns that prevent tobacco use by youth and young adults.

Policies and Programs— Which Work Best?

There are effective policies and programs that prevent young people from using tobacco. Policies and programs that contain several parts working together to make tobacco use more difficult and less accepted are the ones that work best.

Policies

Policies are very effective because they can change the environment so that choosing a tobacco-free life is encouraged and supported. Government and private entities have implemented a number of policies that are effective in preventing youth tobacco use.

Here are some policies proven to work best:

- Make tobacco products less affordable.

- Restrict tobacco marketing.

- Ban smoking in public places—such as workplaces, schools, daycare centers, hospitals, restaurants, hotels, and parks.

- Require tobacco companies to label tobacco packages with large, graphic health warnings.

Programs

Many states and communities have programs to prevent tobacco use by young people. The most effective ones combine several elements, and include evidence-based curricula in secondary schools, working with policies, and influencing people at work, at home, in school, in healthcare settings, and in public places.

Mass Media Campaigns

Mass media campaigns against tobacco use—most often television ads—have proven very effective at helping prevent tobacco use by young people. Studies show that teens respond most to ads that trigger strong negative feelings, such as ads about how smoking and second-hand smoke harm health, and ads that expose the tobacco industry's marketing strategies that target young people. Even ads that are designed for adult audiences help reduce tobacco use among young people.

Every three or four years, new groups of children and teens reach the age where they are vulnerable to influences encouraging them to smoke. To be effective, mass media campaigns must be repeated so they will reach new vulnerable populations.

Parents—How You Can Help

You can help your children make healthy choices about tobacco use. Try these tips.

Tell them:

- Key facts about tobacco

- You don't want anyone—including them—to use tobacco in your house or car

- You expect they will never use tobacco or will stop using it

Help them:

- Cope with their problems

- Refuse tobacco

- Quit if they're current users

Make sure you:

- Know what they're doing and who their friends are

- Network with other parents who can help you encourage children and teens to refuse tobacco

- Encourage your children's schools to enforce tobacco-free policies for students, faculty, staff, and visitors both on campus and at all school-sponsored events off campus

- Enforce movie age restrictions—and discourage teens from playing video games or using other media that feature smoking

- Never give tobacco to children or teens

- Set a good example by not using tobacco yourself

Section 41.2

Preventing and Reducing Teen Tobacco Use

This section includes text excerpted from "Tobacco Use in Adolescence," U.S. Department of Health and Human Services (HHS), July 26, 2016.

More deaths are caused each year by tobacco use than by human immunodeficiency virus (HIV), illegal drug use, alcohol, motor-vehicle injuries, suicides, and murders combined. However, tobacco use by adolescents and young adults has declined substantially over the past 40 years. In 2017, only one in 25 high-school seniors identified as a daily smoker, and fewer than one in 10 had smoked within the past 30 days. Furthermore, adolescents' use of smoking products is evolving. In 2014, and for the first time in history, more teenagers used electronic cigarettes (or e-cigarettes) than smoked tobacco cigarettes. This trend continued in 2017, as more high-school students used e-cigarettes or similar devices to vape (i.e., inhale vapors that include nicotine) than smoked tobacco cigarettes. These products pose a set of new challenges, as they are known to be harmful, but their health impact is not yet fully understood.

Tobacco use remains the number one cause of preventable deaths in the United States. It is critical that efforts to prevent and reduce teen smoking continue, as the stakes could not be higher. On average, smokers die at least 10 years earlier than nonsmokers and, every day, more than 1,200 people in the United States die from smoking-related causes. Almost 90 percent of those who die from smoking-related causes began using tobacco products at or prior to age 18. The Surgeon General estimates that if all the evidence-based youth antitobacco strategies were implemented, smoking among high-school students would decline by more than 50 percent by 2020.

Ways to Reduce Adolescent Tobacco Use

The Surgeon General's report *Preventing Tobacco Use among Youth and Young Adults* summarizes the causes, solutions, and latest findings about the epidemic of tobacco use among these groups.

The U.S. Preventive Services Task Force (USPSTF) recommends that primary-care clinicians provide interventions, including education or brief counseling, to prevent initiation of tobacco use among school-aged children and adolescents (ages 10–17).

The Community Preventive Services Task Force (CPSTF) offers guidance and recommendations on evidence-based approaches to prevent and reduce tobacco use by adolescents. These include:

- Reducing the initiation of tobacco use (such as by increasing the price of tobacco products)

- Increasing tobacco-use cessation for smokers and users of other tobacco products (through mass media campaigns and other interventions)

- Restricting minors' access to tobacco products (through community mobilization and enforcement)

- Reducing exposure to second-hand smoke (primarily through smoking bans)

Section 41.3

Tobacco-Control Interventions

This section includes text excerpted from "Tobacco Control Interventions," Centers for Disease Control and Prevention (CDC), June 8, 2017.

What Are Effective Statewide Tobacco Interventions?

Effective population-based tobacco control interventions include tobacco price increases, high-impact anti-tobacco mass-media campaigns, and comprehensive smoke-free policies. The evidence shows that implementing and enforcing these strategies, both individually and as part of a comprehensive tobacco-prevention-and-control effort, can reduce smoking initiation and use among adults and youths. Comprehensive tobacco-prevention-and-control efforts involve the coordinated implementation of population-based interventions to prevent tobacco initiation among youth and young adults, promote quitting among adults and youth, eliminate exposure to secondhand smoke, and identify and eliminate tobacco-related disparities among population groups. Tobacco products include cigarettes, cigars, pipes, hookah, smokeless tobacco, and others. Programs combine and integrate multiple evidence-based strategies, including educational, regulatory, economic, and social strategies at local, state, or national levels. Evidence-based interventions that are key components of a comprehensive tobacco prevention and control effort include

- Mass-reach health communications campaigns that use multiple media formats and include hard-hitting or graphic images are intended to change knowledge, beliefs, attitudes, and behaviors affecting tobacco use, and provide tobacco users with information on resources about how to quit.

- Increasing the unit price for tobacco products will decrease the number of people using tobacco, reduce the amount of tobacco consumed, and prevent young people from starting to use tobacco.

- Comprehensive smoke-free policies that prohibit smoking in all indoor areas of workplaces and public places, including restaurants and bars, will prevent involuntary exposure to secondhand smoke.

What Is the Public-Health Issue?

Tobacco use is the single most preventable cause of disease, disability, and death in the United States. Cigarette smoking harms nearly all organs of the body; it has been linked to heart disease, multiple cancers, and lung diseases, among others. Smoking during pregnancy also causes harm to the fetus. In addition to adverse effects on individual- and population-level health, smoking imposes an immense financial burden on society, with over 480,000 premature deaths, over $170 billion in lost productivity costs, and at least $133 billion in direct medical-care expenditures in the United States each year. The use of smokeless tobacco, cigars, and pipes can also have deadly consequences, including lung, larynx, esophageal, and oral cancers. Moreover, the effects of tobacco use are not limited to the user. Secondhand smoke exposure can cause death and many serious diseases, including lung cancer, heart disease, and stroke among adults, and respiratory illness, ear infections, asthma attacks, and sudden infant death syndrome (SIDS) among children and infants. An estimated one in four nonsmokers (58 million people), including about two in five children, are exposed to secondhand smoke.

What Is the Evidence of the Health Impact and Cost Effectiveness?

A systematic review of proven population-based tobacco-control interventions found that these programs were associated with:

- Reductions in the prevalence of tobacco use among adults and young people

- Reductions in tobacco-product consumption

- Increased quitting

States that have made larger investments in comprehensive tobacco-control efforts have seen larger declines in cigarettes sales than the United States as a whole, and the prevalence of cigarette smoking among adults and youth has declined faster as spending for tobacco-control programs has increased. Comprehensive tobacco-control efforts have also contributed to reductions in tobacco-related diseases and deaths, and were effective across diverse racial, ethnic, educational, and socioeconomic groups. The review also found that these programs were cost-effective and that healthcare savings were greater than the cost of the intervention.

Additional systematic reviews examining the impact of single interventions that may be implemented individually or included as part of a comprehensive tobacco control program, such as mass-media campaigns, price increases, and smokefree policies, also found strong evidence of their efficacy and cost-effectiveness.

Mass-Reach Communications Campaigns

Mass-media campaigns were associated with lower prevalence of tobacco use, increased cessation and use of available cessation services, and decreased initiation of tobacco use among young people:

- Median decrease of 5.0 percent points in the prevalence of tobacco use among adults

- Median decrease of 3.4 percent points in the prevalence of tobacco use among young people (11 to 24 years of age)

- Median increase of 3.5 percent points in cessation of tobacco use

- Median relative increase of 132 percent in the number of calls to quitlines

- Decrease of 6.7 percent points in tobacco use initiation among young people (11 to 24 years of age)

An economic review of the evidence found the benefit-to-cost ratio (BCR) for mass-reach health-communications campaigns ranged from 7:1 to 74:1, with an estimated cost of $213 (2011 dollars) per life year saved.

Increasing the Price of Tobacco Products

Increases in the price of tobacco products reduce demand for tobacco, thereby prompting quit attempts, reducing consumption among those who do not quit, and preventing youth from starting. Increasing the unit price of tobacco by 20 percent was found to be associated with the following reductions:

- 7.4% reduction in demand among adults ages 30 and older

- 14.8% reduction in demand among young people ages 13 to 29

- 3.6% reduction in the proportion of adults ages 30 and older who use tobacco

- 7.2% reduction in the proportion of young adults ages 19 to 29 who use tobacco

- 8.6% reduction in tobacco use initiation among young people ages 13 to 29

- 6.5% increase in quitting among adults ages 30 and older

- 18.6% increase in quitting among young people ages 13 to 29

An economic review of the evidence estimated that healthcare cost savings from a 20 percent price increase for tobacco products ranged from -$0.14 to $90.02 per person per year (2011 dollars) in addition to averted productivity losses.

Comprehensive Smoke-Free Policies

Comprehensive smoke-free policies have been shown to substantially improve indoor air quality, reduce secondhand smoke exposure, change social norms regarding the acceptability of smoking, prevent smoking initiation by youth and young adults, help smokers quit, and reduce heart attack and asthma hospitalizations among nonsmokers. Comprehensive smoke-free policies were associated with

- Decreased exposure to secondhand smoke (50% reduction in biomarkers)

- Decreased prevalence of tobacco smoking (absolute reduction of 2.7% points)

- Decreased tobacco consumption (absolute reduction of 1.2 cigarettes per day)

- Fewer cardiovascular events (5.1% reduction in hospital admissions)

- Decreased asthma morbidity (20.1% reduction in hospital admissions)

An economic review of the evidence estimated that net savings resulting from a nationwide U.S. smokefree policy would range from $700 to $1,297 per person not currently covered by a smokefree policy (2011 dollars). It also found that smokefree policies did not have an adverse economic impact on the business activity of restaurants, bars, or establishments catering to tourists; some studies found a small positive effect of these policies.

Chapter 42

Comprehensive Tobacco-Control Programs

The Centers for Disease Control and Prevention's (CDC) Office on Smoking and Health (OSH) created the National Tobacco Control Program (NTCP) in 1999 to encourage coordinated, national efforts to reduce tobacco-related diseases and deaths. The program provides funding and technical support to state and territorial health departments. The NTCP funds:

- All 50 states

- The District of Columbia

- Eight U.S. territories/jurisdictions

- Eight national networks

- Twelve tribal support organizations

NTCP-funded programs are working to achieve the objectives outlined in OSH's Best Practices for Comprehensive Tobacco Control Programs.

This chapter contains text excerpted from the following sources: Text in this chapter begins with excerpts from "National Tobacco Control Program," Centers for Disease Control and Prevention (CDC), February 6, 2018; Text beginning with the heading "The Real Cost Campaign" is excerpted from "The Real Cost Campaign," U.S. Food and Drug Administration (FDA), February 5, 2019.

The NTCP's goals are to:

- Eliminate exposure to secondhand smoke

- Promote quitting among adults and youth

- Prevent initiation among youth and young adults

- Identify and eliminate tobacco-related disparities

The NTCP's strategies are:

- Population-based community interventions

- Counter-marketing

- Program policy/regulation

- Surveillance and evaluation

"The Real Cost" Campaign

Given the devastating consequences of cigarette addiction, the U.S. Food and Drug Administration (FDA) launched its first tobacco-prevention campaign, "The Real Cost" in 2014, to educate at-risk teens on the harmful effects of cigarette smoking. In 2018, the campaign expanded to educate teens on the dangers of e-cigarette use and had previously expanded to educate rural boys on the harms of smokeless tobacco in 2016.

- Youth E-Cigarette Prevention Campaign

- Smokeless Tobacco Prevention Campaign

- Smoking Prevention Campaign: A Cost-Effective Approach

- Awards and Recognition

"The Real Cost" Youth E-Cigarette Prevention Campaign

E-cigarettes surpassed combustible cigarettes as the most commonly used tobacco product among U.S. middle- and high-school students in 2014. In 2018, more than 3.6 million middle- and high-school students reported they currently use e-cigarettes, with many parents, teachers, and school administrators raising alarm about pervasive vaping in schools. Additional research shows that about 80 percent of youth do not see great risk of harm from regular use of e-cigarettes, particularly alarming considering that harm perceptions can influence tobacco-use behaviors. To address this "cost-free" mentality, FDA expanded its award-winning "The Real Cost" campaign to educate the nearly 10.7 million youth aged 12 to 17 who have ever used e-cigarettes or are open to trying them about the potential risks of e-cigarette use. Campaign messages focus on educating youth that using e-cigarettes, just like cigarettes, puts them at risk for addiction and other health consequences.

Advertising and other prevention materials are delivered where teens spend most of their time—online and in school—including:

- Online video ads

- Additional content on "The Real Cost" campaign's youth-targeted website

- Digital and social media content

- Materials for use in high schools nationwide (e.g., posters for school bathrooms)

"The Real Cost" Smokeless-Tobacco Prevention Campaign

Each day in the United States, more than 950 male youth under 18 years of age use smokeless tobacco for the first time. Many of these boys are not aware of the negative health consequences of using smokeless tobacco products, or "dipping." "The Real Cost" smokeless tobacco prevention campaign seeks to educate rural, male teenagers about risks of dipping—including loss of control, gum disease, tooth loss, and multiple kinds of cancer.

The campaign's central message is "smokeless doesn't mean harmless," which aims to motivate teens to reconsider what they think they know about smokeless tobacco use. The campaign will have a national digital presence and out-of-home advertisements will appear in select markets across 20 states specifically selected to reach the campaign's at-risk target audience.

"The Real Cost" Smoking Prevention Campaign: A Cost-Effective Approach

The FDA's first smoking prevention campaign, "The Real Cost," seeks to educate the more than 10 million at-risk teens in the United

States about the harmful effects of cigarette smoking. Launched in 2014, the campaign strives to prevent youth who are open to smoking from trying it and to reduce the number of youth who move from experimenting with cigarettes to regular use by messaging on loss of control due to addiction, health consequences, and dangerous chemicals found in cigarettes.

In its first two years, research shows the campaign has done just that: "The Real Cost" prevented an estimated 350,000 teens ages 11 to 18 from initiating smoking between 2014 and 2016, half of whom might have gone on to become established adult smokers. Preventing teens from initiating smoking doesn't just impact their personal health, but also the health of their families and smoking-related costs borne to society. Ultimately, by preventing these kids from becoming established smokers, the campaign has saved them, their families, and the country more than $31 billion by reducing smoking-related costs such as early loss of life, costly medical care, lost wages, lower productivity, and increased disability.

The campaign obtained these impressive results through a carefully executed paid media strategy based on evidence-based best practices for tobacco-prevention campaigns, and by ensuring its messaging was designed to reach and motivate an at-risk teen audience. For example, the FDA conducts research with at-risk teens across the country to develop campaign advertising that resonates. Near-final TV ads are tested with thousands of target audience members for perceived effectiveness and message comprehension prior to being placed in market.

The FDA also hired an independent firm, RTI International, to conduct a multi-year evaluation to measure indicators of success throughout the first two years of the campaign, including advertising awareness and changes in the target audience's tobacco-related beliefs, intentions, and behaviors. Results from this research are impressive: More than 90 percent of the target audience was aware of the first wave of ads less than a year after launch, and the campaign changed teens' perceptions and beliefs about tobacco, ultimately resulting in a 30 percent decrease in youth smoking initiation from 2014 to 2016.

These results not only reinforce the importance of the FDA's public-education efforts in reducing the public-health and financial burden of tobacco use, but also highlight the importance of investing in tobacco-related education campaigns. Investment in tobacco prevention can have huge returns: The campaign had a cost savings of $128 for every dollar of the nearly $250 million invested in the first two years of the campaign. The campaign continues to air nationally across TV, radio, print, web, and social media.

Awards and Recognition

"The Real Cost" campaign has earned two Effie awards to date, including a 2015 gold Effie in the Disease Awareness and Education category and a 2017 bronze Effie in the Youth Marketing category. The Effies are the advertising industry's most prestigious award, recognizing marketing ideas that work and have demonstrated effectiveness. "The Real Cost" campaign was recognized for its insightful communications strategy, outstanding creativity, and success in market.

"The Real Cost" campaign also earned a 2016 Shorty Award for its creative work on Tumblr. Shortys are prestigious, highly coveted awards for the best work in social media.

Chapter 43

Why Quitting Is Hard

One of the first steps is to learn why you feel like you need to smoke. Once you understand why you smoke, you can prepare yourself to find the best ways to quit. Build a Quit Plan to help you identify your smoking triggers, learn about managing cravings, and explore different quit methods.

Withdrawal

One of the main reasons smokers keep smoking is nicotine. Nicotine is a chemical in cigarettes that makes you addicted to smoking. Over time, your body gets used to having nicotine. However, the more you smoke, the more nicotine you need to feel normal. When your body doesn't get nicotine, you may feel uncomfortable and crave cigarettes. This is called withdrawal.

It takes time to get over withdrawal. Most physical symptoms go away after a few days to a week, but cigarette cravings may stick around longer. There are ways you can be prepared for withdrawal.

Triggers

When you smoke, certain activities, feelings, and people become linked to your smoking. These may "trigger" your urge to smoke. Try to anticipate these smoking triggers and develop ways to deal with them:

This chapter includes text excerpted from "Why Quitting Is Hard," Smoke-free.gov, U.S. Department of Health and Human Services (HHS), July 30, 2013. Reviewed February 2019.

- Go to places that don't allow smoking. Shops, movie theaters, and many restaurants are now smoke-free.

- Spend more time with nonsmokers. You won't want to smoke as badly if you are around people who don't smoke.

- Keep your hands busy. Play a game on your phone, eat a healthy snack, or squeeze a stress ball.

- Take a deep breath. Remind yourself why you want to stop smoking. Think of people in your life who will be happier and healthier because you decided to quit.

Consider Using a Quit Smoking Program

Quit smoking programs help smokers understand and cope with problems they have when trying to quit. The programs teach problem-solving and other coping skills. A quit smoking program can help you quit for good by:

- Helping you understand why you smoke

- Teaching you how to handle withdrawal and stress

- Teaching you tips to help resist the urge to smoke

Get started using a quit program today:

- Try a text message program. Sign up for SmokefreeTXT online or text QUIT to 47848.

- Download a smartphone app. The FDA's free apps help you track cravings and understand your smoking patterns.

- Visit Smokefree on social media. Grow your support network and stay connected.

- Talk to an expert at a quitline. Call the National Cancer Institute (NCI) Quitline at 800-QUIT-NOW (800-784-8669) or find your state's quitline by calling 877-44U-QUIT (877-448-7848).

- Chat with a quit smoking counselor. LiveHelp is Monday through Friday, 9:00 a.m. to 9:00 p.m. Eastern time. LiveHelp is also available in Spanish.

Chapter 44

Dealing with Secondhand Smoke

Smoking harms both you and the ones you love. Quitting smoking will benefit you plus help you protect the people in your life.

Quitting will make the people you care about happier and healthier. This may be one of your reasons for quitting.

Dangers of Secondhand Smoke

The main way smoking hurts nonsmokers is through secondhand smoke (SHS). Secondhand smoke is the combination of smoke that comes from a cigarette and smoke breathed out by a smoker. When a nonsmoker is around someone smoking, they breathe in secondhand smoke.

Secondhand smoke is dangerous to anyone who breathes it in. It can stay in the air for several hours after somebody smokes. Breathing secondhand smoke for even a short time can hurt your body.

This chapter contains text excerpted from the following sources: Text in this chapter begins with excerpts from "Secondhand Smoke," Smokefree.gov, U.S. Department of Health and Human Services (HHS), June 5, 2016; Text under the heading "What Can You Do to Reduce Exposure to Secondhand Smoke?" is excerpted from "Secondhand Smoke and Smoke-Free Homes," U.S. Environmental Protection Agency (EPA), December 10, 2018.

Health Effects of Secondhand Smoke

Over time, secondhand smoke has been associated with serious health problems in nonsmokers:

- Lung cancer in people who have never smoked

- More likely that someone will get heart disease, have a heart attack, and die early

- Breathing problems like coughing, extra phlegm, wheezing, and shortness of breath

Secondhand smoke is especially dangerous for children, babies, and women who are pregnant:

- Mothers who breathe secondhand smoke while pregnant are more likely to have babies with low birth weight

- Babies who breathe secondhand smoke after birth have more lung infections than other babies

- Secondhand smoke causes kids who already have asthma to have more frequent and severe attacks

- Children exposed to secondhand smoke are more likely to develop bronchitis, pneumonia, and ear infections and are at increased risk for sudden infant death syndrome (SIDS)

The only way to fully protect nonsmokers from the dangers of secondhand smoke is to not allow smoking indoors. Separating smokers from nonsmokers (with "no smoking" sections in restaurants, and so on), cleaning the air, and airing outbuildings do not get rid of secondhand smoke.

Other Ways Smoking Affects Others

Smoking affects the people in your life in other ways, beyond their health. When you smoke, you may miss out on:

- Spending time with family and friends

- Having more money to spend on the people you love

- Setting a good example for your children. Children who are raised by smokers are more likely to become smokers themselves

Steps You Can Take to Protect Your Loved Ones

The best thing you can do to protect your family from secondhand smoke is to quit smoking. Right away, you get rid of their exposure to secondhand smoke in your home and car, and reduce it anywhere else you go together.

Make sure your house and car remain smokefree. Kids breathe in secondhand smoke at home more than any other place. The same goes for many adults. Don't allow anyone to smoke in your home or car. Setting this rule will:

- Reduce the amount of secondhand smoke your family breathes in

- Help you quit smoking and stay smokefree

- Lower the chance of your child becoming a smoker

When you're on the go, you can still protect your family from secondhand smoke:

- Make sure caretakers such as nannies, babysitters, and daycare staff do not smoke

- Eat at smokefree restaurants

- Avoid indoor public places that allow smoking

- Teach your children to stay away from secondhand smoke

What Can You Do to Reduce Exposure to Secondhand Smoke?

Eliminating secondhand smoke in indoor environments will reduce its harmful health effects, improve the indoor air quality and the comfort or health of occupants. Secondhand-smoke exposure can be reduced through mandated or voluntary smokefree policy implementation. Some workplaces and enclosed public spaces such as bars and restaurants are smokefree by law. People can establish and enforce smokefree rules in their own homes and cars. For multifamily housing, smokefree policy implementation could be mandatory or voluntary, depending on the type of property and location (e.g., ownership and jurisdiction).

- The home is becoming the predominant location for the exposure of children and adults to secondhand smoke.

- Households within buildings with smokefree policies have lower PM2.5 compared to buildings without these policies. PM2.5 is a unit of measure for small particles in the air and is used as one indication of air quality. High levels of fine particles in the air can lead to negative health impacts.

- Prohibiting smoking indoors is the only way to eliminate secondhand smoke from the indoor environment. Ventilation and filtration techniques can reduce, but not eliminate, secondhand smoke.

Chapter 45

Clean Indoor-Air Regulations

Minimal Clinical Interventions

- As reported in 1992 by the U.S. Environmental Protection Agency (EPA), exposure to tobacco smoke in the environment can cause lung cancer in adult nonsmokers. Environmental tobacco smoke (ETS) also has been linked to an increased risk of heart disease among nonsmokers.

- ETS causes about 3,000 lung cancer deaths annually among adult nonsmokers.

- In 1997, the California EPA concluded that ETS causes coronary heart disease and death in nonsmokers. Scientific studies have estimated that ETS accounts for as many as 62,000 deaths from coronary heart disease annually in the United States.

- The 1992 EPA report also concluded that ETS causes serious respiratory problems in children, such as greater number and severity of asthma attacks and lower-respiratory-tract infections. ETS exposure increases children's risk for sudden infant death syndrome (SIDS) and middle-ear infections as well.

- Each year ETS causes 150,000 to 300,000 lower-respiratory-tract infections, such as pneumonia and bronchitis, in children.

This chapter includes text excerpted from "Highlights: Clean Indoor Air Regulations," Centers for Disease Control and Prevention (CDC), July 21, 2015. Reviewed February 2019.

- In a large U.S. study, maternal exposure during pregnancy and postnatal exposure of the newborn to ETS increased the risk for SIDS.

- Comparative risk studies performed by the EPA have consistently found ETS to be a risk to public health. ETS is classified as a group A carcinogen (known to cause cancer in humans) under the EPA's carcinogen-assessment guidelines.

- Several studies have documented the widespread exposure of ETS among nonsmoking adults and children in the United States. Testing nonsmokers' blood for the presence of cotinine, a chemical produced when the body metabolizes nicotine, shows that nearly 9 out of 10 nonsmoking Americans (88%) are exposed to ETS.

- A 1988 National Health Interview Survey (NHIS) reported that an estimated 37 percent of the 79.2 million nonsmoking U.S. workers were employed in places that permitted smoking in designated areas, and that 59 percent of these workers experienced moderate or great discomfort from ETS exposure in the workplace.

- Under common law (laws based on court decisions rather than government laws and regulations), employers must provide a work environment that is reasonably free of recognized hazards. Courts have ruled that common-law duty requires employers to provide nonsmoking employees protection from the proven health hazards of ETS exposure.

- The Occupational Safety and Health Administration (OSHA) is considering regulations that would either prohibit smoking in all workplaces or limit it to separately ventilated areas.

- The federal government has instituted increasingly stringent regulations on smoking in its own facilities. On August 9, 1997, President Clinton signed an Executive Order declaring that Executive Branch federal worksites be smoke-free, thereby protecting nonsmoking federal employees and thousands of citizens who visit federal facilities from the dangers of ETS.

- The Pro-Children's Act of 1994 (Public Law 103–227, secs. 1041–1044) prohibits smoking in facilities where federally funded children's services are provided on a regular or routine basis.

- As of December 31, 1999, at least some degree of smoke-free indoor-air laws were present in 45 states and the District of Columbia. These laws vary widely, from limited smoking restrictions on public transportation to comprehensive restrictions in worksites and public places.

- Twenty states and the District of Columbia limit smoking in private worksites. Of these states, only one (California) meets the nation's Healthy People 2010 objective to eliminate exposure to ETS by either banning indoor smoking or limiting it to separately ventilated areas.

- Forty-one states and the District of Columbia have laws restricting smoking in state government worksites, but only 13 of these states meet the nation's Healthy People 2010 objective.

- Thirty-one states have laws that regulate smoking in restaurants; of these, only Utah and Vermont completely prohibit smoking in restaurants. California requires either a no-smoking area or separate ventilation for smoking areas.

Additional Benefits

- An additional benefit of clean indoor-air regulations may contribute to a reduction in smoking prevalence among workers and the general public. Studies have found that moderate or extensive laws for clean indoor-air are associated with a lower smoking prevalence and higher quit rates.

- The majority of smokers support smoke-free hospitals. Smokers and nonsmokers were in favor of smoke-free workplace six months after a smoke-free policy was implemented.

- Employers are likely to save money by implementing policies for smoke-free workplaces. Savings include costs associated with such things as fire risk, damage to property and furnishings, cleaning, workers compensation, disability, retirement, injuries, and life insurance. Cost savings were estimated at $1,000 per smoking employee based on 1988 dollars.

- The EPA estimates a nationwide, comprehensive policy on clean indoor-air would save $4 to $8 billion per year in building operations and maintenance costs.

Establishing Public Policy

- Involuntary exposure to ETS remains a common public-health hazard that is entirely preventable by adopting appropriate regulatory policies.

- To fight the establishment of such policies, the tobacco industry tries to shift the focus from the science-based evidence on the health hazards of ETS to the controversial social issue of personal freedom. The industry has lobbied extensively against legislation to restrict smoking, and has supported the passage of state laws that preempt stronger local ordinances. (Preemptive legislation is defined as legislation that prevents a local jurisdiction from enacting laws more stringent than, or at a variance with, the state law.)

- A case study conducted in six states found that the existence of an organized smoking-prevention coalition among local citizens was a key determinant in successfully enacting clean indoor-air legislation.

- Smoke-free environments are the most effective method for reducing ETS exposure. Healthy People 2010 objectives address this issue and seek optimal protection of nonsmokers through policies, regulations, and laws requiring smoke-free environments in all schools, work sites, and public places.

Chapter 46

Healthy People 2020: Goals for Reducing Tobacco Use

Scientific knowledge about the health effects of tobacco use has increased greatly since the first Surgeon General's report on tobacco was released in 1964. Since the publication of that report, more than 20 million Americans have died because of smoking.

Tobacco use causes:

- Cancer (oropharynx, larynx, esophagus, trachea, bronchus, lung, acute myeloid leukemia, stomach, liver, pancreas, kidney and ureter, cervix, bladder, and colorectal)

- Heart disease and stroke

- Lung diseases (emphysema, bronchitis, chronic airway obstruction, chronic obstructive pulmonary disease, and pneumonia)

- Reproductive effects (ectopic pregnancy, premature birth, low birth weight, stillbirth, reduced fertility in women, and erectile dysfunction; and birth defects, including cleft-lip and/or cleft palate)

This chapter includes text excerpted from "Tobacco Use," Office of Disease Prevention and Health Promotion (ODPHP), U.S. Department of Health and Human Services (HHS), October 2, 2014. Reviewed February 2019.

- Other effects (Type 2 diabetes, age-related macular degeneration, rheumatoid arthritis, blindness, cataracts, hip fractures, impaired immune function, periodontitis, and overall diminished health)

The harmful effects of tobacco do not end with the user. There is no risk-free level of exposure to secondhand smoke. Since 1964, 2.5 million deaths have occurred among nonsmokers who died from diseases caused by secondhand smoke exposure. Secondhand smoke causes heart disease, lung cancer, and stroke in adults, and can cause a number of health problems in infants and children, including:

- More severe asthma attacks
- Respiratory infections
- Ear infections
- Sudden infant death syndrome (SIDS)

In addition, smokeless tobacco causes a number of serious oral health problems, including cancer of the mouth and gums, periodontitis, and tooth loss.

Why Is Preventing Tobacco Use Important?

Tobacco use is the largest preventable cause of death and disease in the United States. Each year, approximately 480,000 Americans die from tobacco-related illnesses. Further, more than 16 million Americans suffer from at least one disease caused by smoking.

Smoking-related illness in the United States costs more than $300 billion each year, including nearly $170 billion for direct medical care for adults and more than $156 billion in lost productivity.

Healthy People 2020: A Framework for Ending the Tobacco Use Epidemic

Healthy People 2020 provides a framework for action to reduce tobacco use to the point that it is no longer a public-health problem for the nation. Research has identified effective strategies that will contribute to ending the tobacco-use epidemic, including:

- Increasing the price of tobacco products
- Enacting comprehensive smokefree policies
- Expanding cessation treatment in clinical care settings and providing access to proven cessation treatment to all smokers

- Implementing hard-hitting anti-tobacco media campaigns

- Fully funding tobacco-control programs at Centers for Disease Control and Prevention (CDC)-recommended levels

- Controlling access to tobacco products, including e-cigarettes and combustible and noncombustible products

- Reducing tobacco advertising and promotion directed at children

The Healthy People 2020 Tobacco Use objectives are organized into 3 key areas:

1. **Tobacco use prevalence:** Implementing policies to reduce tobacco use and initiation among youth and adults

2. **Health system changes:** Adopting policies and strategies to increase access, affordability, and use of smoking cessation services and treatments

3. **Social and environmental changes:** Establishing policies to reduce exposure to secondhand smoke, increase the cost of tobacco, restrict tobacco advertising, and reduce illegal sales to minors

Preventing tobacco use and helping tobacco users quit can improve the health and quality of life (QOL) for Americans of all ages. People who stop smoking greatly reduce their risk of disease and premature death. Benefits are greater for people who stop at earlier ages, but quitting tobacco use is beneficial at any age.

Understanding Tobacco Use

Many factors influence tobacco use. Risk factors include race/ethnicity, age, education, and socioeconomic status. Significant disparities in tobacco use exist geographically; such disparities typically result from differences among states in smokefree protections, tobacco prices, and program funding for tobacco control.

Emerging Issues in Tobacco Use

Major advances have been made in recent years to address the tobacco epidemic:

- In 2009, the U.S. Food and Drug Administration (FDA) was granted the authority to regulate the sales, marketing, and manufacture of all tobacco products marketed in the United States.

- In 2009, the federal tobacco excise tax was increased by $0.61 to $1.01 per pack.

- As of 2015, 26 states and the District of Columbia (DC) had comprehensive smokefree laws prohibiting smoking in workplaces, restaurants, and bars. Moreover, 15 states and DC had cigarette excise tax rates of at least $2 per pack.

- In 2010, the Affordable Care Act (ACA) was passed, which provided the opportunity for access to barrier-free proven tobacco use cessation treatment including counseling and medication to all smokers.

- In 2012, the CDC launched the first-ever paid national tobacco education campaign—Tips From Former Smokers (Tips). The initial campaign resulted in 1.6 million additional smokers making a quit attempt and over 100,000 sustained quitters. Similarly, in 2015, Tips ads had an immediate and strong impact, showing that the campaign's success has continued since its launch in 2012.

However, there are several emerging issues in tobacco use:

1. The tobacco product landscape is rapidly changing, and use of emerging tobacco products is increasing, particularly among youth. In 2014, e-cigarettes became the most commonly used tobacco product among U.S. middle- and high-school students, surpassing cigarettes.

2. The Family Smoking Prevention and Tobacco Control Act prohibits characterizing flavors other than tobacco and menthol in cigarettes; however, characterizing flavors are not currently prohibited in other tobacco products, including for instance, cigars, cigarillos, e-cigarettes, and hookahs. In 2014, an estimated 70 percent (3.26 million) of all current youth tobacco users had used at least one flavored tobacco product in the past 30 days.

3. From 2002 to 2012, the number of youth and young adults who tried cigarette smoking during the past year increased from 1.9 to 2.3 million. Raising the minimum age of sale for tobacco products to 21 years has emerged as a potential strategy for addressing use among this population.

4. An estimated 58 million Americans remain exposed to secondhand smoke each year. The home is the primary source of secondhand smoke exposure for children, and multiunit housing residents are particularly vulnerable to involuntary exposure in their homes.

Chapter 47

Preventing Tobacco Use during Pregnancy

What Are the Health Effects of Tobacco Use on Pregnancy?

Smoking during pregnancy remains one of the most common preventable causes of pregnancy complications illness, and death among infants. Women who quit smoking before or during pregnancy reduce their risk for poor pregnancy outcomes.

Compared with nonsmokers, women who smoke before pregnancy are about twice as likely to experience the following conditions:

- Delay in conception
- Infertility
- Ectopic pregnancy
- Premature rupture of the membranes
- Placental abruption
- Placenta previa

Compared with babies born to nonsmokers, babies born to women who smoke during pregnancy are more likely to have one of the following conditions:

This chapter includes text excerpted from "Information for Healthcare Providers and Public Health Professionals: Preventing Tobacco Use during Pregnancy," Centers for Disease Control and Prevention (CDC), July 20, 2016.

- Premature birth

- Low birth weight

- Small for gestational age or fetal growth restricted

- Born with a cleft lip, or cleft palate

- Higher risks of SIDS (sudden infant death syndrome)

All tobacco products that are burned contain nicotine and carbon monoxide. These are harmful during pregnancy. These products include cigarettes, little cigars, cigarillos, and hookah.

What Is the Prevalence of Smoking before, during, and after Pregnancy?

The Centers for Disease Control and Prevention's (CDC) Pregnancy Risk Assessment Monitoring System (PRAMS) monitors the prevalence of smoking before, during, and after pregnancy based on a mother's self-report. In 2011, data from 24 states (representing about 40% of U.S. live births) showed

Before Pregnancy

- About 23 percent of women smoked during the 3 months before pregnancy

During Pregnancy

- About 10 percent of women smoked during the last three months of pregnancy

- Groups who reported the highest prevalence of smoking during pregnancy included (see Figure 47.1)

  - American Indians/Alaska Natives

  - Those younger than 25 years of age

  - Those with 12 years of education or less

- Women enrolled in Medicaid were three times more likely to smoke than women with private insurance (see Figure 47.2)

- About 55 percent of women who smoked before pregnancy reported they quit smoking by the last three months of pregnancy

After Pregnancy

- Of those who quit smoking during pregnancy, 40 percent relapsed within six months after delivery

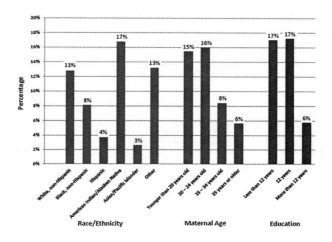

Figure 47.1. *Prevalence of Smoking during the Last Three Months of Pregnancy by Demographic Characteristics, 24 PRAMS States, 2011.*

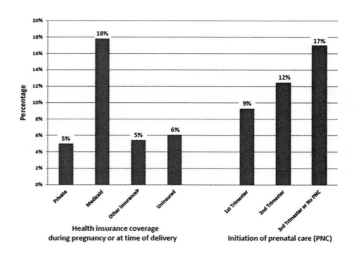

Figure 47.2. *Prevalence of Smoking during the Last Three Months of Pregnancy by Insurance Status and Prenatal Care Initiation, 24 PRAMS States, 2011.*

Trends from 2000 to 2011 (Data from Nine States)

- The prevalence of smoking in the three months before pregnancy did not change (see Figure 47.3). About one of four women smoked before pregnancy.

- The prevalence of smoking declined in the last three months of pregnancy (13.2% to 11.6%) and after delivery (17.8% to 16.6%).

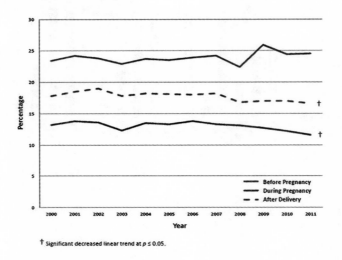

† Significant decreased linear trend at $p \leq 0.05$.

Figure 47.3. *Trends of Smoking before Pregnancy, during Pregnancy, and after Delivery, PRAMS States, 2000–2011*

What about Products That Don't Burn, Such as Electronic Cigarettes and Smokeless Tobacco?

Women may perceive tobacco products that don't burn to be safer than smoking cigarettes. In addition, the use of electronic cigarettes— also referred to as e-pens, e-hookah, tanks, or vape pens—is increasing rapidly among youth and adults.

- All tobacco products contain nicotine, which is a reproductive toxicant and has adverse effects on fetal brain development.

- Pregnant women and women of reproductive age should be cautioned about the use of nicotine-containing products, such as electronic cigarettes, as alternatives to smoking. The health effects of using electronic cigarettes before or during pregnancy have not been studied.

- Electronic cigarettes are not regulated and have not been shown to be a safe and effective cessation aid in smokers.

- The use of smokeless tobacco products, such as snus, during pregnancy has been associated with preterm delivery, stillbirth, and infant apnea.

- There are a number of FDA-approved cessation products, including nicotine replacement therapies (NRT), that are available for the general population of smokers to use to reduce their dependence on nicotine.

- Pregnant women who haven't been able to quit smoking on their own or with counseling can discuss the risks and benefits of using cessation products, such as NRT, with their healthcare provider.

What Works to Help Pregnant Women Quit Smoking?
Counseling by Healthcare Providers

The majority of pregnant women receive prenatal care. Prenatal care visits provide a valuable opportunity to address women's smoking behavior.

- Pregnancy-specific counseling (e.g., counseling based on the 5A's model) increases smoking cessation in pregnant women. Steps of the 5A's include the following:

  - Ask the patient about smoking status at first prenatal visit and follow up at subsequent visits

  - Advise the patient to quit

  - Assess the patient's willingness to quit

  - Assist the patient by providing resources

  - Arrange follow-up visits to track the progress of the patient's attempt to quit

- If women are unable to quit on their own or with counseling, the American College of Obstetricians and Gynecologists (ACOG) recommends that nicotine replacement therapies be considered under the close supervision of a provider.

- Quitlines can be used to support pregnant smokers in their goal to quit. Quitline counseling is available in every state, easy to use, and generally provided at no cost to the user.

- Healthcare system changes, such as provider reminders and documentation of smoking status and cessation intervention, can increase the number of patients who quit.

Population-Based Interventions

State and community tobacco-control interventions that promote tobacco cessation, prevent tobacco initiation, and reduce secondhand smoke not only reduce smoking prevalence in the general population, but also decrease prevalence in pregnant women.

- A $1.00 increase in cigarette taxes increased quit rates among pregnant women by five percentage points. Higher cigarette prices also reduced the number of women who start smoking again after delivery.

- Full smoking bans in private work sites can increase the number of women who quit during pregnancy by about five percentage points.

- Expanded Medicaid tobacco-cessation coverage increased quitting by almost two percentage points in women who smoked before pregnancy.

What about Cutting Back the Number of Cigarettes Smoked without Quitting?

Pregnant women should be advised that complete cessation has the most health benefits by far, and any amount of smoking can be harmful to the fetus. Studies support that cutting down without quitting before the third trimester of pregnancy may improve fetal growth. However, smoking has many other health effects and the potential benefits of simply reducing the number of cigarettes smoked without quitting should be weighed against the following:

- Nicotine is a reproductive toxicant and smoking has been found to contribute to adverse health effects on pregnancy, including preterm birth and stillbirth.

- Nicotine has lasting adverse effects on fetal brain development.

- Nicotine is believed to affect fetal lung development and contribute to the risk of SIDS.

- Smoking most likely affects fetal growth through products of combustion, such as carbon monoxide (CO). There are more than

7,000 other chemicals in tobacco smoke, many of which could also affect fetal health.

- Fetal growth cannot be viewed as a measure of other health effects. It is unknown whether reducing the number of cigarettes smoked improves outcomes other than fetal growth.

What Can Be Done?
Doctors, Midwives, Nurses, and Other Healthcare Providers Can . . .

- Ask all pregnant women about their tobacco use (cigarettes, cigars, little cigars, cigarillos, hookah, smokeless, and electronic cigarettes) and provide nonjudgmental support for women who want to quit. An interactive Web-based program teaches best-practice approaches to help pregnant smokers and women of reproductive age to quit. This program is endorsed by the CDC and ACOG and is available for continuing education credits.

- Refer pregnant women to their state quitline, 800-QUIT-NOW (800-784-8669). Quitlines provide special services and counseling for pregnant and postpartum women. Follow up with pregnant women to make sure they have initiated counseling.

- Share and use resources from the CDC Tips from Former Smokers campaign such as posters, videos, and factsheets.

Public Health Professionals Can . . .

- Link health systems and organizations that serve women who are at high risk for smoking during pregnancy with available resources. An interactive Web-based program teaches best-practice approaches to help pregnant smokers and women of reproductive age to quit.

- Educate providers and pregnant women on tobacco-cessation coverage benefits and services available in your state. As of October 2010, states are to provide tobacco cessation counseling and medication without cost sharing for pregnant Medicaid beneficiaries.

- Monitor your state's prenatal smoking prevalence by using the CDC's PRAMStat.

Women Can . . .

- Quit smoking before you get pregnant, which is best.

- If you are pregnant, quit smoking to help reduce your and your baby's risk of health problems. It's never too late to quit smoking. Don't start smoking again after your baby is born.

- Talk with your doctor, nurse, or healthcare professional about quitting. For additional support, call the quitline at 800-QUIT-NOW (800-784-8669). The Quitline provides special resources for pregnant women.

- Watch or read real stories from mothers who quit smoking or whose children are affected by tobacco smoke.

- Learn more online about the effects of smoking and health risks of smoking.

Family and Friends Can . . .

- Be supportive and nonjudgmental.

- If you smoke, don't smoke around the expectant mother. Better yet, show your support and quit smoking yourself. It will benefit your health as well as hers.

- Let the expectant mother know about support from the quitline 800-QUIT-NOW (800-784-8669). The Quitline provides special resources for pregnant women.

- Tell her about real stories from mothers who quit smoking or whose children are affected by tobacco smoke.

- Learn more online about the effects of smoking and health risks of smoking.

Chapter 48

Talking to Your Teens about Tobacco

Young bodies are more sensitive to nicotine, and youth become addicted more quickly than adults. Even social smoking once or twice a month puts teens at serious risk for nicotine addiction that will keep them smoking longer and increase their chances of getting a serious disease. Most young smokers already show signs of damage to their hearts and blood vessels. One out of three teens who continue to smoke regularly will die prematurely—an average of 13 years earlier than their peers—because of smoking. Smokeless tobacco products also cause nicotine addiction.

What Can You Do?

You can influence your child's decision about whether to smoke. Even if you use tobacco yourself, your child will listen if you discuss

---

This chapter contains text excerpted from the following sources: Text in this chapter begins with excerpts from "What You Need to Know about Tobacco to Talk to Your Teens," Centers for Disease Control and Prevention (CDC), April 22, 2012. Reviewed February 2019; Text under the heading "Before the Talk" is excerpted from "Talk with Your Teen about E-cigarettes: A Tip Sheet for Parents," Office of the Surgeon General (OSG), December 8, 2016; Text under the heading "Adolescents and Tobacco: Tips for Parents" is excerpted from "Adolescents and Tobacco: Tips for Parents," U.S. Department of Health and Human Services (HHS), April 11, 2016.

your struggles with nicotine addiction and your regrets about starting in the first place. Be clear that you don't approve of smoking and that you expect your child to live tobacco-free.

- Tell your children that most kids don't smoke.

- Make your home and your car tobacco-free zones for everyone—family, friends, and visitors—and ban the use of all tobacco products.

- Set a good example and don't use tobacco yourself—it's the best thing you can do.

- Talk with your kids often about what nicotine addiction can cost them as they get older—and tell them you expect them to say no to tobacco.

- Make sure your children's schools enforce tobacco-free policies on campus and at all school-sponsored events.

- If your child is using tobacco, get help to quit for him or her right away. Nicotine is a highly addictive drug, and even experimenting with cigarettes one time increases your child's chance of being hooked for life. You can start by talking with your child's doctor.

Talk with Your Kids about Tobacco Use—Every Chance You Get!

Kids are less likely to smoke if they know you disapprove of it. They also respond if you share your own struggles with tobacco. If you've never smoked, tell them about family or friends who regret starting in the first place—or who have died from a smoking-related disease. The important thing is to talk with your children every chance you get and to get help right away if your child is using tobacco.

Try these conversation starters:

- Do you smoke or use tobacco? Do you think you ever would smoke?
  There are lots of reasons most teenagers don't smoke.

  - It's expensive

  - It smells bad

  - It makes your clothes and breath stink

  - It gives you wrinkles

- It's hard to quit once you start

- It doesn't make you slim

- Most teens think smoking is a dirty habit

- Most high school seniors would rather date nonsmokers

- What would you do if your best friend offered you a cigarette?

You can reduce your child's risk by helping her or him be ready to say no. For some kids, the direct approach is best ["Gross! Those things stink!"]. For others, a more low-key approach works ("No thanks—I'm good."). The important thing is to help your kids be prepared to resist peer pressure to use tobacco.

- Tell me what you've heard about smoking and health. How long do you think you have to smoke for it to affect your health?

You can tell your kids they don't have to smoke a lot or smoke a long time to get a disease from it. Smoking can affect athletic performance, activity level, and endurance in a very short time. Worst of all, smoking is addictive. Like heroin and cocaine, nicotine changes the way your brain works. It makes you crave more nicotine.

Before the Talk
Know the Facts

- Get credible information about e-cigarettes and young people at E-cigarettes.SurgeonGeneral.gov.

Be Patient and Ready to Listen

- Avoid criticism and encourage an open dialogue.

- Remember, your goal is to have a conversation, not to deliver a lecture.

- It's okay for your conversation to take place over time, in bits and pieces.

Set a Positive Example by Being Tobacco-Free

- If you use tobacco, it's never too late to quit. For free help, visit smokefree.gov or call 800-QUIT-NOW (800-7848-669).

Start the Conversation
Find the Right Moment

A more natural discussion will increase the likelihood that your teen will listen. Rather than saying "we need to talk," you might ask your teen what he or she thinks about a situation you witness together, such as:

- Seeing someone use an e-cigarette in person or in a video
- Passing an e-cigarette shop when you are walking or driving
- Seeing an e-cigarette advertisement in a store or magazine or on the Internet

Ask for Support

- Not sure where to begin? Ask your healthcare provider to talk to your teen about the risks of e-cigarettes
- You might also suggest that your teen talk with other trusted adults, such as relatives, teachers, faith leaders, coaches, or counselors whom you know are aware of the risks of e-cigarettes
- These supportive adults can help reinforce your message as a parent

Answer Their Questions

Here are some questions and comments you might get from your teen about e-cigarettes and some ideas about how you can answer them.

Why Don't You Want Me to Use E-Cigarettes?

- Science shows that e-cigarettes contain ingredients that are addictive and could harm different parts of your body.
- Right now, your brain is still developing, which means you are more vulnerable to addiction. Many e-cigarettes contain nicotine, and using nicotine can change your brain to make you crave more nicotine. It can also affect your memory and concentration. I don't want that for you!
- E-cigarettes contain chemicals that are harmful. When people use e-cigarettes, they breathe in tiny particles that can harm their lungs.

458

- The cloud that people exhale from e-cigarettes can expose you to chemicals that are not safe to breathe.

What's the Big Deal about Nicotine?

- Your brain is still developing until about age 25. The Surgeon General reported that nicotine is addictive and can harm your brain development.

- Using nicotine at your age may make it harder for you to concentrate, learn, or control your impulses.

- Nicotine can even train your brain to be more easily addicted to other drugs like meth and cocaine.

- I don't say this to scare you, but I want you to have the facts because nothing is more important to me than your health and safety.

Aren't E-Cigarettes Safer Than Conventional Cigarettes?

- Because your brain is still developing, scientific studies show that it isn't safe for you to use any tobacco product that contains nicotine, including e-cigarettes.

- Whether you get nicotine from an e-cigarette or a cigarette, it's still risky.

- Some e-cigarette batteries have even exploded and hurt people.

I Thought E-Cigarettes Didn't Have Nicotine—Just Water and Flavoring?

- I used to think that too. But many e-cigarettes have nicotine. There are also other chemicals in them that can be harmful.

- Let's look at the Surgeon General's website on e-cigarettes (E-cigarettes.SurgeonGeneral.gov) together so you can see for yourself.

I (Or My Friends) Have Tried E-Cigarettes and It Was No Big Deal

- I appreciate your honesty. In the future, I hope you (or your friends) will stay away from e-cigarettes and other tobacco products, including cigarettes. Science shows that e-cigarettes contain ingredients that are addictive and could harm different parts of your body.

- Next time we go to the doctor, let's ask about the risks of nicotine, e-cigarettes, and other tobacco products.

You Used Tobacco, So Why Shouldn't I?

- If I could live my life over again, I never would have started smoking. I learned that people who smoke cigarettes are much more likely to develop, and die from, certain diseases than people who don't smoke. This was really scary, so I quit smoking.

- Quitting was really hard, and I don't want you to go through that. The best thing is to not start at all.

Keep the Conversation Going

Many parents find that texting is a great way to reach their teens. Here are some suggestions for text messages that might catch your teen's attention. And, you can easily share pages of the website (e-cigarettes.surgeongeneral.gov) with your teen.

Adolescents and Tobacco: Tips for Parents

When it comes to preventing teens from using tobacco or helping them quit, parents and other caring adults can take several steps.

Set an example. Choose not to smoke or use other tobacco products. In addition to modeling desirable behavior, this would prevent adolescents' exposure to second-hand smoke, which can cause many of the same negative health effects that adolescents would experience if they smoked themselves.

Don't be shy. You should speak up before adolescents begin smoking or if tobacco use of any kind is suspected. Youth who do not use tobacco before the age of 26 are likely to never start.

Go the distance to prevent secondhand smoke exposure. In addition to not smoking yourself, you can prevent adolescents' exposure to secondhand smoke by not allowing anyone to smoke anywhere in or near an adolescents' home; not allowing smoking in the cars they ride in, even with a window down; ensuring that adolescents' schools are tobacco-free; and—if your state still allows smoking in public areas—frequenting restaurants and other places that do not allow smoking (note that "no-smoking sections" are not enough to shield a person from secondhand smoke).

Monitor. The amount of monitoring you do (such as having expectations about when adolescents will be home and checking on their plans) can lessen your teen risks of nicotine dependence.

Strongly disapprove. Adolescents whose parents or other adults in their lives strongly disapprove of their smoking—even if the adults themselves smoke—are less likely to take up smoking. Your disapproval has even been found to counteract the influence of peers on smoking.

Know what children watch. Setting limits on adolescents' movie choices may help prevent them from starting to smoke; many adult-oriented movies include depictions of smoking that may glamorize the habit.

Enlist allies. Other adults in adolescents' lives, such as teachers, grandparents, aunts, and uncles, influence whether teens start using tobacco and whether they stop. These adults can be important allies in communicating a no-smoking message to teens.

Chapter 49

Regulating Tobacco Advertising and Promotion

Chapter Contents

463

Section 49.1

Tobacco Industry Marketing

This section includes text excerpted from "Tobacco
Industry Marketing," Centers for Disease Control
and Prevention (CDC), May 4, 2018.

Tobacco Industry
Cigarette and Smokeless Tobacco Companies Spend Billions of Dollars Each Year to Market Their Products

- In 2016, cigarette and smokeless tobacco companies spent $9.5 billion on advertising and promotional expenses in the United States alone.

  - Cigarette companies spent $8.7 billion on cigarette advertising and promotion in 2016, an increase from $8.3 billion in 2015.

  - The five major U.S. smokeless tobacco manufacturers spent $759.3 million on smokeless tobacco advertising and promotion in 2016, an increase from $684.9 million spent in 2015.

The Money Cigarette and Smokeless Tobacco Companies Spent in 2016 on U.S. Marketing Amounted To

- About $26 million each day

- About $29 for every person (adults and children) in the United States per year (according to 2016 population estimate of 323,000,000)

- About $250 per year for each U.S. adult smoker (based on 37.8 million adult smokers in 2016)

The Following Three Categories Totaled Approximately $7.87 Billion and Accounted for 90.4 Percent of All Cigarette Company Marketing Expenditures in 2016

- Price discounts paid to retailers and wholesalers to reduce the price of cigarettes to consumers—about $7.2 billion

- Promotional allowances paid to cigarette retailers, such as payments for stocking, shelving, displaying, and merchandising particular brands—$228.8 million

- Promotional allowances paid to cigarette wholesalers, such as payments for volume rebates, incentive payments, value-added services, and promotions—$395.9 million

Marketing to Specific Populations
Youth and Young Adults

Scientific evidence shows that tobacco company advertising and promotion influences young people to start using tobacco.

- Adolescents who are exposed to cigarette advertising often find the ads appealing.

- Tobacco ads make smoking appear to be appealing, which can increase adolescents' desire to smoke.

The three most heavily advertised brands—Marlboro, Newport, and Camel—were the preferred brands of cigarettes smoked by middle- and high-school students in 2016.

Cigarette Brand Preferences among U.S. Middle School Students

- 38.3% preferred Marlboro

- 21.4% preferred Newport

- 13.4% preferred Camel

Cigarette Brand Preferences among U.S. High School Students

- 48.8% preferred Marlboro

- 16.6% preferred Newport

- 13.3% preferred Camel

Women

Women are also targeted by the tobacco industry, and tobacco companies continue to produce brands specifically for women. Marketing toward women is dominated by themes of social desirability, empowerment, and independence, which are conveyed by advertisements featuring slim, attractive, and athletic models.

Racial / Ethnic Communities

Advertisement and promotion of certain tobacco products appear to be targeted to members of racial/minority communities.

- Marketing to Hispanics and American Indians/Alaska Natives has included advertising and promotion of cigarette brands with names such as Rio, Dorado, and American Spirit®.

- The tobacco industry has targeted African American communities in its advertisements and promotional efforts for menthol cigarettes. Strategies include:

  - Campaigns that use urban culture and language to promote menthol cigarettes

  - Tobacco-sponsored hip-hop bar nights with samples of specialty menthol cigarettes

  - Targeted direct-mail promotions

- Tobacco companies' marketing to Asian Americans has included:

  - Sponsorship of Chinese and Vietnamese New Year festivals and other activities related to Asian/Pacific American Heritage Month

  - Heavy billboard and in-store advertisements in predominantly urban Asian American communities

  - Financial and in-kind contributions to community organizations

  - Support of Asian American business associations

Section 49.2

Facts about Tobacco Advertising, Promotion, and Sponsorship

This section includes text excerpted from "Advertising and Promotion," U.S. Food and Drug Administration (FDA), August 13, 2018.

Sponsoring an Event

Manufacturers, retailers, and distributors are prohibited from sponsoring or causing to be sponsored any athletic, musical, artistic or other social or cultural event, or any entry or team in any event, with the brand name, logo, symbol, motto, selling message, recognizable color or pattern of colors, or any other indicia of product identification identical or similar to, or identifiable with those used for any brand of cigarettes or smokeless tobacco, per 21 CFR§ 1140.34(c).

Use of Brand Name, Logo, Symbol, Motto, etc. in Items and Services

Manufacturers and distributors of imported cigarettes and smokeless tobacco, are prohibited from marketing, licensing, distributing, selling, or causing to be marketed, licensed, distributed, or sold any item (other than cigarettes or smokeless tobacco or roll-your-own paper) or service that bears the brand name, logo, symbol, motto, selling message, recognizable color or pattern of colors, or any other indicia of product identification identical or similar to, or identifiable with those used for any brand of cigarettes or smokeless tobacco, per 21 CFR§ 1140.34(a).

Free Sample Restrictions

The prohibition on the distribution of free samples applies to all "tobacco products," (except for smokeless tobacco products distributed in a "qualified adult-only facility" and in limited quantities as specified in the law, per section 102 of the Tobacco Control Act) including newly-deemed tobacco products, per 21 CFR § 1140.16(d).

Format and Display Requirements for Required Warning Statements on Advertisements

For cigarette tobacco, roll-your-own tobacco, and other covered tobacco products, it is unlawful for any such tobacco product manufacturer, packager, importer, distributor, or retailer of the tobacco product to advertise or cause to be advertised within the United States any tobacco product unless each advertisement bears the required warning statement. The required warning statement(s) on print advertisements or other advertisements with a visual component (including, for example, advertisements on signs, shelf-talkers, Web pages, and email) for cigarette tobacco, roll-your-own tobacco, and covered tobacco products must:

- Appear on the upper portion of the advertisement within the trim area;

- Occupy at least 20 percent of the area of the advertisement (warning area);

- Be printed in at least 12-point font size and ensure that the required warning statement occupies the greatest possible proportion of the warning area set aside for the required warning statement;

- Be printed in conspicuous and legible Helvetica bold or Arial bold type or other similar sans serif fonts and in black text on a white background or white text on a black background in a manner that contrasts by typography, layout, or color, with all other material on the advertisement;

- Be capitalized and punctuated, as described in the regulations (21 CFR §§ 1143.3(a)(1), 1143.3(c), and 1143.5(a)(1));

- Be centered in the warning area in which the text is required to be printed and positioned such that the text of the required warning statement and the other textual information in the advertisement have the same orientation; and

- Be surrounded by a rectangular border that is the same color as the text of the required warning statement and that is not less than 3 millimeters (mm) or more than 4 mm.

The warning statement requirement for advertisements outlined in 21 CFR §§ 1143.3(b) and 1143.5(b) apply to a retailer only if that retailer is responsible for or directs the health warning required under the paragraph. However, this does not relieve a retailer of liability if the retailer displays, in a location open to the public, an advertisement that:

- Does not contain a health warning; or

- Contains a health warning that has been altered by the retailer in a way that is material to the requirements of sections 1143.3(b) and 1143.5(b) (21 CFR §§ 1143.3(b)(3) and 1143.5(b) (3)).

Section 49.3

Cigar Labeling and Warning Statement Requirements

This section includes text excerpted from "Cigar Labeling and Warning Statement Requirements," U.S. Food and Drug Administration (FDA), August 13, 2018.

The deeming regulation requires that all cigar packages that are manufactured, packaged, sold, offered for sale, distributed, or imported for sale or distribution within the United States must bear one of the following required warning statements, per 21 CFR § 1143.5(a), on the package label:

- WARNING: Cigar smoking can cause cancers of the mouth and throat, even if you do not inhale.

- WARNING: Cigar smoking can cause lung cancer and heart disease.

- WARNING: Cigars are not a safe alternative to cigarettes.

- WARNING: Tobacco smoke increases the risk of lung cancer and heart disease, even in nonsmokers.

- WARNING: Cigar use while pregnant can harm you and your baby.

- SURGEON GENERAL WARNING: Tobacco Use Increases the Risk of Infertility, Stillbirth, and Low Birth Weight.

- WARNING: This product contains nicotine. Nicotine is an addictive chemical.

Each required warning statement must appear directly on the package and must be clearly visible underneath any cellophane or other clear wrappings as follows, per 21 CFR § 1143.5(a):

i. Be located in a conspicuous and prominent place on the two "principal display panels" of the package and the warning area must comprise at least 30 percent of each of the principal display panels;

ii. Appear in at least 12-point font size and ensure that the required warning statement occupies the greatest possible proportion of the warning area set aside for the required text;

iii. Be printed in conspicuous and legible Helvetica bold or Arial bold type (or other similar sans serif fonts) and in black text on a white background or white text on a black background in a manner that contrasts by typography, layout, or color, with all other printed material on the package;

iv. Be capitalized and punctuated as indicated in 21 CFR § 1143.5(a); and

v. Be centered in the warning area in which the text is required to be printed and positioned such that the text of the required warning statement and the other information on that principal display panel have the same orientation.

A cigar retailer will not be in violation of this section for packaging that:

- Contains a health warning

- Is supplied to the retailer by the tobacco product manufacturer, importer, or distributor, who has the required state, local, or Alcohol and Tobacco Tax and Trade Bureau (TTB)-issued license or permit, if applicable, and

- Is not altered by the retailer in a way that is material to the requirements of this section

Cigars in Small Packages

The U.S. Food and Drug Administration (FDA) issued the *Guidance: Compliance Policy for Required Warning Statements on Small-Packaged Cigars*. This guidance document is intended to assist any person who manufactures, packages, sells, offers to sell, distributes, or imports cigars in small packages with respect to the warning statement requirements in 21 CFR § 1143 for product packaging.

Cigar Warning Plans for Packaging

You must submit a proposed warning plan to the FDA no later than August 10, 2017, or 12 months before advertising or commercially marketing a cigar product, whichever is later (21 CFR § 1143.5(c)). Except for cigars sold individually and not in a product package, the six required warning statements must be randomly displayed in each 12-month period, in as equal a number of times as is possible on each brand of cigar sold in product packaging, in accordance with an FDA-approved warning plan, per 21 CFR § 1143.5(c).

The six warning statements must also be randomly distributed in all areas of the United States where the product is marketed, in accordance with the aforementioned plan submitted by the responsible cigar manufacturer, importer, distributor, or retailer to, and approved by the FDA.

Cigars Sold Individually without Packaging

For cigars that are sold individually, and not in a product package, the retailer must display all six of the required warning statements on a sign posted at the point-of-sale, per 21 CFR § 1143.5 (a).

The sign at point-of-sale must be:

- At least 8.5 × 11 inches in size;

- Clear, legible, and conspicuous;

- Printed in black Helvetica bold or Arial bold type or other similar sans serif fonts against a solid white background, in at least 17-point font size with appropriate spacing between the warning statements;

- Posted on or within 3 inches of each cash register where payment may be made; and

- Unobstructed in its entirety and read easily by each consumer making a purchase.

- The required warning statements on the sign must be:
  - Printed in a manner that contrasts, by typography, layout, or color, with all other printed material; and
  - Capitalized and punctuated as indicated in 21 CFR § 1143.5(a).

Other Cigar-Labeling Requirements

There are other labeling requirements. For example, tobacco products deemed under the deeming final rule to be subject to the FDA's authority, if in package form, must bear a label containing:

- The name and place of business of the tobacco product manufacturer, packer, or distributor;

- An accurate statement of the quantity of the contents in terms of weight, measure, or numerical count;

- An accurate statement of the percentage of the tobacco used in the product that is domestically grown tobacco and the percentage that is foreign grown tobacco; and

- The statement required under section 920(a)...(section 903(a) of the Federal Food, Drug, and Cosmetic Act (FD&C Act)).

Section 920(a) of the FD&C Act provides that the label, packaging, and shipping containers of tobacco products shall bear the statement "Sale only allowed in the United States."

Section 49.4

Federal Rules for Tobacco Sales

This section contains text excerpted from the following sources: Text in this section begins with excerpts from "Summary of Federal Rules for Tobacco Retailers," U.S. Food and Drug Administration (FDA), February 7, 2019; Text under the heading "Higher Cost of Tobacco Products, Cigarettes Increases Quit Attempts" is excerpted from "Smoking and Tobacco Use—Federal Tax Increase," Centers for Disease Control and Prevention (CDC), February 2, 2016.

The U.S. Food and Drug Administration (FDA) regulates all tobacco products, including e-cigarettes, hookah tobacco, and cigars. If you sell tobacco products, you must comply with all applicable federal laws and regulations for retailers.

How Do I Comply?

These rules apply to all "covered tobacco products" beginning August 8, 2016:

- Check photo identification (ID) of everyone under age 27 who attempts to purchase any tobacco product.

- Only sell tobacco products to customers age 18 or older.

- Do NOT sell tobacco products in a vending machine unless in an adult-only facility.

- Do NOT give away free samples of tobacco products to consumers, including any of their components or parts.

These rules, along with rules specific to each tobacco product, are listed below.

"This Is Our Watch" Program

The "This Is Our Watch" program helps retailers comply with federal tobacco law and regulations and protect minors. "This is Our Watch" program materials include a mix of educational pieces for owners, managers, and clerks, as well as a variety of point-of-purchase tools to inform customers of the law and emphasize the retailer's role.

Rules for Cigarettes, Cigarette Tobacco, and Roll-Your-Own Tobacco Sales

These rules have been in place since 2010:

- Check photo ID of everyone under age 27 who attempts to purchase cigarettes, cigarette tobacco, or roll-your-own tobacco.

- Only sell cigarettes, cigarette tobacco, and roll-your-own-tobacco to customers age 18 or older.

- Do NOT sell cigarettes, cigarette tobacco, or roll-your-own tobacco in a vending machine or self-service display unless in an adult-only facility.

- Do NOT give away free samples of cigarettes, cigarette tobacco, or roll-your-own tobacco to consumers, including any of their components or parts.

- Do NOT sell cigarettes, cigarette tobacco, or roll-your-own tobacco that contain a characterizing flavor (except menthol or tobacco flavor).

- Do NOT sell cigarette packages containing fewer than 20 cigarettes, including single cigarettes, known as "loosies."

- Do NOT break open packages of cigarettes, cigarette tobacco, or roll-your-own tobacco to sell products in smaller amounts.

Beginning August 10, 2018:

- Do NOT sell or distribute cigarette tobacco or roll-your-own tobacco products without a warning statement on the package.

- Do NOT display advertisements for cigarette tobacco or roll-your-own tobacco products without a warning statement.

Rules for Smokeless Tobacco Sales

- Check photo ID of everyone under age 27 who attempts to purchase smokeless tobacco.

- Only sell smokeless tobacco to customers age 18 and older.

- Do NOT sell smokeless tobacco in a vending machine or self-service display unless in an adult-only facility.

- Do NOT give away free samples of smokeless tobacco unless in a "qualified adult-only facility" and in limited quantities as specified in the law.

- Do NOT break open smokeless tobacco packages to sell products in smaller amounts.

- Do NOT sell smokeless tobacco without a health warning statement displayed on the package.

- Do NOT display advertisements for smokeless tobacco products without a warning statement.

Rules for Cigar Sales

- Check photo ID of everyone under age 27 who attempts to purchase cigars.

- Only sell cigars to customers age 18 and older.

- Do NOT sell cigars in a vending machine unless in an adult-only facility.

- Do NOT give away free samples of cigars to consumers, including any of their components or parts.

Beginning August 10, 2018:

- Do NOT sell or distribute cigars without a health warning statement displayed on the package.

- Do NOT display advertisements for cigars without a health warning statement.

- If you sell cigars individually, and not in a product package, you must post a sign with six required warning statements within 3 inches of each cash register.

Rules for Hookah and Pipe Tobacco Sales

- Check photo ID of everyone under age 27 who attempts to purchase hookah tobacco or pipe tobacco.

- Only sell hookah or pipe tobacco to customers age 18 and older.

- Do NOT sell hookah and pipe tobacco in a vending machine unless in an adult-only facility.

- Do NOT give away free samples of hookah or pipe tobacco to consumers, including any of their components or parts.

Beginning August 10, 2018:

- Do NOT sell or distribute hookah or pipe tobacco without a health warning statement displayed on the package.

- Do NOT display advertisements for hookah or pipe tobacco without a health warning statement.

Rules for Sales of E-Cigarettes and Other Electronic Nicotine Delivery Systems

Some examples of ENDS include e-cigarettes, vape pens, e-hookahs, e-cigars, personal vaporizers, and electronic pipes.

- Check photo ID of everyone under age 27 who attempts to purchase e-cigarettes or other ENDS.

- Only sell e-cigarettes and other ENDS to customers age 18 and older.

- Do NOT sell e-cigarettes or other ENDS in a vending machine unless in an adult-only facility.

- Do NOT give away free samples of e-cigarettes or other ENDS to consumers, including any of their components or parts.

Beginning August 10, 2018, these rules apply to all "covered tobacco products":

- Do NOT sell or distribute e-cigarettes or other ENDS without a health warning statement on the package.

- Do NOT display advertisements for e-cigarettes or other ENDS without a health warning statement.

Rules for Sales of Nicotine Gels

- Check photo ID of everyone under age 27 who attempts to purchase nicotine gel.

- Only sell nicotine gel to customers age 18 and older.

- Do NOT sell nicotine gel in a vending machine unless in an adult-only facility.

- Do NOT give away free samples of nicotine gel to consumers.

Beginning August 10, 2018:

- Do NOT sell or distribute nicotine gel without a health warning statement on the package.
- Do NOT display advertisements for nicotine gel without a health warning statement.

Rules for Sales of Dissolvables

These rules apply to dissolvable tobacco products that are not already regulated as smokeless tobacco.

- Check photo ID of everyone under age 27 who attempts to purchase dissolvable tobacco products.
- Only sell dissolvable tobacco products to customers age 18 and older.
- Do NOT sell dissolvable tobacco products in a vending machine unless in an adult-only facility.
- Do NOT give away free samples of dissolvable tobacco products to consumers.

Beginning August 10, 2018:

- Do NOT sell or distribute a dissolvable tobacco product without a health warning statement on the package.
- Do NOT display advertisements for dissolvable tobacco products without a health warning statement.

Higher Cost of Tobacco Products, Cigarettes Increases Quit Attempts

The largest federal tobacco tax increase in history took effect on April 1, 2009—a move expected to prompt one million smokers to quit and prevent 2 million young people from ever starting.

The 62-cent increase raises federal excise taxes on cigarettes to $1.01 per pack, up from 39 cents. The tax increase, which will fund the State Children's Health Insurance Program (SCHIP) for lower-income children, covers cigarettes, smokeless tobacco, and cigars.

Research has proven that raising cigarette prices, through excise taxes or other methods, increases the quit rate among adult smokers and is especially effective in discouraging children and young people from ever starting to smoke.

Quitting smoking has immediate, as well as long-term, health benefits, including reducing risks for smoking-related diseases such as cancer, heart disease, stroke, and some respiratory illnesses.

For support in quitting, including free confidential coaching, educational materials, and referrals to local resources, call 800-QUIT-NOW (800-784-8669).

Chapter 50

Health Warnings for Cigarettes: What You Need to Know

The Family Smoking Prevention and Tobacco Control Act (TCA) was signed by President Barack Obama on June 22, 2009, to help reduce the toll of death and disease from tobacco use. In compliance with this act, the U.S. Food and Drug Administration (FDA) issued a proposed rule on November 10, 2010, that requires new graphic health warnings for cigarette packages and advertisements.

For cigarette packages, the rule requires that graphic health warnings cover at least one half of the package, including the top front and back of each package. The rule also requires that graphic warnings appear in each cigarette advertisement and occupy at least 20 percent of the ad. The proposed rule includes nine new textual warning statements and 36 proposed images designed to help the public better understand the negative health consequences of smoking.

The FDA sought public comments on the proposed rule through January 9, 2011. They considered all comments, the latest scientific literature, and other research before selecting nine of the 36 images

This chapter includes text excerpted from "Graphic Health Warnings for Cigarettes: What You Need to Know," Centers for Disease Control and Prevention (CDC), February 7, 2017.

with warning statements. They issued the final regulations about these graphic warning labels in June 2011.

The warnings display facts such as:

- Cigarettes are addictive

- Tobacco smoke can harm your children

- Cigarettes cause fatal lung disease

- Smoking during pregnancy can harm your baby

- Quitting smoking now greatly reduces serious risks to your health

These and four other messages are accompanied by color graphics showing the negative health consequences of smoking. Nine different warning labels appear on cigarette packages and advertisements on a rotating basis.

After the regulations were finalized, all cigarette packages and advertisements had to include these graphic warning labels starting September 2012. In addition, tobacco manufacturers could no longer distribute for sale packages of cigarettes without these labels after October 2012.

Do your part to reduce the negative health consequences of tobacco use. Get help to quit and discourage others, especially youth, from starting.

Part Six

Additional Help and Information

Chapter 51

Glossary of Terms Related to Tobacco Use and Smoking Cessation

absorption: The process of taking in. For a person or an animal, absorption is the process of a substance getting into the body through the eyes, skin, stomach, intestines, or lungs.

ACE inhibitor: A drug that is used to lower blood pressure. An ACE inhibitor is a type of antihypertensive agent. Also called angiotensin-converting enzyme inhibitor.

acquired immunodeficiency syndrome (AIDS): A disease caused by the human immunodeficiency virus (HIV). People with AIDS are at an increased risk for developing certain cancers and for infections that usually occur only in individuals with a weak immune system.

addiction: A chronic, relapsing disease characterized by compulsive drug seeking and use despite serious adverse consequences, and by long-lasting changes in the brain.

adverse effect: An unexpected medical problem that happens during treatment with a drug or other therapy. Adverse effects may be mild,

---

This glossary contains terms excerpted from documents produced by several sources deemed reliable.

moderate, or severe, and may be caused by something other than the drug or therapy being given. Also called adverse event.

alcohol: A chemical substance found in drinks such as beer, wine, and liquor. It is also found in some medicines, mouthwashes, household products, and essential oils (scented liquid taken from certain plants). It is made by a chemical process called fermentation that uses sugars and yeast.

amenorrhea: The abnormal absence of menstrual periods. Early amenorrhea caused by overtraining can cause bones to become brittle and break.

analgesics: A group of medications that reduce pain.

angioplasty: A procedure to enlarge the opening in a blood vessel that has become narrowed or blocked by plaque (a buildup of fat and cholesterol on the inner wall of the blood vessel).

antibiotic: A drug used to treat infections caused by bacteria and other microorganisms.

anxiety: Feelings of fear, dread, and uneasiness that may occur as a reaction to stress. A person with anxiety may sweat, feel restless and tense, and have a rapid heartbeat.

arthritis: A general term for conditions that cause inflammation (swelling) of the joints and surrounding tissues. Some forms of arthritis may occur simultaneously with osteoporosis and Paget disease.

aspiration: The removal of fluid or tissue through a needle. Also, the accidental breathing in of food or fluid into the lungs.

assessment: The process of gathering evidence and documentation of a student's learning.

asthma: A chronic disease in which the bronchial airways in the lungs become narrowed and swollen, making it difficult to breathe. Symptoms include wheezing, coughing, tightness in the chest, shortness of breath, and rapid breathing.

backbone: The bones, muscles, tendons, and other tissues that reach from the base of the skull to the tailbone. The backbone encloses the spinal cord and the fluid surrounding the spinal cord. Also called spinal column, spine, and vertebral column.

bacteria: A large group of single-cell microorganisms. Some cause infections and disease in animals and humans.

biomedical testing: Testing of persons to find out whether a change in a body function might have occurred because of exposure to a hazardous substance.

blood: A tissue with red blood cells (RBCs), white blood cells (WBCs), platelets, and other substances suspended in fluid called plasma. Blood takes oxygen and nutrients to the tissues, and carries away wastes.

bone: A living, growing tissue made mostly of collagen.

breast cancer: Cancer that forms in tissues of the breast. The most common type of breast cancer is ductal carcinoma, which begins in the lining of the milk ducts (thin tubes that carry milk from the lobules of the breast to the nipple).

bypass: A surgical procedure in which the doctor creates a new pathway for the flow of body fluids.

calcium: A mineral that is an essential nutrient for bone health. It is also needed for the heart, muscles, and nerves to function properly and for blood to clot.

calorie: A measurement of the energy content of food. The body needs calories as to perform its functions, such as breathing, circulating the blood, and physical activity. When a person is sick, their body may need extra calories to fight fever or other problems.

cancer: A term for diseases in which abnormal cells divide without control and can invade nearby tissues. Cancer cells can also spread to other parts of the body through the blood and lymph systems.

cannabis: The dried leaves and flowering tops of the *Cannabis sativa* or *Cannabis indica plant*. *Cannabis* contains active chemicals called cannabinoids that cause drug-like effects all through the body, including the central nervous system and the immune system.

carbohydrate: A sugar molecule. Carbohydrates can be small and simple (for example, glucose) or they can be large and complex (for example, polysaccharides such as starch, chitin, or cellulose).

carcinogen: A substance that can cause cancer.

chewing tobacco: A type of shredded or twisted smokeless tobacco that the user keeps in his or her mouth, between the cheek and gum.

chromosome: A chromosome is an organized package of deoxyribonucleic acid (DNA) found in the nucleus of the cell. Different organisms have different numbers of chromosomes. Humans have 23 pairs of chromosomes—22 pairs of numbered chromosomes, called autosomes, and one pair of sex chromosomes, X and Y.

chronic disease: A disease that has one or more of the following characteristics: is permanent; leaves residual disability; is caused by nonreversible pathological alternation; requires special training of the patient for rehabilitation; or may be expected to require a long period of supervision, observation, or care.

chronic obstructive pulmonary disease (COPD): A type of lung disease marked by permanent damage to tissues in the lungs, making it hard to breathe. COPD includes chronic bronchitis, in which the bronchi (large air passages) are inflamed and scarred, and emphysema, in which the alveoli (tiny air sacs) are damaged. It develops over many years and is usually caused by cigarette smoking.

chronic pain: Pain that can range from mild to severe, and persists or progresses over a long period of time.

cigar: A tube of tobacco that is thicker than a cigarette, wrapped in tobacco leaf, lit, and smoked. Cigars include regular cigars, cigarillos, and little filtered cigars.

cigarette: A tube-shaped tobacco product that is made of finely cut, cured tobacco leaves wrapped in thin paper. It may also have other ingredients, including substances to add different flavors.

computed tomography (CT): A procedure for taking X-ray images from many different angles and then assembling them into a cross-section of the body. This technique is generally used to visualize bone.

constipation: A decrease in frequency of stools or bowel movements with hardening of the stool. Some forms of osteogenesis imperfecta are associated with increased risk for constipation caused by increased perspiration, growth impairment, pelvic malformation, and diminished physical activity.

coronary heart disease: A disease in which there is a narrowing or blockage of the coronary arteries (blood vessels that carry blood and oxygen to the heart). Coronary heart disease is usually caused by atherosclerosis (a buildup of fatty material and plaque inside the coronary arteries).

diabetes: A disease in which the body does not produce or properly use insulin. Insulin is a hormone that is needed to convert sugar, starches, and other food into energy. Having diabetes may increase osteoporosis risk.

diet: What a person eats and drinks. Any type of eating plan.

dip: Dipping tobacco (so called because users dip their fingers into the package to pinch a portion to insert into the mouth) is moist ground tobacco placed between the lower lip or cheek and the gums; it is not used nasally.

dissolvables: Dissolvables are finely ground tobacco pressed into shapes such as tablets, orbs, sticks, or strips and slowly dissolve in the mouth.

dopamine: A brain chemical, classified as a neurotransmitter, found in regions that regulate movement, emotion, motivation, and pleasure.

e-cigarette: A device that has the shape of a cigarette, cigar, or pen and does not contain tobacco. It uses a battery and contains a solution of nicotine, flavorings, and other chemicals, some of which may be harmful.

enzyme: A protein that speeds up chemical reactions in the body.

estrogen: A type of hormone made by the body that helps develop and maintain female sex characteristics and the growth of long bones.

exercise: A type of physical activity that involves planned, structured, and repetitive bodily movement done to maintain or improve one or more components of physical fitness.

fracture: Broken bone. People with osteoporosis, osteogenesis imperfecta, and Paget disease are at greater risk for bone fracture.

gynecologist: A doctor who diagnoses and treats conditions of the female reproductive system and associated disorders.

head and neck cancer: Cancer that arises in the head or neck region (in the nasal cavity, sinuses, lips, mouth, salivary glands, throat, or larynx [voice box]).

hookah: Hookahs are water pipes that are used to smoke specially made tobacco that comes in different flavors, such as apple, mint, cherry, chocolate, coconut, licorice, cappuccino, and watermelon. Also called water pipes.

immune system: A complex system of cellular and molecular components having the primary function of distinguishing self from not self and defense against foreign organisms or substances.

inflammatory bowel disease (IBD): Diseases, including ulcerative colitis (UC) and Crohn disease, that cause swelling in the intestine and/or digestive tract, which may result in diarrhea, abdominal pain, fever, and weight loss. People with IBD are at an increased risk for osteoporosis.

inhalation: The act of breathing. A hazardous substance can enter the body this way.

lesion: An area of abnormal tissue. A lesion may be benign (not cancer) or malignant (cancer).

lung cancer: Cancer that forms in tissues of the lung, usually in the cells lining air passages. The two main types are small cell lung cancer and nonsmall cell lung cancer. These types are diagnosed based on how the cells look under a microscope.

lupus: A chronic inflammatory disease that occurs when the body's immune system attacks its own tissues and organs. Also, called systemic lupus erythematosus (SLE). Inflammation caused by lupus can affect many different body systems including joints, skin, kidneys, blood cells, heart, and lungs. People with lupus are at increased risk for osteoporosis.

magnetic resonance imaging (MRI): A noninvasive procedure that uses magnetic fields and radio waves to produce three-dimensional computerized images of areas inside the body.

marijuana: The dried leaves and flowering tops of the *Cannabis sativa* or *Cannabis indica plant*. Marijuana contains active chemicals called cannabinoids that cause drug-like effects all through the body, including the central nervous system and the immune system.

menopause: The cessation of menstruation in women. Bone health in women often deteriorates after menopause due to a decrease in the female hormone estrogen.

metabolism: The chemical changes that take place in a cell or an organism. These changes make energy and the materials cells and organisms need to grow, reproduce, and stay healthy. Metabolism also helps get rid of toxic substances.

neurotransmitter: A chemical produced by neurons that carry messages from one nerve cell to another.

nicotine: Chemical in tobacco that causes and maintains the powerful addicting effects of tobacco products.

nutrition: The taking in and use of food and other nourishing material by the body. Nutrition is a 3-part process. First, food or drink is consumed. Second, the body breaks down the food or drink into nutrients.

oral cancer: Cancer that forms in tissues of the oral cavity (the mouth) or the oropharynx (the part of the throat at the back of the mouth).

organ: A part of the body that performs a specific function. For example, the heart is an organ.

osteoporosis: Literally means "porous bone." This disease is characterized by too little bone formation, excessive bone loss, or a combination of both, leading to bone fragility and an increased risk of fractures of the hip, spine and wrist.

over-the-counter (OTC): Refers to a medicine that can be bought without a prescription (doctor's order). Examples include analgesics (pain relievers), such as aspirin and acetaminophen. Also called nonprescription and OTC.

overweight: Overweight refers to an excessive amount of body weight that includes muscle, bone, fat, and water. A person who has a body mass index (BMI) of 25 to 29.9 is considered overweight.

periodontitis: A chronic infection that affects the gums and the bones that support the teeth. Bacteria and the body's own immune system break down the bone and connective tissue that hold teeth in place. Teeth may eventually become loose, fall out, or have to be removed.

physical activity: Any bodily movement that is produced by the contraction of skeletal muscle and that substantially increases energy expenditure.

pipe: A tube with a small bowl at one end that is filled with tobacco, lit, and smoked.

portion size: The amount of a food served or eaten in one occasion. A portion is not a standard amount.

pregnancy: The condition between conception (fertilization of an egg by a sperm) and birth, during which the fertilized egg develops in the uterus. In humans, pregnancy lasts about 288 days.

prevention: Actions that reduce exposure or other risks, keep people from getting sick, or keep disease from getting worse.

prognosis: The likely outcome or course of a disease; the chance of recovery or recurrence.

prostate cancer: A disease in which abnormal tumor cells develop in the prostate gland. Men who receive hormone deprivation therapy for prostate cancer have an increased risk of developing osteoporosis and broken bones.

protein: A molecule made up of amino acids. Proteins are needed for the body to function properly. They are the basis of body structures, such as skin and hair, and of other substances such as enzymes, cytokines, and antibodies.

psychosis: A mental disorder characterized by delusional or disordered thinking detached from reality; symptoms often include hallucinations.

radiation: Energy moving in the form of particles or waves. Familiar radiations are heat, light, radio, and microwaves.

rheumatoid arthritis (RA): An inflammatory disease that causes pain, swelling, stiffness, and loss of function in the joints. It occurs when the immune system, which normally defends the body from invading organisms, attacks the membrane lining the joints. Studies have found an increased risk of bone loss and fracture in individuals with RA.

saliva: The watery fluid in the mouth made by the salivary glands. Saliva moistens food to help digestion and it helps protect the mouth against infections.

salivary gland cancer: A rare cancer that forms in tissues of a salivary gland (gland in the mouth that makes saliva). Most salivary gland cancers occur in older people. A rare cancer that forms in tissues of a salivary gland (gland in the mouth that makes saliva). Most salivary gland cancers occur in older people.

secondhand smoke: Smoke that comes from the burning of a tobacco product and smoke that is exhaled by smokers. Inhaling secondhand smoke is called involuntary or passive smoking. Also, called environmental tobacco smoke and ETS.

serum: The liquid part of blood that remains after clotting proteins and blood cells are removed.

smokeless tobacco: A type of tobacco that is not smoked or burned. It may be used as chewing tobacco or moist snuff, or inhaled through the nose as dry snuff. Smokeless tobacco contains nicotine and many harmful, cancer-causing chemicals. Using it can lead to nicotine addiction and can cause cancers of the mouth, esophagus, and pancreas. It may also cause heart disease, gum disease, and other health problems.

smoking cessation: To quit smoking. Smoking cessation lowers the risk of cancer and other serious health problems. Counseling, behavior therapy, medicines, and nicotine-containing products, such as nicotine patches, gum, lozenges, inhalers, and nasal sprays, may be used to help a person quit smoking.

snuff: Finely ground smokeless tobacco usually sold in round cans. Can be placed between the cheek and gum or may be sniffed.

snus: Moist snuff packaged in ready-to-use pouches that resemble small tea bags Pouch is placed between cheek or teeth and gums; does not require spitting.

sodium: A mineral and an essential nutrient needed by the human body in relatively small amounts (provided that substantial sweating does not occur).

steroid: Any of a group of lipids (fats) that have a certain chemical structure. Steroids occur naturally in plants and animals or they may be made in the laboratory.

tobacco: A plant with leaves that have high levels of the addictive chemical nicotine. After harvesting, tobacco leaves are cured, aged, and processed in various ways. The resulting products may be smoked (in cigarettes, cigars, and pipes), applied to the gums (as dipping and chewing tobacco), or inhaled (as snuff).

toxic: Causing temporary or permanent effects detrimental to the functioning of a body organ or group of organs.

virus: In medicine, a very simple microorganism that infects cells and may cause disease. Because viruses can multiply only inside infected cells, they are not considered to be alive.

vitamin D: A nutrient that the body needs to absorb calcium.

weight-bearing exercise: Exercise that forces you to work against gravity, such as walking, hiking, jogging, climbing stairs, playing tennis, dancing, and lifting weights. This type of exercise is best for strengthening bone.

withdrawal: Symptoms that occur after chronic use of a drug is reduced abruptly or stopped.

X-ray: A type of radiation used in the diagnosis and treatment of cancer and other diseases. In low doses, X-rays are used to diagnose diseases by making pictures of the inside of the body.

Chapter 52

Tobacco Information and Control Resources

Government Agencies That Provide Information about Smoking

Agency for Healthcare Research and Quality (AHRQ)
5600 Fishers Ln.
Seventh Fl.
Rockville, MD 20857
Phone: 301-427-1364
Website: www.ahrq.gov

Bureau of Alcohol, Tobacco, Firearms and Explosives (ATF)
99 New York Ave. N.E.
Rm. 5S 144
Washington, DC 20226
Toll-Free: 800-800-3855
Phone: 202-648-7777
Website: www.atf.gov
E-mail: ATFTips@atf.gov

Center for Substance Abuse Prevention (CSAP)
5600 Fishers Ln.
Rockville, MD 20857
Phone: 240-276-2420
Website: www.samhsa.gov

Centers For Disease Control and Prevention (CDC)
1600 Clifton Rd.
Atlanta, GA 30329-4027
Toll-Free: 800-CDC-INFO (800-232-4636)
Toll-Free TTY: 888-232-6348
Website: www.cdc.gov

Resources in this chapter were compiled from several sources deemed reliable; all contact information was verified and updated in February 2019.

Federal Trade Commission (FTC)

Public Reference Branch
600 Pennsylvania Ave. N.W.
Washington, DC 20580
Toll-Free: 877-FTC-HELP
(877-382-4357)
Phone: 202-326-2222
Toll-Free TTY: 866-653-4261
Website: www.ftc.gov

National Cancer Institute (NCI)

9609 Medical Center Dr.
BG 9609, MSC 9760
Bethesda, MD 20892-9760
Toll-Free: 800-4-CANCER
(800-422-6237)
Phone: 301-435-3848
Website: www.cancer.gov
E-mail: cancergovstaff@mail.nih.gov

National Health Information Center (NHIC)

Office of Disease Prevention and
health promotion (ODPHP)
1101 Wootton Pkwy.
Ste. LL100
Rockville, MD 20852
Fax: 240-453-8281
Website: www.health.gov/nhic
E-mail: nhic@hhs.gov

National Heart, Lung, and Blood Institute (NHLBI)

Bldg. 31, 31 Center Dr.
Bethesda, MD 20892
Website: www.nhlbi.nih.gov
E-mail: nhlbiinfo@nhlbi.nih.gov

National Institute for Dental and Craniofacial Research (NIDCR)

31 Center Dr. MSC 2190
Bldg. 31, Rm. 5B55
Bethesda, MD 20892-2190
Toll-Free: 866-232-4528
Phone: 301-496-4261
Fax: 301-496-9988
Website: www.nidcr.nih.gov
E-mail: nidcrinfo@mail.nih.gov

National Institute on Drug Abuse (NIDA)

Office of Science Policy and
Communications (OSPC)
6001 Executive Blvd.
Rm. 5213, MSC 9561
Bethesda, MD 20892-9561
Phone: 301-443-1124
Website: www.nida.nih.gov

NIH News in Health

NIH Office of Communications
and Public Liaison (OCPL)
9000 Rockville Pike
Bldg. 31, Rm. 5B52
Bethesda, MD 20892
Phone: 301-451-8224
Website: newsinhealth.nih.gov
E-mail: nihnewsinhealth@od.nih.gov

NIH Osteoporosis and Related Bone Diseases— National Resource Center (NIH OBRD—NRC)
2 AMS Cir.
Bethesda, MD 20892-3676
Toll-Free: 800-624-BONE
(800-624-2663)
Phone: 202-223-0344
TTY: 202-466-4315
Fax: 202-293-2356
Website: www.bones.nih.gov
E-mail: NIHBoneInfo@mail.nih.gov

Office of the Surgeon General (OSG)
U.S. Department of Health and Human Services (HHS)
200 Independence Ave. S.W.
Humphrey Bldg.
Ste. 701H
Washington, DC 20201
Website: www.surgeongeneral.gov

Office on Women's Health (OWH)
U.S. Department of Health and Human Services (HHS)
200 Independence Ave. S.W.
Rm. 712E
Washington, DC 20201
Toll-Free: 800-994-9662
Phone: 202-690-7650
Fax: 202-205-2631
Website: www.womenshealth.gov
E-mail: womenshealth@hhs.gov

Substance Abuse and Mental Health Services Administration (SAMHSA)
5600 Fishers Ln.
Rockville, MD 20857
Toll-Free: 877-SAMHSA-7
(877-726-4727)
Toll-Free TTY: 800-487-4889
Fax: 240-221-4292
Website: www.samhsa.gov

Tobacco Control Research Branch (TCRB)
National Cancer Institute (NCI)
9609 Medical Center Dr.
BG 9609, MSC 9760
Bethesda, MD 20892-9760
Phone: 301-496-8584
Fax: 301-496-8675
Website: smokefree.gov
E-mail: NCISmokefreeTeam@mail.nih.gov

U.S. Department of Agriculture (USDA)
1400 Independence Ave. S.W.
Washington, DC 20250
Phone: 202-720-2791
Website: www.usda.gov

U.S. Department of Health and Human Services (HHS)
200 Independence Ave. S.W.
Washington, D.C. 20201
Toll-Free: 877-696-6775
Website: www.hhs.gov

*U.S. Environmental
Protection Agency (EPA)*
1200 Pennsylvania Ave. N.W.
Washington, DC 20460
Toll-Free: 800-424-8802
Phone: 202-564-4700
Website:.www.epa.gov

*U.S. Food and Drug
Administration (FDA)*
10903 New Hampshire Ave.
Silver Spring, MD 20993-0002
Toll-Free: 888-INFO-FDA
(888-463-6332)
Website: www.fda.gov

*U.S. National Library of
Medicine (NLM)*
8600 Rockville Pike
Bethesda, MD 20894
Website: www.nlm.nih.gov

Private Agencies That Provide Information about Smoking

Advocacy Institute
Phone: 540-364-0051
Website: www.advocacyinstitute.
org

*American Cancer Society
(ACS)*
250 Williams St. N.W.
Atlanta, GA 30303
Toll-Free: 800-ACS-2345
(800-227-2345)
Website: www.cancer.org

*The American College
of Obstetricians and
Gynecologists (ACOG)*
409 12th St. S.W.
P.O. Box 70620
Washington, DC 20024-9998
Toll-Free: 800-673-8444
Phone: 202-638-5577
Website: www.acog.org

*American Council on Science
and Health (ACSH)*
110 E. 42nd St.
Ste. 1300
New York, NY 10017
Phone: 212-362-7044
Website: www.acsh.org
E-mail: acsh@acsh.org

*American Heart Association
National Center*
7272 Greenville Ave.
Dallas, TX 75231
Toll-Free: 800-AHA-USA1
(800-242-8721)
Website: www.heart.org

American Lung Association (ALA)
55 W. Wacker Dr.
Ste. 1150
Chicago, IL 60601
Toll-Free: 800-LUNGUSA
(800-586-4872)
Website: www.lung.org

American Medical Association (AMA)
330 N. Wabash Ave.
Ste. 39300
Chicago, IL 60611-5885
Toll-Free: 800-621-8335
Website: www.ama-assn.org

Americans for Nonsmokers' Rights (ANR)
2530 San Pablo Ave.
Ste. J
Berkeley, CA 94702
Phone: 510-841-3032
Website: www.no-smoke.org

Association of State and Territorial Health Officials (ASTHO)
2231 Crystal Dr.
Ste. 450
Arlington, VA 22202
Phone: 202-371-9090
Fax: 571-527-3189
Website: www.astho.org

Cancer Research Foundation of America (CRFA)
1600 Duke St.
Ste. 500
Alexandria, VA 22314
Toll-Free: 800-227-2732
Phone: 703-836-4412
Fax: 703-836-4413
Website: www.preventcancer.org
E-mail: PCF@preventcancer.org

Center for Tobacco Research and Intervention (CTRI)
University of Wisconsin Medical School
1930 Monroe St.
Ste. 200
Madison, WI 53711
Phone: 608-262-8673
Fax: 608-265-3102
Website: www.ctri.wisc.edu
E-mail: infoctri@ctri.wisc.edu

The Foundation For a Smoke Free America
8117 W. Manchester Ave.
Ste. 500
Playa del Rey, CA 90293-8745
Phone: 310-577-9828
Fax: 310-388-1350
Website: anti-smoking.org

Robert Wood Johnson Foundation (RWJF)
50 College Rd. E.
Princeton, NJ 08540-6614
Toll-Free: 877-843-7953
Phone: 609-627-6000
Website: www.rwjf.org

March of Dimes Birth Defects Foundation
1550 Crystal Dr.
Ste. 1300
Arlington, VA 22202
Toll-Free: 888-MODIMES
(888-663-4637)
Website: www.marchofdimes.org

National Center for Tobacco-Free Kids
1400 I St. N.W.
Ste. 1200
Washington, DC 20005
Phone: 202-296-5469
Fax: 202-296-5427
Website: www.tobaccofreekids.org
E-mail: info@tobaccofreekids.org

National Families in Action (NFIA)
2957 Clairmont Rd. N.E.
P.O. Box 133136
Atlanta, GA 30333-3136
Phone: 404-248-9676
Website: www.nationalfamilies.org
E-mail: nfia@nationalfamilies.org

National Federation of State High School Associations
P.O. Box 690
Indianapolis, IN 46206
Phone: 317-972-6900
Fax: 317-822-5700
Website: www.nfhs.org

National Governors Association (NGA)
444 N. Capitol
Ste. 267
Washington, DC 20001-1512
Phone: 202-624-5300
Fax: 202-624-5313
Website: www.nga.org

SmokeFree.net
Website: www.smokefree.net
E-mail: info@smokefree.net

Truth Initiative
900 G St. N.W.
Fourth Fl.
Washington, DC 20001
Phone: 202-454-5555
Website: truthinitiative.org

Chapter 53

Smoking Cessation: Hotlines, Helplines, and Internet Resources

Nationwide Smoking Cessation Hotlines and Helplines

American Cancer Society (ACS)
Toll-Free: 800-ACS-2345
(800-227-2345)
Toll-Free TDD/TTY:
866-228-4327
Website: www.cancer.org

American Lung Association
Toll-Free: 800-LUNGUSA
(800-586-4872)
Website: www.lung.org

National Cancer Institute's Smoking Quitline
Toll-Free: 877-44U-QUIT
(877-448-7848)
Website: www.cancer.gov

State-by-State Help for Smoking Cessation

Alaska
Toll-Free: 800-QUIT-NOW
(800-784-8669)

Arizona
Toll-Free: 800-556-6222
Website: www.ashline.org

Resources in this chapter were compiled from several sources deemed reliable; all contact information was verified and updated in February 2019.

California
Toll-Free: 800-NO-BUTTS
(800-662-8887)
Spanish: 800-45-NO-FUME
(800-456-6386)
Vietnamese: 800-778-8440
Mandarin and Cantonese:
800-838-8917
Korean: 800-556-5564
Tobacco Chewers: 800-844-
CHEW (800-844-2439)
Website: www.nobutts.org

Colorado
Toll-Free: 800-QUIT-NOW
(800-784-8669)
Website: www.coquitline.org

Connecticut
Toll-Free: 800-QUIT-NOW
(800-784-8669)
Website: portal.ct.gov

Delaware
Toll-Free: 866-409-1858
Website: dhss.delaware.gov/
dhss/dph/dpc/quitline.html

Florida
Toll-Free: 877-U-CAN-NOW
(877-822-6669)
Toll-Free TDD/TTY:
877-777-6534
Website: www.quitnow.net/florida

Georgia
Toll-Free: 877-270-STOP (877-
270-7867; in-state calls only)
Toll-Free TDD/TTY:
877-777-6534
Website: www.quitnow.net/
georgia

Illinois
Toll-Free: 800-QUIT-YES
(800-784-8937)
Website: www.smoke-free.
illinois.gov/sf_quit.htm

Iowa
Toll-Free: 800-QUIT-NOW
(800-784-8669)
Website: iowa.quitlogix.org

Maine
Toll-Free: 800-207-1230

Maryland
Toll-Free: 800-QUIT-NOW
(800-784-8669)
Website: mdquit.org/quitline

Massachusetts
Toll-Free: 800-QUIT-NOW
(800-784-8669)
Spanish and Portuguese:
800-8-Déjalo (800-833-5256)
TDD/TTY: 888-229-2182
Website: makesmokinghistory.
org/quit-now/
what-is-the-helpline

Michigan
Toll-Free: 800-QUIT-NOW
(800-784-8669)
Website: www.michigan.gov/
mdhhs/0,5885,7-339--210613--
,00.html

Mississippi
Toll-Free: 800-QUIT-NOW
(800-784-8669)
Website: quitlinems.com/
contact-us

Montana
Toll-Free: 800-QUIT-NOW
(800-784-8669)
Website: dphhs.mt.gov/
publichealth/mtupp/quitline

Nevada
Toll-Free: 800-QUIT-NOW
(800-784-8669)
Website: nevada.quitlogix.org/
en-US

New Jersey
Toll-Free: 866-NJSTOPS
(866-657-8677)
Website: njquitline.org

New Mexico
Toll-Free: 800-QUIT-NOW
(800-784-8669)
TTY: 877-777-6534
Website: quitnownm.com

New York
Toll-Free: 866-NY-QUITS
(866-697-8487)
Website: www.nysmokefree.com

Ohio
Toll-Free: 800-QUIT-NOW
(800-784-8669)
Website: odh.ohio.gov/wps/portal/
gov/odh/know-our-programs/
tobacco-use-prevention-and-
cessation/cessation

Oklahoma
Toll-Free: 800-QUIT-NOW
(800-784-8669)
Spanish: 855-DÉJELO-YA
TDD/TTY: 877-777-6534
Website: okhelpline.com

Oregon
Toll-Free: 800-QUIT-NOW
(800-784-8669)
Website: www.quitnow.net/oregon

Pennsylvania
Toll-Free: 800-QUIT-NOW
(800-784-8669)
Spanish: 855-DEJELO-YA
(855-335-3569)
Website: www.health.pa.gov/
topics/programs/tobacco/Pages/
Quitline.aspx

Rhode Island
Toll-Free: 800-QUIT-NOW
(800-784-8669
Website: www.health.ri.gov/
healthrisks/tobacco/about/
quitsmoking

South Dakota
Toll-Free: 866-SD-QUITS
(866-737-8487)
Website: www.sdquitline.com

Virginia
Toll-Free: 800-QUIT-NOW
(800-784-8669)
Spanish: 855-DEJELO-YA
(855-335-3569)
TTY: 877-777-6534
Website: www.vdh.virginia.
gov/tobacco-free-living/
quit-now-virginia

West Virginia
Toll-Free: 800-QUIT-NOW
(800-784-8669)
Website: dhhr.wv.gov/wvdtp/
cessation/Quitline/Pages/default.
aspx

Wisconsin
Toll-Free: 800-QUIT-NOW
(800-784-8669)
Website: centralwitobaccofree.
org/quit-tobacco/
wisconsin-quit-line

Additional Online Resources

**Clinical Guidelines
for Prescribing
Pharmacotherapy for
Smoking Cessation**
Website: www.ahrq.gov/
professionals/clinicians-
providers/guidelines-
recommendations/tobacco/
prescrib.html

Committed Quitters
Website: www.quit.com

Guía para Dejar de Fumar
Website: www.cdc.gov/tobacco/
campaign/tips/spanish/dejar-
fumar/guia/index.html

Kick Butts Day
Website: kickbuttsday.org

**National Lung Health
Education Program**
Website: www.nlhep.org

Nicotine Anonymous
Website: www.nicotine-
anonymous.org

Pathways to Freedom
Website: pathways.thinkport.org

QuitNet
Website: quitnet.meyouhealth.
com

Smokefree.gov
Website: www.smokefree.gov

**Smoking: It's Never Too Late
to Stop**
Website: www.niapublications.
org/engagepages/smoking.asp

Tackling Tobacco
Website: quit.org.au/
resources/community-services/
tackling-tobacco-program

Index

Index

cavities, tobacco 26

CDC *see* Centers for Disease Control and Prevention

Centers for Disease Control and Prevention (CDC)
contact 493
publications
American Indians/Alaska Natives and tobacco use 140n
Asian Americans, Native Hawaiians, or Pacific Islanders and tobacco use 136n
asthma and secondhand smoke 203n
Buerger disease 222n
burden of tobacco use 149n
cancer 176n
cancer and tobacco use 176n
chronic obstructive pulmonary disease (COPD) 200n
cigarette smoking among adults 125n
clean indoor air regulations 437n
dual use of tobacco products 170n
e-cigarettes 29n
e-cigarettes and pregnancy 250n
economic trends in tobacco 97n
federal tax increase on tobacco 473n
first 20 minutes of quitting 294n
gum (periodontal) disease 272n
health effects of cigarette smoking 164n
health effects of smoking 164n
health warnings for cigarettes 479n
heart disease and stroke 223n
Hispanics/Latinx and tobacco use 132n
lesbian, gay, bisexual, and transgender (LGBT) persons and tobacco use 129n
manage your quit day 294n
marijuana medicine 101n

Centers for Disease Control and Prevention (CDC)
publications, *continued*
minors' access to tobacco 143n
people living with HIV 274n
preventing tobacco use among youth and young adults 416n
preventing tobacco use during pregnancy 447n
quitting smoking 284n
smokeless tobacco products and marketing 89n
smokeless tobacco use 109n
smoking and diabetes 258n
smoking and reproduction 269n
smoking and tobacco use—data and statistics 109n
smoking during pregnancy 367n
smoking, pregnancy, and babies 246n
steps to quit smoking 287n
talking to your teens about tobacco 455n
tobacco brand preferences 84n
tobacco-control interventions 421n
tobacco-control program 425n
tobacco industry marketing 464n
tobacco ingredient and nicotine reporting 57n
tobacco products 42n
tobacco timeline 4n
tobacco use among adults with mental illness and substance-use disorders (SUDs) 228n
tobacco use and pregnancy 246n
tobacco-related mortality 149n
vision loss and blindness 256n
women and smoking 121n
youth and tobacco use 116n
Center for Substance Abuse Prevention (CSAP), contact 493
Center for Tobacco Research and Intervention (CTRI), contact 497
central nervous system (CNS)
pregnancy 270
tobacco 9
cerebral palsy, premature birth 249

510

WITHDRAWN

$76.50

LONGWOOD PUBLIC LIBRARY
800 Middle Country Road
Middle Island, NY 11953
(631) 924-6400
longwoodlibrary.org

LIBRARY HOURS

Monday-Friday	9:30 a.m. - 9:00 p.m.
Saturday	9:30 a.m. - 5:00 p.m.
Sunday (Sept-June)	1:00 p.m. - 5:00 p.m.